D0846110

Bio-Fluid Mechanics

International Series on Advances in Fluid Mechanics

Objectives:

The field of fluid mechanics is rich in exceptional researchers worldwide who have advanced the science and brought a greater technical understanding of the subject to their institutions, colleagues and students.

It is to bring such advances to the attention of the broad international community, that this book series has been established. Its aims are achieved by contributions to volumes from leading researchers by invitation only. This is backed by an illustrious Editorial Board for the Series who represent much of the active research in fluid mechanics worldwide.

Volumes in the series cover areas of current interest or active research and will include contributions by leaders in the field.

Topics for the series include: Bio-Fluid Mechanics, Biophysics and Chemical Physics, Boundary Element Methods for Fluids, Experimental & Theoretical Fluid Mechanics, Fluids with Solids in Suspension, Fluid-Structure Interaction, Geophysics, Groundwater Flow, Heat and Mass Transfer, Hydrodynamics, Hydronautics, Magnetohydrodynamics, Marine Engineering, Material Sciences, Meteorology, Ocean Engineering, Physical Oceanography, Potential Flow of Fluids, River amd Lakes Hydrodynamics, Slow Viscous Fluids, Stratified Fluids, High Performance Computing in Fluid Mechanics, Tidal Dynamics, Viscous Fluids, and Wave Propagation and Scattering.

C.C. Hsiung
Department of Naval Architecture
Technical University of Nova Scotia
PO Box 1000
Halifax, Nova Scotia
Canada B3J 2X4

D.B. Ingham
Department of Applied Mathematical Studies
School of Mathematics
The University of Leeds
Leeds LS2 9JT
UK

M. Isaacson
Department of Civil Engineering
University of British Columbia
Vancouver, BC
Canada V6T 1Z4

S. Kim
Department of Chemical Engineering
University of Wisconsin-Madison
1415 Johnson Drive
Madison, Wisconsin 53706-1691
USA

T. Matsui
Department of Architecture
Nagoya University
Furo-cho, Chikusa-ku
Nagoya 464
Japan

T.B. Moodie
Department of Mathematics
University of Alberta
632 Central Academic Building
Edmonton
Canada T6G 2G1

M.A. Noor
Department of Mathematics
King Saud University
PO Box 2455
Riyadh 11451
Saudi Arabia

A.J. Nowak
Institute of Thermal Technology
Technical University of Silesia
44-101 Gliwice
Konarskiego 22
Poland

J. Noye
Department of Applied Mathematics
GPO Box 498
Adelaide
South Australia 5001

M. Ohkusu
Research Institute for Applied Mechanics
Kyushu University
Kasuga-koen 6, Kasuga
Fukuoka
Japan

W. Perrie
Bedford Institute of Oceanography
PO Box 1006
Dartmouth, Nova Scotia
Canada B2Y 4A2

H. Pina
Instituto Superior Tecnico
Av. Rovisco Pais
1096 Lisboa Codex
Portugal

H. Power
Wessex Institute of Technology
Ashurst Lodge, Ashurst
Southampton SO40 7AA
UK

D. Prandle
Proudman Oceanographic Laboratory
Bidston Observatory
Birkenhead
Merseyside L43 7RA
UK

S.K. Sarma
Department of Civil Engineering
Imperial College
Imperial College Road
London SW7 2BU
UK

H. Schmitt
Department of Fluid Mechanics Research
Institute of Theoretical Fluid Mechanics
Bunsenstrasse 10
D-3400 Göttingen
Germany

M.P. Singh
Department of Mathematics
Indian Institute of Technology
New Delhi
India

P. Skerget
Faculty of Technical Sciences
University of Maribor
PO Box 224, Smetanova 17
62000 Maribor
Slovenia

P.A. Tyvand
Department of Agricultural Engineering
Agricultural University of Norway
PO Box 5065
N-1432 AS-NLH
Norway

L.C. Wrobel
Wessex Institute of Technology
Ashurst Lodge, Ashurst
Southampton SO40 7AA
UK

M. Zamir
Department of Applied Mathematics
The University of Western Ontario
Western Science Centre
London, Ontario
Canada N6A 5B7

Advances in Fluid Mechanics
Volume 3
Series Editor: M. Rahman

Bio-Fluid Mechanics

Editor:
H. Power
Wessex Institute of Technology, Southampton, UK

Computational Mechanics Publications
Southampton Boston

Series Editor
M. Rahman

Editor
H. Power
Wessex Institute of Technology
Ashurst Lodge
Ashurst
Southampton SO40 7AA
UK

Published by

Computational Mechanics Publications
Ashurst Lodge, Ashurst, Southampton, SO40 7AA, UK
Tel: 44(0)1703 293223; Fax: 44(0)1703 292853
Email: CMI@uk.ac.rl.ib; Intl Email: CMI@ib.rl.ac.uk

For USA, Canada and Mexico

Computational Mechanics Inc
25 Bridge Street, Billerica, MA 01821, USA
Tel: 508 667 5841; Fax: 508 667 7582
Email: CMINA@netcom.com

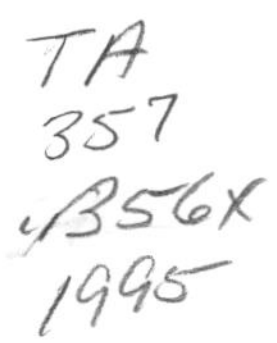

British Library Cataloguing-in-Publication Data

A Catalogue record for this book is available
from the British Library

ISBN 1 85312 286 6 Computational Mechanics Publications, Southampton
ISBN 1 56252 210 8 Computational Mechanics Publications, Boston
ISSN 1353 808X Series

Library of Congress Catalog Card Number 94-69712

Printed and bound in Great Britain by Bell & Bain Ltd, Glasgow.

CONTENTS

PREFACE

Bioengineering applies the methods of engineering, applied mathematics and physics to the study of biological phenomena, and the use of their concepts to describe these phenomena. This approach describes biological processes in terms consistent with their function. In that regard, numerical analysis has been helpful to the understanding of biological phenomena by means of their computer simulation. The increasing speed and expanded storage capacity of modern computers, together with new advanced numerical methods and programming techniques, have greatly improved the ability to study complex biological processes.

Fluid mechanics is one of the main topics of applied mathematics and physics, and some of the more relevant theories of these sciences have been found during the study of fluid flow problems. Also, fluid mechanics is one of the most challenging areas of computational mechanics; the simulation of fluid motion is always a serious test for any numerical method, in particular the non-linear problems. Since fluids are one of the major components of a living organism, it is logical that fluid mechanics play a major role in Bioengineering, by analyzing and simulating the fluid flow problem associated with physiological processes.

This book contains a selection of invited review articles covering some application of fluid mechanics in Bioengineering (Bio-fluid Mechanics). The nine chapters in the book have been contributed by leading authors in the field. The first chapter has been written by M. Karlsson, J. Engvall, P. Ask, B. Wranne and D. Loyd on *Orifice Flow in Stenotic and Regurgitant Valve Lesions: Modelling and Computer Simulation.* This chapter is concerned with the fluid mechanics of abnormal heart valves as well as the flow phenomena associated with these conditions. Disturbances in the blood flow may cause a prolonged suffering to patients and even cause their death, as several of the most notorious heart problems are associated with valves suffering from reduced efficiency-stenosis and/or regurgitation.

Chapter 2 is written by A.P. Yoganathan, J.D. Lemmon, R.A. Levine and C.C. Vesier on *A Three-Dimensional Simulation of Intraventricular Fluid Dynamics: Examination of Left Ventricular Early Systolic Function.* In this chapter a computational model is developed to conduct simulations of blood flow in the left ventricle during the contraction stage of the cardiac cycle called systole. The modelling involves the solution of the full three-dimensional non-permanent Navier-Stokes equations in a compliant thin-walled, anatomically correct left ventricle, and calculating the forces within the thin-walled structure by anatomical relations of stress-strain for cardiac muscle. The model produced endocardial motion which was consistent with normal human heart data, and velocity fields consistent with those occurring in a normal contracting left ventricle.

Chapter 3 entitled *Numerical Modelling of Blood Flow in Compliant Arteries and Arterial Bifurcations*, written by X.Y. Xu and M.W. Collins, describes a numerical model for the coupled solution of the equations of motion for the flow and wall in the larger arteries. The numerical model is able to solve for three-dimensional, unsteady flows with complex moving boundaries, as well as the full, time-dependent displacement and stress fields. Analysis of flow in a compliant artery with various EDRF-related interventions, and flow in a distensible bifurcation are presented.

The fourth chapter, *Numerical Simulation of Arterial Hemodynamics* has been written by S. Cavalcanti and G. Gnudi. This chapter describes a mathematical model for the simulation of arterial hemodynamics, as well as discussing several results obtained by the simulation, including the effects on axial velocity profiles, shear-stress and their time variation during a cardiac cycle. The numerical simulation is aimed at gaining a deeper insight into the effects caused by the mechanical and geometrical properties of arteries, as well as pressure and flow conditions on various hemodynamics variables such as velocity profiles, parietal shear-stress, etc.

In Chapter 5, *A Numerical Heart and Circulation Model to Simulate Hemodynamics for Rate-Responsive Pacing* written by A. Urbaszek and M. Schaldach, a computer model of the human cardiovascular system is presented, for simulating the short-term regulatory processes that adapt the cardiac output of the heart to changing circulatory demands. Both peripheral and cardiac control mechanics are considered. The model includes Starling's law as well as chronotropic and inotropic response, controlled by the autonomic nervous system. The goal of the simulations is to develop effective control algorithms for rate-responsive pacemakers based on measured cardiac parameters, especially for those systems providing closed loop control.

Chapter 6, *Evaluation of Haemodialysis Systems Using Computer Simulation,* is written by D. De Wachter, P. Verdonck and R. Verhoeven. In this chapter a numerical model is developed to calculate extra corporeal blood flow to describe haemodialysis systems and therapies. The results are used to determine the kinematic transport parameters in an artificial kidney. The model is employed to compute dialysis adequacy; a comparison is made between different systems and between different dialysis strategies.

In Chapter 7, *Folding Motifs, Kinetics, and Function in the Proximal Convoluted Tubule* written by Ch.J. Lumsden, simulation is used to study the dynamics of indicators interacting with the renal proximal convoluted tubule (PCT). The structural model of the PCT emulates the tubule's complex pattern of three-dimensional folding and vascular perfusion. Diffusional shunting among nearby folds strongly influences the washout and regional retention of secreted and re-absorbed indicators. Visualization of the tracer clouds in the blood, interstitial, and urinary compartments allows the regional pattern of indicator progression through the system to be followed in detail.

The numerical modelling of the complex mechanical system of the lung respiration has been the subject of Chapter 8, *Coupled Behavior of Lung Respiration: Computational Respiratory Mechanics Approach* written by S. Wada and M. Tanaka. The respiration model is made up from the mechanical elements of the respiratory system, such as gas flow, the tissue defamation, the blood flow, the gas diffusion, etc. The mechanical phenomena are directly expressed by the model taking account of the interaction among them. The breathing dynamics is simulated to exhibit the surface tension characteristics which could not be observed directly in the lung. The pulmonary blood flow affected by the breathing is investigated by means of the simulation supported by experimental observations. On the basis of the coupled behavior of the pulmonary circulation with the breathing and ventilation, the authors attempt to seek an appropriate condition in artificial ventilation.

Finally, in Chapter 9, *Micropolar Fluid Model for the Brain Fluid Dynamics* written by H. Power, the problem of determining the low Reynolds number flow of the cerebrospinal fluid through the subarachnoid space passing around the brain and spinal cord is formulated exactly as a system of linear Fredholm integral equations of the second kind. The formulation is based upon the assumption that the fluid belongs to the type described by the non-Newtonian theory of micropolar fluid. The resulting system of integral equation possesses a unique continuous solution, and it shows the different mechanisms that the brain have to control the flow in the subarachnoid space.

Chapter 1

Orifice flow in stenotic and regurgitant valve lesions - modelling and computer simulations

M. Karlsson, J. Engvall, P. Ask, B. Wranne & D. Loyd
Cardiovascular Dynamics Group, Departments of Mechanical Engineering, Clinical Physiology and Biomedical Engineering, Linköping University and University Hospital, Linköping, Sweden

1 Introduction

Many heart problems are associated with valves suffering from a reduced efficiency - stenosis, i.e. constriction, regurgitation, i.e. leakage, or both. In this chapter the basic fluid mechanics and typical flow phenomena of valve disease are discussed.

In the analysis of biological flow system two major aspects are of interest: (a) the bulk relationships between input and output and (b) the details of the flow. The former is best described by a "black-box" approach (i.e. a one-dimensional analysis), while for the latter the differential equations which describe the characteristics of the complete flow field are analysed. These equations are derived to describe the flow characteristics of an infinitesimal fluid element using the continuum assumption. Most models describing the flow within the human heart are, however, derived from empirical observations. The available measurement techniques are therefore important and have evolved in parallel with an increased understanding of the flow phenomena.

1.1 (a) The "black-box" approach

The first quantitative measurement methods were invasive, i.e. data were obtained from catheters introduced into the vessels and the heart. Flow was calculated using the Fick principle or applying dilution techniques, Guyton[1]. Pressure was measured with manometers connected to fluid filled catheters. The relationship between pressure and flow velocity in a one-dimensional description was described by the Bernoulli equation. Using two pressure catheters and combining the Bernoulli and the continuity equations flow and valve area could be estimated. With the introduction of Continuous Wave Doppler (CWD) the reversed procedure was applied and the pressure gradient across a stenotic valve could be estimated from the measured maximum velocity.

1.2 (b) Analysis of the velocity field

With the introduction of echocardiography, anatomical details and later on, in combination with Doppler technique, flow velocities within the heart could be imaged. The Colour-Doppler technique gives a colour coded map of the velocity vectors in the direction of the Doppler beam superimposed on a tissue image. For some applications this introduces a need for signal processing, e.g. the velocity vectors obtained by Colour-Doppler must be angle corrected in order to get the velocity vector in the desired direction. Due to the sweep time of the Doppler beam it is also necessary to introduce a time correction in order to get the instantaneous velocity field. The time correction may e.g. be obtained by controlling aquisition from the ECG. The time-corrected velocity field is then constructed from data originating from several consecutive heart beats.

Magnetic Resonance Imaging (MRI) estimates the instantaneous velocity field from averages of several consecutive heart beats. Thus, the velocity field obtained from MRI is also constructed using an averaging procedure.

Non-invasive techniques have opened a new window of information regarding cardiac flow. The extensive signal processing necessitates a thorough understanding of basic fluid mechanics in order to interpret the information correctly. During the last twenty years the development of non-invasive measurement techniques has been very fast. Image quality has improved due to the development of faster computers allowing more elaborate post-processing. Unfortunalely, post processing gives less possibilities for the physician to evaluate the raw data.

A qualitative diagnosis of obstructive or regurgitant flow lesions is often easy to obtain. Quantification, however, is necessary for risk assessment and for decisions regarding intervention. For this task, invasive pressure and flow measurements have to a large extent been replaced by echo Doppler.

Fast and reliable protocols, based on basic principles of fluid mechanics without too many restraining assumptions are therefore needed in order to provide the clinician with necessary and sufficient information for the diagnosis. The remaining part of this chapter deals with basic fluid mechanics applicable to biological flow and a deeper analysis of some different methods for estimating stenotic valve area and regurgitant volume.

1.3 An integrated approach

Computer simulations are normally validated using model experiments, which in some sense represent the real situation. In bio-fluid mechanics, however, the difference between the *in-vitro* setting and the *in-vivo* situation can be large. The difference between a computer model and the *in-vitro* setting is, on the other hand, normally small. In order to make the best progress, a "unified" approach is taken, i.e. the method best suited for a specific type of investigation is used. In bio-fluid mechanics it is not always possible to predict which method should be used. Thus, a multi-disciplinary approach is often advantageous. It is desirable to compare measurements with computer simulations, which may increase the understanding of artefacts in the velocity field.

Modelling the human cardiovascular system is very complex. Over the years several contributions concerning the fluid mechanics of the heart have been presented, e.g.: Peskin *et al.* [2] computed flow within a model of the heart. Vesier *et al.* [3] studied computer models of the left ventricle in order to simulate the behaviour of the flow in the immediate neighbourhood of the valve. Rogers *et al.* [4] studied the non-stationary three-dimensional flow field inside a model of an artificial heart in order to investigate turbulence induced shear stresses, which may damage the blood cells.

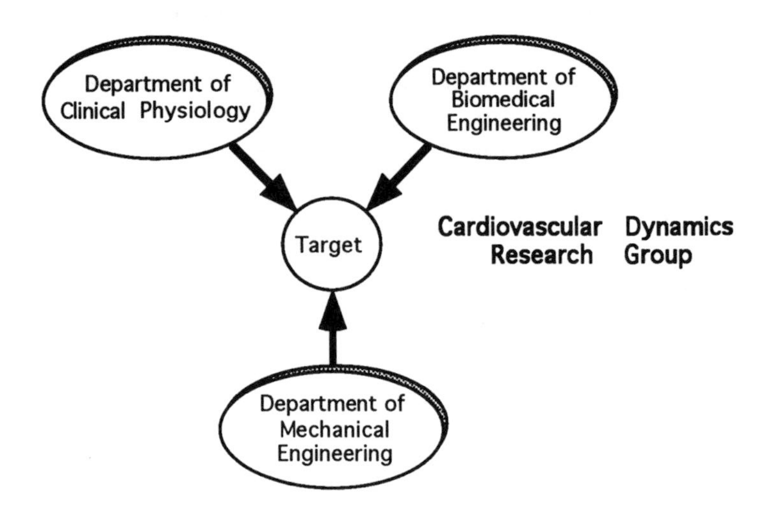

Figure 1: A Multi-Disciplinary Approach

2 Mathematical background

2.1 Basic assumptions

Any model well suited for studies of phenomena related to biological flow contains several assumptions and simplifications. However, the model must be "complex enough", i.e. all significant features must be resolved.

Some basic assumptions are applied throughout the the whole chapter. The fluid is assumed to be incompressible as we are dealing with the flow of a liquid at moderate velocities. Furthermore, the fluid is generally assumed to be Newtonian, i.e. the shear-stress is proportional to the velocity gradient.

2.2 The governing equations

The equations governing the flow are based on two conservation equations: conservation of mass (the continuity equation) and conservation of momentum (the Navier-Stokes equations). The governing equations for incompressible, laminar flow with constant viscosity may in a cartesian co-ordinate system be written as, Panton[5]

$$\frac{\partial u}{\partial t}+u\frac{\partial u}{\partial x}+v\frac{\partial u}{\partial y}+w\frac{\partial u}{\partial z}=X-\frac{1}{\rho}\frac{\partial p}{\partial x}+\nu(\frac{\partial^2 u}{\partial x^2}+\frac{\partial^2 u}{\partial y^2}+\frac{\partial^2 u}{\partial z^2}) \quad (1)$$

$$\frac{\partial v}{\partial t}+u\frac{\partial v}{\partial x}+v\frac{\partial v}{\partial y}+w\frac{\partial v}{\partial z}=Y-\frac{1}{\rho}\frac{\partial p}{\partial y}+\nu(\frac{\partial^2 v}{\partial x^2}+\frac{\partial^2 v}{\partial y^2}+\frac{\partial^2 v}{\partial z^2}) \quad (2)$$

$$\frac{\partial w}{\partial t}+u\frac{\partial w}{\partial x}+v\frac{\partial w}{\partial y}+w\frac{\partial w}{\partial z}=Z-\frac{1}{\rho}\frac{\partial p}{\partial z}+\nu(\frac{\partial^2 w}{\partial x^2}+\frac{\partial^2 w}{\partial y^2}+\frac{\partial^2 w}{\partial z^2}) \quad (3)$$

where t is the time and u, v and w are the velocity components in the x, y and z directions, respectively. p is pressure, ρ density and ν kinematic viscosity. X, Y and Z are used for body forces. The continuity equation can be written as

$$\frac{\partial u}{\partial x}+\frac{\partial v}{\partial y}+\frac{\partial w}{\partial z}=0 \quad (4)$$

This set of equations and appropriate boundary and initial conditions determine velocities and pressure in the analysed domain. In many physiological flow situations the geometry can be regarded as axi-symmetric, resulting in a simplification of the equations. Introducing v_z and v_r as the velocity components in the z (axial) and r (radial) direction respectively, we obtain

$$\frac{\partial v_r}{\partial t}+v_r\frac{\partial v_r}{\partial r}+v_z\frac{\partial v_r}{\partial z}=X_r-\frac{1}{\rho}\frac{\partial p}{\partial r}+\nu[\frac{\partial}{\partial r}(\frac{1}{r}\frac{\partial}{\partial r}(rv_r))+\frac{\partial^2 v_r}{\partial z^2}] \quad (5)$$

$$\frac{\partial v_z}{\partial t}+v_r\frac{\partial v_z}{\partial r}+v_z\frac{\partial v_z}{\partial z}=X_z-\frac{1}{\rho}\frac{\partial p}{\partial z}+\nu[\frac{1}{r}\frac{\partial}{\partial r}(r\frac{\partial v_z}{\partial r})+\frac{\partial^2 v_z}{\partial z^2}] \quad (6)$$

where X_r and X_z are body forces. The continuity equation has the following form

$$\frac{1}{r}\frac{\partial}{\partial r}(rv_r)+\frac{\partial}{\partial z}v_z=0 \quad (7)$$

Neglecting friction in Equations (5) - (7) gives the Euler equations

$$\frac{\partial v_r}{\partial t} + v_r \frac{\partial v_r}{\partial r} + v_z \frac{\partial v_r}{\partial z} = X_r - \frac{1}{\rho} \frac{\partial p}{\partial y} \tag{8}$$

$$\frac{\partial v_z}{\partial t} + v_r \frac{\partial v_z}{\partial r} + v_z \frac{\partial v_z}{\partial z} = X_z - \frac{1}{\rho} \frac{\partial p}{\partial z} \tag{9}$$

$$\frac{1}{r} \frac{\partial}{\partial r}(r v_r) + \frac{\partial}{\partial z} v_z = 0 \tag{10}$$

The geometry of complex problems is as a rule difficult to define in terms of cartesian or cylindrical co-ordinates. The two-dimensional Euler equations in streamline co-ordinates give the following expressions, Massey[6]

$$\frac{\partial w_s}{\partial t} + w_s \frac{\partial w_s}{\partial s} = S - \frac{1}{\rho} \frac{\partial p}{\partial s} \tag{11}$$

$$\frac{\partial w_n}{\partial t} - \frac{{w_s}^2}{r} = N - \frac{1}{\rho} \frac{\partial p}{\partial n} \tag{12}$$

where s and n are the co-ordinates parallell and perpendicular to the streamlines, respectively. S and N are body forces. Equation (11) along the streamline can be formulated as

$$\int \frac{\partial w_s}{\partial t} ds + \frac{{w_s}^2}{2} + gh + \frac{p}{\rho} = constant \tag{13}$$

For stationary flow, Equation (13) can be written as

$$\frac{{w_s}^2}{2} + gh + \frac{p}{\rho} = constant \tag{14}$$

known as the Bernoulli equation. The Bernoulli equation states that the sum of static pressure, dynamic pressure and height pressure is constant along one particular streamline. For some situations, however, the constant has the same value for the whole flow field, Massey[6]. The simplified Bernoulli equation for blood flow is derived for use in conjunction with ultra-sound Doppler

$$\Delta p = 4V^2 \tag{15}$$

where Δp is the pressure difference in [mmHg], V the maximal velocity measured by Continuous wave Doppler in [m/s] and 4 is a constant that includes density, viscosity and mercury conversion.

2.3 Modelling turbulent flow

Several turbulence models exist representing different degrees of complexity, from the decomposition of the velocities into separate terms for mean and fluctuating components to the simpler approach, where an effective viscosity is introduced into the governing equations. The effective viscosity is calculated from

$$\mu_{eff} = \mu_{lam} + \mu_{turb} \tag{16}$$

The simplest turbulence model is the mixing length model, Hinze[7], where the turbulent diffusivity is assumed to be proportional to the squared velocity gradient. The mixing length approach, however, depends on an unknown parameter, the mixing length, l_m. The mixing length may be thought of as the mean free path for the collision or mixing of globules of turbulent fluid. The mixing length approach is widely and successfully used for boundary layer and thin shear layer flow. In general, however, the model is difficult to apply in complex flows because of the difficulty to specify the mixing length.

A more elaborate model, however still assuming a homogenous turbulence, for calculation of the effective viscosity is the k-ϵ model. It is based on two equations, the first describes production of turbulent kinetic energy, k, and the second equation describes dissipation of turbulent kinetic energy, ϵ. The turbulent flow is characterised by a turbulent eddy velocity, u_t, and a turbulent length scale, l_t. The turbulent viscosity $\mu_t \propto \rho_0 u_t l_t$. As $u_t \propto \sqrt{k}$ and $l_t \propto k^{1.5}/\epsilon$, the turbulent viscosity can be written as, Hinze[7]

$$\mu_{turb} \propto \frac{\rho_0 k^2}{\epsilon} \tag{17}$$

The necessary transport equations for k and ϵ are derived from the Navier-Stokes equations, Launder & Spalding[8].

2.4 Boundary and initial conditions

The computational domain is defined by walls/symmetry lines, inlet and outlet. The velocity components at a wall are generally zero and the transverse velocity component at a symmetry line is zero. The inflow to the domain is modelled with prescribed profiles (e.g. velocity, turbulent kinetic energy and dissipation of turbulent kinetic energy). The outflow is normally assumed to be free.

Initial condition for the calculation is prescribed for each quantity, e.g. velocity, turbulent kinetic energy and dissipation.

3 The necessity of validation

The numerical simulation must be able to describe the cardiovascular flow properly. The obvious way in order to of validate the computer simulation would be to compare it with measurements in humans, which is frequently done. However, in several occasions the complex biological flow in humans cannot be exactly described in a numerical model. In other cases the blood flow in humans is not measurable with the required precision.

The solution in the first case would be to simplify the flow, study flow phenomena of main interest and ignore other flow contributions. This can be done in an *in-vitro* model describing the cardiovascular flow of interest. This model is also more accessable for flow measurements and therefore a solution of the second problem. In an animal model the heart can be freely exposed and furthermore it is possible to manipulate physiological parameters more freely than in humans by pharmacological agents and other types of stimulus.

A powerful instrument in both *in-vivo* and *in-vitro* investigations of flow is a duplex ultrasound Doppler scanner giving structural as well as flow information from the investigated area. The information is two-dimensional, but from this information a 3D image can be reconstructed. A limitation with the ultrasound technique is the requirement of an "ultrasound window" without bone or air. Another limitation is that the measured maximal flow velocity is limited by aliasing.

The phase coded MRI is a more complex technique that also has the problem of velocity aliasing but no "window" is needed. The laser-Doppler technique is highly accurate for velocity measurements and serves therefore as a "reference method". However, the method is only possible for *in-vitro* studies since the test object must be optically transparent.

Examples of competing techniques are pressure measurement with catheter as well as the use of transit time ultrasound probe. The latter is mainly used for measurements of flow through an exposed vessel or through a plastic tube. A system with water-filled catheters measures in relation to the fix level of the transducer and is insensitive to a varying hydrostatic component due to change of catheter orifice location. For pressure measurements catheter tip transducers have the advantage of a wide bandwidth.

4 Clinical applications I: Valve stenosis

The assessment of valve stenosis is of great clinical significance. Measurement of a transvalvular pressure difference (pressure gradient) across a stenotic valve gives some information but may be misleading since the difference varies with i.a. blood flow. For this reason it is important to determine the difference as well as the valve area. Ultrasound 2D-echocardiography may trace the valve orifice directly but is hampered by shadowing from calcification. The continuity equation assesses the area in relation to flow. The Gorlin equation is a special case of the continuity equation, where the flow velocity is replaced by the pressure difference from the Bernoulli equation. The different approaches to area estimation are schematically shown in Figure (2).

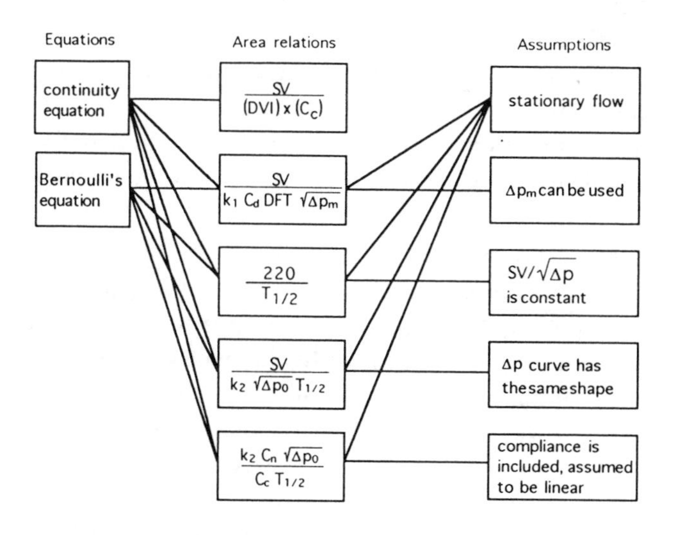

Figure 2: Different approaches to area estimation

Resistance to flow is another, however less validated, approach to assess the severity of the stenosis. The pressure half-time, defined as the time needed for the pressure difference across the stenosis to reach half its initial value, is widely used for the assessment of mitral stenosis.

4.1 Pressure differences

The general equation describing the relation between flow, Q, and pressure difference, Δp, for moderate to severe constrictions, Massey[6] and Merrit[9]

$$Q = K\sqrt{\Delta p} \tag{18}$$

For mild constrictions or constrictions at low flow, however, a linear relationship between flow and pressure difference is applicable. A problem, however, with Equation (18) is the determination of the coefficient K. In general it is written as, Merrit[9]

$$K = C_d A_a \sqrt{\frac{2}{\rho}} \tag{19}$$

where C_d is the discharge coefficient. $C_d = C_c C_v$, C_c accounts for the area correction between the anatomical area, A_a, and A_{vc}, the area at vena contracta. C_c varies generally between 0.6 and 1.00 depending on the shape of the orifice. C_v accounts for frictional losses and varies as a rule between 0.8 and 0.99 for biological orifices.

Besides the connection between pressure difference and flow, there are several problems of technical nature that need to be discussed.

4.1.1 Choice of transducers

Fluid filled catheter systems have a low resonance frequency which may cause pressure ringings. One should demand 10 × base frequency, which is 10 Hz for a heart rate of 60 BPM, 20 Hz for 120 BPM and so on. Most fluid filled catheters have a natural frequency < 10 Hz. The advantage with this type of catheter system, however, is that the location of the catheter tip is not very critical and that they are relatively cheap.

High fidelity catheter tip manometers have a high natural frequency, but they are expensive. The base location of the catheter tip becomes important, i.e. the pressure is not measured relative to a fix level. In the pulmonary artery such difference can be about 15 mmHg, which is large when the normal systolic pulmonary artery pressure is around 20 mmHg.

4.1.2 Influence of dynamic pressure

As seen in Equation (14), the total pressure consists of three components: static pressure, dynamic pressure and height pressure. Ignoring the influence from the height pressure still two components of pressure remain. Using a side-hole catheter the static pressure is in principle measured directly, while the tip-manometer catheter gives the stagnation pressure, i.e. the sum of static and dynamic pressure. The location and direction of the catheter tip becomes sometimes important. A normal aortic pressure of 120 mmHg and a velocity of 1 m/s gives a dynamic component of about 4 mmHg, i.e. the dynamic component is only about 3% of the total pressure. In the pulmonary artery during exercise, however, the pressure is about 30 mmHg and the velocity 2 m/s. The dynamic component is then 16 mmHg, i.e. the dynamic component is slightly more than 50% of the total pressure.

4.1.3 Pressure recovery

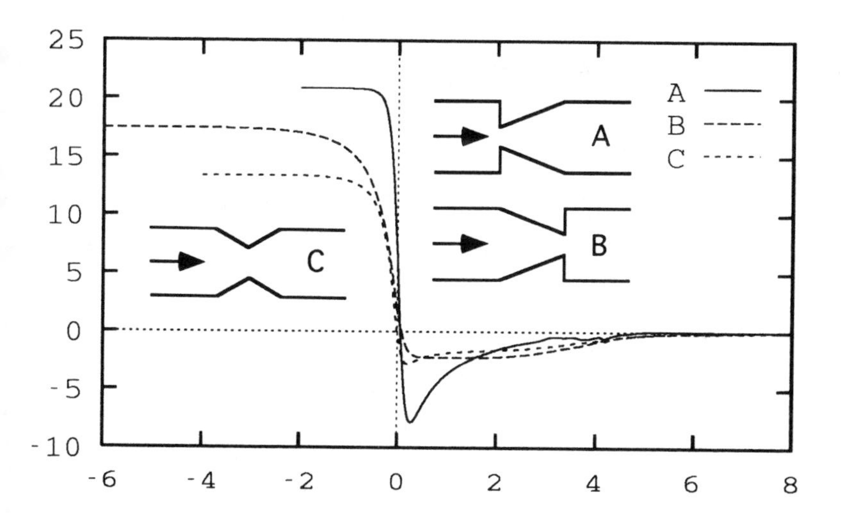

Figure 3: Centre line pressure curves for different geometries

The fluid is accelerated through the constriction and the pressure decreases. The pressure decrease occurs in order to conserve the momentum. When the fluid deccelerates distal to the constriction the velocity decreases and the pressure increases slightly. This *pressure recovery* is an important feature of stenotic flow. If, for example, the pressure difference derived from Continuous Wave Doppler is compared with invasive catheter recordings a discrepancy may occur. Figure (3) shows the pressure curve for three different geometries of stenosis. The size of the pressure recovery depends on the severity *and* the shape of the constriction.

4.1.4 Difference between Doppler and catheter measurement

The Doppler derived pressure difference is calculated from the simplified Bernoulli equation (15). The measured peak velocity is the maximal velocity. The Doppler derived difference is thus the peak instantaneous pressure difference. Using catheter transducers different combinations of pressure differences can be obtained. The peak instantaneous as well as the instantaneous difference can only be obtained using two transducers, while peak-to-peak and mean pressure difference can be obtined from one transducer, utilizing a pullback procedure.

4.2 Valve area

4.2.1 Continuity equation

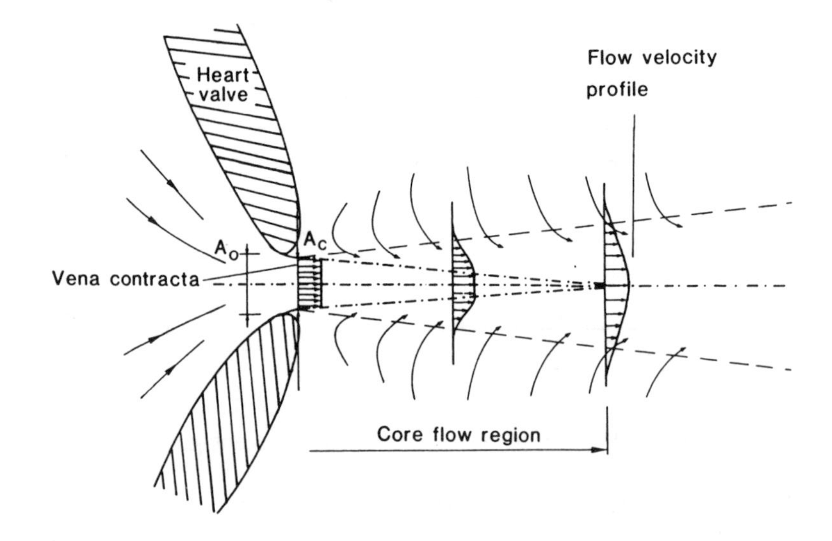

Figure 4: Flow through a stenotic valve

Flow through a stenotic valve is illustrated in Figure (4). The easiest and most obvious way to calculate mitral valve area, A_a, is from the continuity equation with measured volume flow and flow velocity or from non-invasively obtained stroke volume and Doppler velocity integral

$$A_{vc} = \frac{Q}{v} = \frac{SV}{DVI} \tag{20}$$

where Q is the volume flow, v the velocity, SV the stroke volume and DVI the time integral of the diastolic flow velocity through the valve. The area obtained in this way is the area at vena contracta, A_{vc}, where the fluid jet has its minimal diameter. The relation between A_a and A_{vc} is discussed in section 4.1.

4.3 The Gorlin equation

The location of vena contracta is normally about half a diameter downstream the anatomical orifice, A_a. The anatomical area can be obtained from the area at the vena contracta divided by the contraction coefficient, C_c, section 4.1

$$A_a = \frac{A_{vc}}{C_c} \tag{21}$$

The anatomical area can thus be estimeted from the following relation

$$A_a = \frac{Q}{vC_c} = \frac{SV}{DVI\, C_c} \tag{22}$$

This formula was derived by Gorlin & Gorlin[10], who, however, had problems to measure the flow velocity. They therefore calculated the velocity from the Bernoulli relationship and took friction into account via a velocity coefficient, C_v

$$v = C_v \sqrt{\frac{2}{\rho}} \sqrt{\Delta p} \tag{23}$$

where ρ is the density and Δp is the pressure difference. Designating $\sqrt{2/\rho}$ with k_1, setting $C_v C_c$ equal to the discharge coefficient C_d, and taking the average of the square root of the pressure difference (subscript m) one gets

$$A_a = \frac{Q}{k_1 C_d \overline{\sqrt{\Delta p_m}}} = \frac{SV}{DFT k_1 C_d \overline{\sqrt{\Delta p_m}}} \tag{24}$$

which is the Gorlin equation. The Gorlin constant is therefore equal to $k_1 C_d$ and the value depends on the units used, Cohen & Gorlin[11]. DFT is the diastolic flow time and $k_1 DFT\sqrt{\Delta p_m} \approx DVI$, Wranne *et al.*[12].

4.4 Pressure half-time, $T_{1/2}$

The pressure half-time, $T_{1/2}$, is the time it takes to reduce the transmitral pressure difference to half its initial maximal value. The method was introduced in 1966 by Libanoff & Rodbardt[13]. The main purpose was to find an easier way than the Gorlin equation to obtain a measure of the severity of a mitral stenosis. Later on this concept was introduced for use with ultrasound Doppler by Hatle *et al.*[14]. They suggested a simplified equation for estimation of valve area, A_a $[cm^2]$

$$A_a = \frac{220}{T_{1/2}} \tag{25}$$

where $T_{1/2}$ is measured in $[ms]$. This equation does not include flow and both Libanoff & Rodbardt and Hatle *et al.* claimed that flow hardly influence the area calculation.

When analysing the pressure half time concept, Loyd *et al.*[15] on theoretical and experimental grounds predicted that the half-time besides area also was influenced by the transported volume and the initial pressure difference according to

$$A_a = \frac{SV}{k_2\sqrt{\Delta p_0}T_{1/2}} \tag{26}$$

where Δp_0 is the initial pressure difference and k_2 is a coefficient given by

$$k_2 = \frac{C_d I_o}{a}\sqrt{\frac{2}{\rho}} \tag{27}$$

where I_o and a are coefficients related to the shape of the pressure difference curve, Loyd *et al.*[15]. The pressure difference time curve in Equation (26) was assumed to have the same shape for all patients but the time scale and the maximal pressure could vary.

Equations (24) and (26) can be compared with Equation (22) which is based on the continuity equation. As all these expressions have SV in the numerator we can see that the denominators of Equation (24) and (26) are both estimates of the diastolic time velocity integral. In the Gorlin equation a slight error is obtained by estimating the velocity integral by the term $DFT\sqrt{\Delta p_m}$ instead of $\int\sqrt{\Delta p(t)}dt$. For severe stenoses this error is negligible, but for a mild stenosis the area may be overestimated up to 15%, Wranne *et al.* [12]. In Equation (26) a certain shape of the pressure decay curve was assumed. With this assumption the area is determined from $\sqrt{\Delta p_0}$, the half-time and a time constant.

Comparing the pressure half-time relation of Equation (25) with Equation (26) it is evident that Equation (25) is only a special case of Equation (26) and applicable only when the relation $SV/\sqrt{\Delta p_0}$ is approximately constant. Both equations, however, have the disadvantage that the calculations involve errors when the pressure difference curve is truncated (as in short diastoles) and when atrial contraction in sinus rhythm causes the pressure decline to deviate from the normally expected type of time function.

In the analysis of mitral valve flow, Thomas & Weyman[17] and Thomas *et al.* [16], used Newtons second law, the continuity equation and the Bernoulli equation. The relation between volume and pressure in the atrium and ventricle was descibed by the atrial and ventricular compliances, C_{at} and C_{ve}, respectively. A schematic representation of the system is shown in Figure (5). In the derivation of the pressure half-time it was shown that non-stationary and viscous terms could be ignored. Assuming a constant combined atrial and ventricular compliance $C_n = C_{at}C_{ve}/(C_{at} + C_{ve})$, the differential equation can be solved analytically. The following expression for the pressure difference as a function of time is obtained

$$\Delta p(t) = (\sqrt{\Delta p_0} - \frac{C_c A_a}{C_n\sqrt{2\rho}}t)^2 \tag{28}$$

For $\Delta p(t) = \Delta p_0/2$, t will be equal to the pressure half-time

$$T_{1/2} = \frac{k_3 C_n\sqrt{\Delta p_0}}{C_c A_a} \tag{29}$$

where k_3 is a coefficient. The anatomical area is thus

$$A_a = \frac{k_4 C_n\sqrt{\Delta p_0}}{T_{1/2}} \tag{30}$$

where k_4 is a coefficient. The need to consider the compliance of the atrium and the ventricle has been suggested, Thomas and Weyman [17]

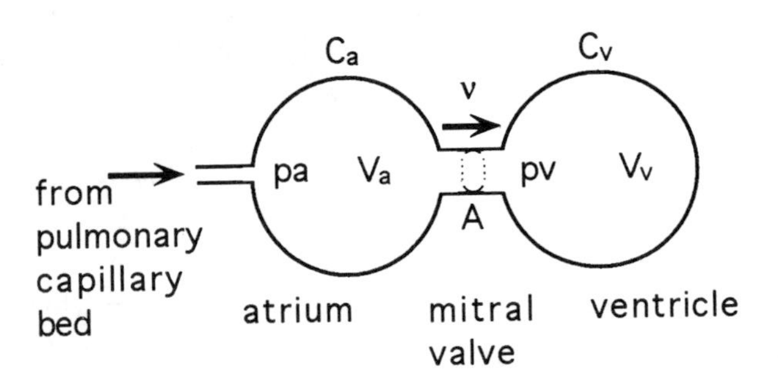

Figure 5: Schematic representation of the system

and Thomas *et al.* [16], and refuted by Loyd *et al.* [15]. At a first glance Equations (26) and (30) seem to be contradictory, but given a more thorough examination, Wranne *et al.* [12], it can be concluded that the $T_{1/2}$ concept with or without inclusion of chamber compliance leads, with certain assumptions, to the same result.

5 Clinical applications II: Regurgitation

Assessment of regurgitation is one of the remaining challenges in cardiology. Generally there is no problem in defining severe or mild leakage, while the grey zone in between is very difficult to assess. Leakage volume, leakage volume in relation to total ventricular stroke volume and the myocardial component - all need to be estimated.

From a clinical point of view, a technique for non-invasive quantification of valvular regurgitation is needed. Angiography has been used as a gold standard for many years, but opacification of a chamber is a very crude measure of the regurgitant volume. The extra volume load imposed on the ventricle may be calculated from the difference between anatomic and effective stroke volume. This volume load is sometimes compounded by a reduction of left ventricular systolic function. Anatomic stroke volume may be determined from planimetry of ventricular volumes determined with angiography, echocardiography or magnetic resonance imaging. Effective stroke volume can be measured with direct Fick technique at catheterisation, with indicator dilution technique or with rebreathing methods. With Doppler echocardiography the velocity field on either side of the valvular plane has been used for assessing mitral regurgitation.

5.1 Distal methods

The Colour-Doppler has been used to study the distal regurgitant jet size, Miyatake *et al.* [18], Helmcke *et al.* [19] and Otsuji *et al.* [20]. The limitations of this technique has been investigated and an alternative relation was suggested by Wranne *et al.* [21] involving flow velocity and orifice diameter. The intrusion of the jet is not only dependent on the flow rate, the driving pressure will also influence. The regurgitant jet may also be affected by the adjacent wall and the jet may also attach to the wall, Grimes *et al.* [22]. In such a case the jet can not be considered as a free jet and the basic equations for free jets can not be applied. Presently, the methods are subjected to several limitations and therefore not regarded as quantitative, Simpson *et al.* [23] and Bolger *et al.* [24]. Magnetic resonance jet velocity mapping is a promising method for the calculation of anatomic as well as effective stroke volume, Søndergaard *et al.* [25]. Flux of momentum is another distal side technique which is based on conservation of momentum of the distal jet, Cape *et al.* [26].

5.1.1 Flux of momentum

As the fluid enters the atrium a shear layer develops between the jet flow and the near stationary fluid of the atrium. The surrounding fluid moves along with the jet flow and some fluid is entrained into the jet. Due to the confined volume of the atrium the fluid starts to move in a circular pattern. The velocity of the jet is slowed due to the entrainment and the jet spreads radially. The jet must, however, travel a certain distance before the centre line velocity starts to decrease. The length of the central core of a *free* jet is about 5-6 orifice diameters, Grimes *et al.* [22]. Downstream the central core the centre line velocity decays with increasing distance from the orifice, Wranne *et al.* [21]

$$v_x = v_o \frac{k_1 d}{k_2 + x} \tag{31}$$

where v_x is the velocity at distance x downstream the orifice, v_o the velocity in the orifice and k_1 and k_2 are constants. d is the orifice diameter. Due to conservation of momentum, the axial momentum J remains constant along the centreline. The axial momentum in the orifice, J_o, is calculated as

$$J_o = \rho Q_o v_o = \rho v_o^2 A_o \tag{32}$$

where v_o is the orifice velocity, A_o the orifice area and Q_o the orifice flux. The rate of transfer of axial momentum at some distal position, downstream the central core, is

$$J_x = \int_A \rho v_x^2 dA \tag{33}$$

Equating $J_o = J_x$ by conservation of momentum gives

$$A_o v_o^2 = \int_A v_x^2 dA \tag{34}$$

According to Cape *et al.*[26], Equation (34), can be rearranged as

$$A_o v_o^2 = v_m^2 x^2 \int_{r=0}^{\infty} 2\pi (\frac{v}{v_m})^2 \frac{r}{x} d(r/x) \tag{35}$$

where v_m is the distal jet centre line velocity. The integral is a constant as a consequence of self-preservation, Cape *et al.*[26]. Thus Equation (35) can be expressed as

$$\frac{v_m}{v_o} = C_1 \frac{d}{x} \tag{36}$$

which essentially is equal to Equation (31). The jet centre line velocity v_x decays from the orifice velocity as

$$\frac{v_x}{v_o} = 6.3 \frac{d}{x} \tag{37}$$

Given the general expression for velocity decay of a free turbulent jet, Cape *et al.* [26], derives an equation which is supposed to be used for Continous Wave Doppler estimation of regurgitant flux, given as

$$Q_o = \frac{\pi v_m^2 x^2}{160 v_o} \tag{38}$$

which includes only v_m and v_o, i.e. the velocity at some distal location and the velocity at vena contracta. The equation, however, is strongly influensed by how accurately the vena contracta can be located and the velocity measured.

Furthermore, as the distal jet issues into a relativly small chamber the jet may impact the distal atrial wall and swirl, given that the jet has sufficient momentum. In low momentum mitral regurgitation the jet may instead be opposed by the inflow from the pulmonary veins. Skew orifices increase the risk for wall jets due to the Coanda effect, Grimes *et al.*[22].

5.2 Proximal methods

Determination of regurgitant volume with Doppler-echocardiography using the proximal isovelocity surface area (PISA) method shows promise but may be difficult for clinical use. Other availabe techniques include integration over proximal hemispheres (IPROV), which is a development of the original PISA method. A recent alternative technique is surface integration of velocity vectors (SIVV), Sun *et al.*[27]. With this method velocity vectors are integrated over a spherical surface with centre at the ultrasound transducer centre. This approach eliminates effects of angle errors.

5.3 The PISA-method

The PISA-method is based on the interpretation of the *Proximal Isovelocity Surface Area*, Recusani *et al.*[28] and Utsunomiya *et al.*[29]. The flow in the left ventricle is accelerated into the left atrium due to the higher pressure in the left ventricle. The regurgitant flux is calculated as

$$Q = 2\pi r^2 v \tag{39}$$

where r is the radius from orifice to the measured velocity, v, Figure (6). Various investigators have used a finite-difference approach to solve the Navier-Stokes equations for the proximal velocity field. The impact from the size of the orifice was studied, Rodriguez *et al.*[30], in a model which only included the the proximal velocity field.

Barclay *et al.*[31] showed through model experiments and computer simulations that the PISA method only estimated the flow rate correctly at one particular distance from the orifice. The flow was overestimated

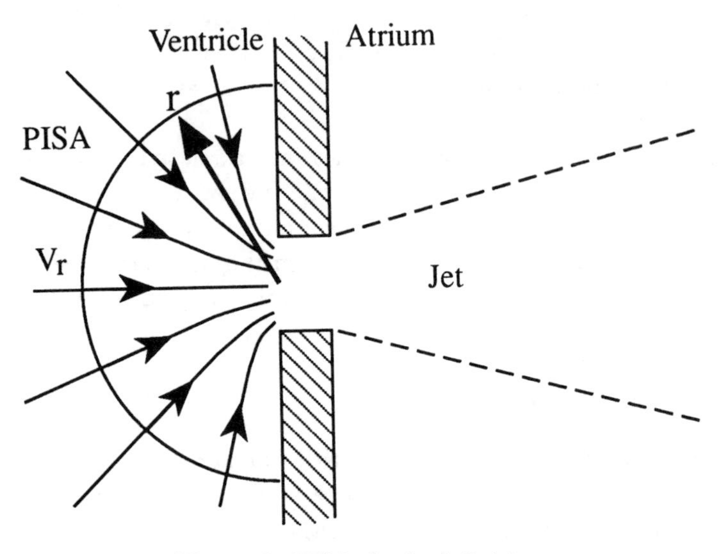

Figure 6: PISA: basic definitions

if the surface area was located at too short a distance from the orifice and the flow underestimated if the surface area was located too far from the orifice. Close to the orifice the velocity gradients are high and the velocity based on the Doppler signal may be influensed by aliasing. Far away from the orifice the gradients are too small and the Doppler signal very weak.

Karlsson & Loyd[32] showed that the difference between potential flow and laminar viscous flow was small, except in the immediate neighbourhood of the orifice. Their examination of different proximal flow field models is described briefly in section 5.3.1.

5.3.1 Fluid mechanics of proximal acceleration

The original PISA-method is based on the idea that the velocity vectors proximal to an orifice are directed towards the centre of the orifice. In the calculation of volumetric flow, the velocity at a certian distance from the orifice is measured and multiplied by the appropriate hemispherical area.

Friction will in principle affect the velocity field very close to the wall and in the extreme neighbourhood of the orifice. Further away from the wall and the orifice, potential flow theory will predict the velocity field with sufficient accuracy. When the fluid is viscous a wall must be modelled with a zero velocity and a zero flux through the boundary.

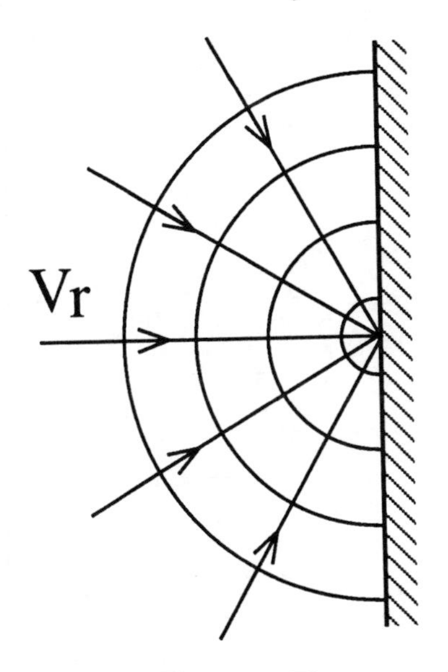

Figure 7: Flow towards a point sink

The flow field for the point sink has a simple analytical solution. For a point sink, Figure (6) the velocity components, using cylindrical co-ordinates, are

$$v_r = \frac{\Gamma}{2\pi r} \qquad v_\varphi = 0 \qquad v_z = 0 \tag{40}$$

where Γ is the strength of the sink and r is the radius, Panton[5], Fig (6). The flow field is given by the the potential function, ϕ

$$\phi = \frac{\Gamma}{2\pi} \ln r + C_1 \tag{41}$$

where C_1 is a constant. The iso-lines for the potential function are concentric hemispheres. In Figure (7) the potential function is plotted for a point sink located at a wall. The centre line velocity is shown in Figure (8). The streamlines, obtained from

$$\frac{\partial \psi}{\partial y} = \frac{\partial \phi}{\partial x} \qquad -\frac{\partial \psi}{\partial x} = \frac{\partial \phi}{\partial y} \tag{42}$$

are in every point perpendicular to the iso-lines of the potential function. For a point sink the streamlines are straight lines directed towards the centre of the orifice.

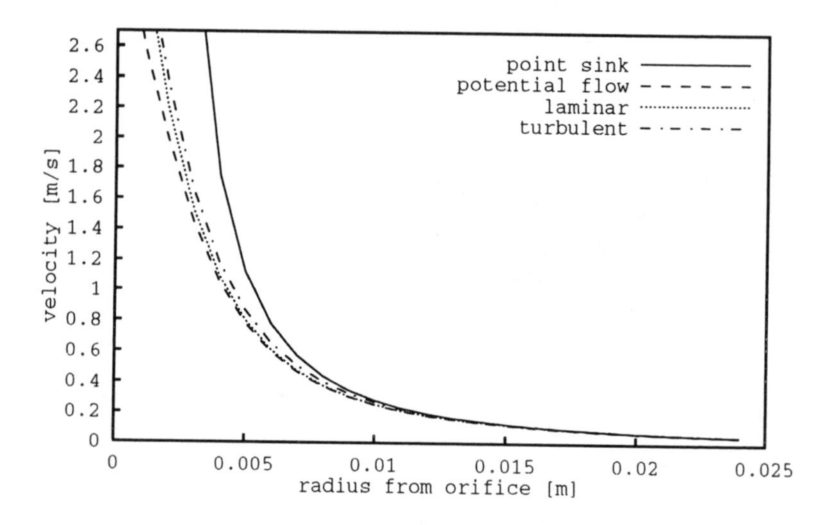

Figure 8: Centre line velocity

As the radius of the orifice has a finite value the point sink technique is not sufficient. The flow field may be found analytically using e.g. a source panel method, Panton[5]. The calculations will, however, be cumbersome and a numerical technique is preferable. Potential flow theory is used to calculate the velocity field for ideal flow, Panton[5]. The velocities can be calculated from the stream function, ψ as

$$u = \frac{\partial \psi}{\partial y} \qquad v = -\frac{\partial \psi}{\partial x} \tag{43}$$

The continuity equation will be identically satisfied and the mathematical treatment of the governing equations will be significantly simplified. The Laplace equation, Panton[5],

$$\frac{\partial^2 \psi}{\partial x^2} + \frac{\partial^2 \psi}{\partial y^2} = 0 \tag{44}$$

describes the flow of an ideal fluid in the limiting case of zero viscosity. The iso-lines for the potential function can be obtained from Equations

42). The centre line velocity for an orifice with a finite radius is shown n Fig (8).

The flow of a viscous fluid is described by the Navier-Stokes equations, Equations (1) - (3) and discussed in section 2.2. The centre line velocity for laminar flow through an orifice of finite size is shown in Fig (8).

If the proximal flow field cannot be assumed to be completely laminar a suitable turbulence model must be used. The k-ϵ model is well suited for simulations using an axi-symmetrical geometry as discussed in section 2.3. In Figure (9) the proximal velocity field is shown and in Figure (8) the centre line velocity for turbulent flow is shown.

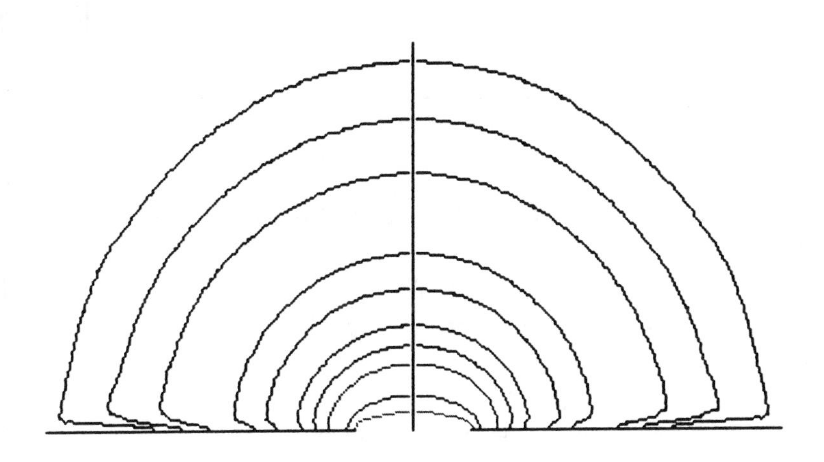

Figure 9: PISA for turbulent flow

5.3.2 Calculated flux from the simulations

The calculated flux through an orifice is shown in Fig (10). The flux is calculated from

$$Q = 2\pi r^2 v_{r=R} \tag{45}$$

Only a small difference between the four curves is seen. The size of the orifice is the prime determinant of the shaping of the velocity curve, while viscosity only plays a minor role at the centre line. Close to the walls,

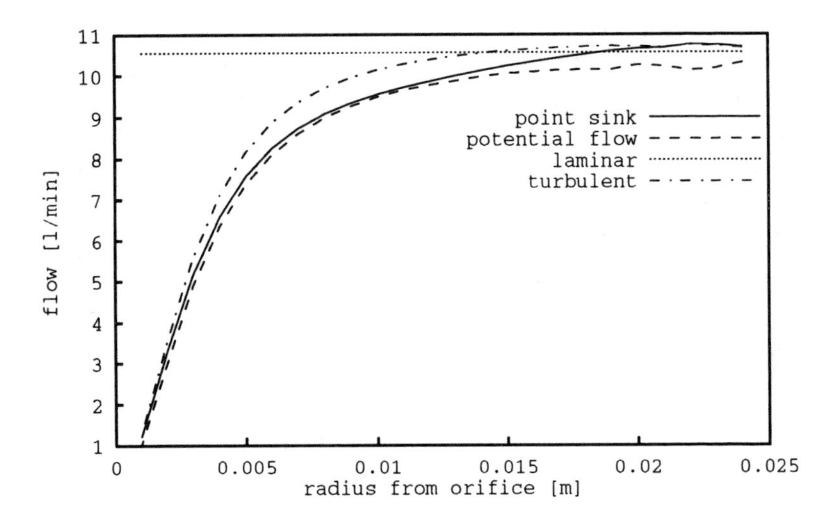

Figure 10: Calculated flux using the PISA method

however, the viscosity distorts the flow field considerably. If the flow is assumed to be turbulent, steeper gradients are possible, which changes the calculated flow curve.

5.3.3 Derivation of a correction factor

If the isovelocity surface has a non-hemispheric shape a correction factor can be defined as, Barclay *et al.*[31]

$$C_{corr} = \frac{Q_{true}}{Q_c} \tag{46}$$

where Q_{true} is the true flow and Q_c is the calculated flow from Equation (45). The correction factor changes with orifice geometry and varies i.a. with the distance from the orifice. The corrected regurgitant flow Q_{cc} is an estimation of Q_{true} and can be calculated as

$$Q_{cc} = C_{corr} 2\pi r^2 v \tag{47}$$

The relative error of the flow calculation can be expressed as

$$\frac{\Delta Q_{cc}}{Q_{cc}} = \sqrt{(\frac{\Delta C_{corr}}{C_{corr}})^2 + (Q_c\frac{d}{dr}(\frac{1}{Q_c})\Delta r + 2\frac{\Delta r}{r} + \frac{1}{v}\frac{dv}{dr}\Delta r)^2 + (\frac{\Delta v}{v})^2} \tag{48}$$

where ΔC_{corr} is the uncertinity in determining the correction factor, Δr the error in determining the radial distance to the orifice and Δv the error in the flow velocity. Barclay *et al.*[31], estimated the error in determining the correction factor to be about 3 %, which is related to the accuracy in measuring the true flow (in this case using a bucket and a stop-watch). The radial error can be estimated to be equal to the sample size volume, i.e. 1.5 mm. At a zone 8 to 9 mm from the orifice the error showed a minimum and the correction factor was less than 15 %.

5.3.4 Distorsion of the flow field

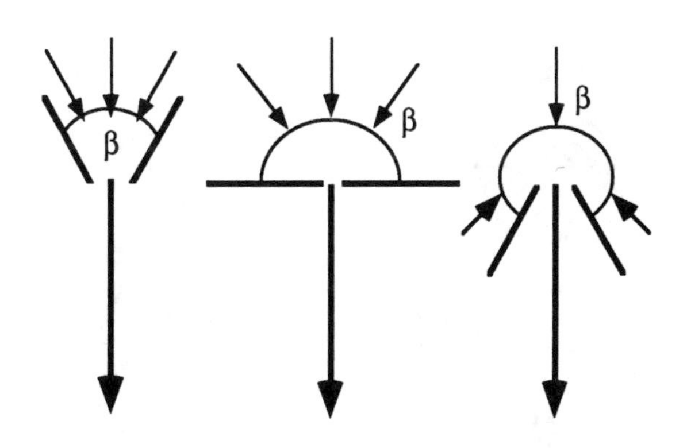

Figure 11: PISA at a non-flat surface

In many cases the orifice plane is not flat. Figure (11) shows three different types of geometry which may occur in mitral regurgitation. It has been argued that the PISA-approximation from centre line velocity can be corrected due to the angle β using, Grimes *et al.*[22],

$$Q = 2\pi r^2(1 - \cos\frac{\beta}{2})v \tag{49}$$

The correction using the angle β is simple to perform in an *in-vitro* setting but remains difficult in an *in-vivo* situation due to difficulties in visualising the valvular plane.

Some *in-vitro* studies of the PISA-method have been conducted in large tanks during steady-state flow conditions. The ultrasound Doppler measurements have been made in the proximity of the orifice. *In-vivo*, however, other flows may distort the flow field, thus reducing the symmetry of the converging flow. When using the PISA method, the main distorsion of the proximal velocity field originates from the outflow from the ventricle into the aorta.

5.4 Integration over proximal hemispheres

To use the available proximal velocity informtion outside the centre line angle correction is needed.

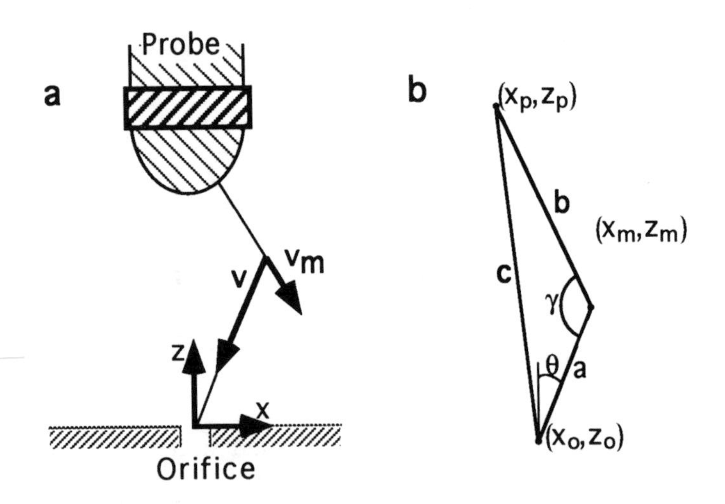

Figure 12: The measuring situation

All velocity vectors displayed in an ultrasound Doppler image are directed along the beam. When analysing the proximal velocity field the velocities of interest are directed towards the orifice and an angle correction is needed for calculations. Eidenvall *et al.* [33] calculated the velocity towards the orifice as

$$v = \frac{v_m}{-\cos\gamma} \tag{50}$$

where γ is the angle between the reflected ultrasound beam and v the velocity vector directed towards the centre of the orifice, Figure (12). v_m is the measured velocity. The angle γ can be calculated from

$$\gamma = \arccos(\frac{a^2 + b^2 - c^2}{2ab}) \tag{51}$$

where a is the distance from the centre of the orifice inlet and point of measurement, b is the distance between the transducer and the point of measurement and c is the distance between the transducer and the centre of the inlet, Figure (12). Combining the two equations, we obtain

$$v = \frac{2abv_m}{c^2 - a^2 - b^2} \tag{52}$$

A fix hemispherical proximal control volume combined with a calcula-

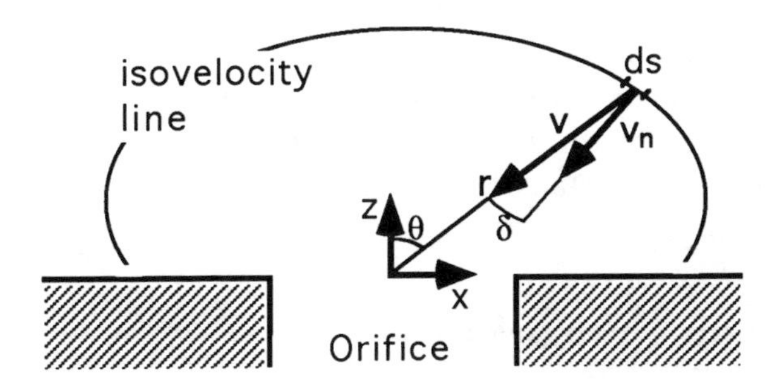

Figure 13: Flow through a proximal hemispherical control volume

tion of the flow from integration of measured actual velocities over that hemisphere is suggested, Eidenvall *et al.*[33], Figure (13). The area A_θ of a spherical segment up to an angle θ from the centre line can be written as

$$A_\theta = 2\pi r^2(1 - \cos\theta) \tag{53}$$

The area dA of a spherical sector with width ds varies with the angle θ as

$$dA(\theta) = 2\pi r^2 \sin\theta d\theta \tag{54}$$

where r is the radius from the orifice centre to the isovelocity segment. Assuming rotational symmetry and that $v = v(\theta)$ the flow through the area up to the angle θ can be calculated as

$$Q_\theta = 2\pi r^2 \int_0^\theta \sin\theta v(\theta) d\theta \tag{55}$$

The total flow through the total hemispherical area, A, is

$$Q = \frac{A}{A_\theta} Q_\theta \tag{56}$$

5.5 Time-corrections for pulsatile flow

A timing error occurs when the ultrasound Doppler beam sweeps through the domain of interest. The acquisition time may be quite long (up to 200 ms per image), thus the data must be time corrected when analysing a non-stationary flow field. Eidenvall *et al.*[34] suggests an external trigger device that enables a varying preset delay in the data acquisition. If, for example, the sweep time is 60 ms and the delay time is 20 ms, a time corrected image can be obtained using information from three consecutive images.

6 Concluding remarks

Non-invasive quantification of stenotic and especially regurgitant valve flow remains a challenging problem in cardiology. The PISA method (and its derivatives) and momentum flux are two promising techniques for assessing mitral regurgitation. Both methods are derived from the basic principles of fluid mechanics but are based on assumptions of idealised flow conditions which may not prevail *in-vivo*. The adoption of these techniques in clinical investigations is still limited. Realistic *in-vitro* experiments and *in-vivo* testing are still needed.

7 Acknowledgements

The authors wish to thank Mr Gunnar Andersson, Department of Mechanical Engineering, for excellent assistance with computer simulations and program development, Ms Elisabeth Forslund, Department of Mechanical Engineering, for excellent assistance with manuscript preparation and Dr Lars Eidenvall for valuable discussions concerning Doppler techniques.

References

1. Guyton A.C., *Textbook of Medical Physiology*, 7th Edition, W.B. Saunders Company, Philadelphia, 1986

2. Peskin C.S., Numerical analysis of numerical blood flow in the heart, *J Comp Physics*, **25**:220-252, 1977

3. Vesier C.C., Levine R.A. & Yoganathan A.P., Simulation of blood flow in the left ventricle: the effect of papillary muscle on mitral valve function, In: *Computers in Biomedicine*, Eds: K.D. Held, C.A. Brebbia and R.D. Ciskowski, Computational Mechanics Publications, Southampton, 1991

4. Rogers S., Kutler P., Kwak D. & Kiris C., Numerical simulation of flow through an artificial heart, *CRAY Channels*, Winter, 1990

5. Panton R.L., *Incompressible flow*, John Wiley & Sons, New York, 1984

6. Massey B.S., *Mechanics of Fluids*, 6th Edition, Chapman & Hall, London, 1991

7. Hinze J.O., *Turbulence*, 2nd Edition, McGraw-Hill Book Company, New York, 1975

8. Launder B.E. & Spalding D.B., The numerical computation of turbulent flows, *Computer Methods in Applied Mechanics and Engineering*, **3**:269-289, 1974

9. Merrit H.E., *Hydraulic control systems*, John Wiley & Sons, New York, 1967

10. Gorlin R. & Gorlin S.J., Hydraulic formula for calculation of the stenotic mitral valve, other cardiac valves, and central circulatory shunts, *Am Heart J*, **41**:1-29, 1951

11. Cohen M.V. & Gorlin R., Modified orifice equation for the calculation of mitral valve area, *Am Heart J*, **84**:839-840, 1972

12. Wranne B., Ask P. & Loyd D., Assessment of mitral stenosis. A theoretical end experimental study on the validity of the gradient half time, Proc. of The 2nd International Conference on Cardiac Doppler, Kyoto, Japan, Nov 25-29, 1986

13. Libanoff A.J. & Rodbard S., Evaluation of the severity of mitral stenosis and regurgitation, *Circulation*, **33**:218-227, 1966

14. Hatle L., Angelsen B. & Tromsdal A., A non-invasive assessment of atrioventricular pressure half-time by Doppler ultrasound, *Circulation*, **60**:1096-1104, 1979

15. Loyd D., Ask P. & Wranne B., Pressure half time does not always predict mitral valve area correctly, *J Am Soc Echo*, **1**:313-321, 1988

16. Thomas J.D., Wilkins G.T., Choong C.Y.P., Abscal V.M., Palacios I.F., Block B.C. & Weyman A.E., Inaccuracy of mitral pressure half-time immediately after percutangeous mitral valvotomy. Dependence on transmitral gradient and left atrial and ventricular compliance, *Circulation*, **78**:980-993, 1988

17. Thomas J.D. & Weyman A.E., Fluid dynamics model of mitral valve flow: determination with in-vivo validation. *J Am Coll Cardiol*, **13**:221-233, 1989

18. Miyatake K., Izumi S., Okamoto M., Kinoshita N., Asonuma H., Nakagawa H., Yamamoto K., Takayima M., Sakakibara H. & Nimura Y., Semiquanitative grading of severity of mitral regurgitation by real-time two-dimensional Doppler flow imaging technique, *J Am Coll Cardiol*, **7**:82-88, 1986

19. Helmcke F., Nanda N.C., Hsiung M.C., Soto B., Adey C.K., Goyal R.G. & Gatewood R.P., Color Doppler assessment of mitral regurgitation with orthogonal planes, *Circulation*, **75**:175-183, 1987

20. Otsuji Y., Tei C., Kisanuki A., Natsugoe K. & Kawazoe Y., Color Doppler echocardiography of the change in mitral regurgitant volume, *Am Heart J*, **114**:349-354, 1987

21. Wranne B., Ask P. & Loyd D., Quantification of heart valve regurgitation: a critical analysis from a theoretical and experimental point of view, *Clin Physiol*, **5**:81-88, 1985

22. Grimes R.Y., Burleson A., Levine R.A. & Yoganathan A.P., Quantification of cardiac jets: theory and limitations, *Echocardiography*, **11**:267-280, 1994

23. Simpson I.A., Valdes-Cruz L.M., Sahn D.J., Murillo A., Tamura T. & Chung K.J., Doppler color flow mapping of simulated in vitro regurgitant jets: evaluation of the effects of orifice size and hemodynamic variables, *J Am Coll Cardiol*, **13**:1195-1207, 1989

24. Bolger A.F., Eigler N.L. & Maurer G., Quantifying valvular regurgitation: limitations and inherent assumptions of Doppler techniques, *Circulation*, **78**:1316-1318, 1988

25. Søndergaard L., Thomsen C., Ståhlberg F., Gymoese E., Lindvig K., Hildebrandt P. & Henriksen O., Mitral and aortic valvular flow: quantification with MR phase mapping, *JMRI*, **2**:295-302, 1992

26. Cape E.G., Yoganathan A.P. & Levine R.A., A new theoretical model for noninvasive quantification of mitral regurgitation, *J Biomechanics*, **23**:27-33, 1990

27. Sun Y., Ask P., Janerot Sjöberg B., Eidenvall L., Loyd D. & Wranne B., Estimation of volume flow rate by surface integration of velocity vectors in color Doppler images from two orthogonal planes, Submitted for publication, 1994

28. Recusani F., Bargiggia G.S., Yoganathan A.P., Raisaro A., Valdes-Cruz L.M., Sung H.W., Bertucci C., Gallati M., Moises V.A., Simpson I.A., Tronconi L. & Sahn D.J., A new method for quantification

of regurgitant flow rate using color Doppler flow imaging of the flow convergence region to a discrete orifice: an in-vitro study, *Circulation*, **83**:594-604, 1991

29. Utsunomiya T., Ogawa T., Doshi R., Patel D., Quan M., Henry W.L. & Gardin J.M., Doppler color flow "Proximal Isovelocity Surface Area" metod for estimating volume flow rate: effects of orifice shape and machining factors, *J Am Coll Cardio*, **17**:1103-1111, 1991

30. Rodriguez L., Anconina J., Flachkampf F.A., Weyman A.E., Levine R.A. & Thomas J.D., Impact of finite orifice size on proximal flow convergence: implications for Doppler quantification of valvular regugitation, *Circulation Research*, **70**:923-930, 1992

31. Barclay S.A., Eidenvall L., Karlsson M., Andersson G., Changsheng X., Ask P., Loyd D. & Wranne B., The shape of the proximal isovelocity surface area varies with regurgitant orifice size and distance from orifice: Computer simulation and model experiments using color M-mode Technique, *J Am Soc Echocard*, **6**:433-45, 1992

32. Karlsson M. & Loyd D., A fluid mechanics analysis of the PISA-method, In: *Computational Biomedicine*, Eds: K.D. Held, C.A. Brebbia, R.D. Ciskowski and H. Power, Computational Mechanics Publications, Southampton, 1993

33. Eidenvall L., Barclay S., Loyd D., Wranne B. & Ask P., Regurgitant heart valve flow from 2D proximal velocity field: continued search for the ideal method, Accepted for publication in *Med & Biol Eng & Comp*, 1994

34. Eidenvall L., Janerot Sjöberg B., Ask P., Loyd D. & Wranne B., Two-dimensional color Doppler flow velocity profiles can be time corrected with an external ECG-delay device, *J Am Soc Echocard*, **5**:405-413, 1992

Chapter 2

A three-dimensional simulation of intraventricular fluid dynamics: examination of left ventricular early systolic function

A.P. Yoganathan, J.D. Lemmon, Jr.,
R.A. Levine & C.C. Vesier
Cardiovascular Fluid Mechanics Laboratory,
Georgia Institute of Technology, Atlanta, GA 30332, USA

Abstract

Computational methods are emerging as extremely powerful tools in the area of Biomedical Engineering because they provide detailed information about the systems modeled. In particular, computational research can be used to gain an elaborate picture of cardiac function by not only providing the three-dimensionality of blood flow, but also the motion of the heart walls interacting with the blood. By using Peskin's Immersed Boundary Method, a computational model was designed to conduct simulations of blood flow in the left ventricle during the contraction stage of the cardiac cycle called systole. This involved the solution of the full, three-dimensional Navier-Stokes equations for time-dependent flow in a compliant thin-walled, anatomically correct left ventricle, and calculating the forces within the thin-walled structure by anatomical relations of stress-strain for cardiac muscle. The structure was permitted to move freely within a fixed fluid domain by use of an interpolation function developed by Peskin which transfers forces between the two systems. The model produced endocardial wall motion which was consistent with normal human heart data, and velocity fields consistent with those occurring in a normally-contracting left ventricle. The results were also consistent with clinical measurements of mitral valve-papillary muscle apparatus displacement, and changes in the mitral valve annular area. This study shows that the thin-walled, three-dimensional left ventricular model simulates observed normal heart phenomena. This model was also used to show what changes occur to the ventricle flow fields and wall motion when it is associated with disease. In particular, the mitral valve disease Systolic Anterior Motion (SAM) of the Mitral Valve Leaflets was examined to investigate its causation.

Nomenclature

SAM -	Systolic Anterior Motion of the Mitral Valve Leaflets
HCM -	Hypertrophic Cardiomyopathy
PM-MV -	Altered Papillary Muscle - Mitral Valve Geometry theory
AML -	Anterior Mitral Leaflet
AMRL-	Anterior Mitral Residual Leaflet
PML -	Posterior Mitral Leaflet
IVS -	Intraventricular Septum

1 Introduction

For the past two decades, computational fluid dynamics has been a useful research tool in the study of cardiovascular fluid mechanics. Several groups developed models to simulate steady laminar flow through a two-dimensional rigid heart valve.[1,2] By approximating the time-varying geometry, Hung & Schuessler[3] simulated the effect of valve motions on an inviscid flow field. As the available technology expanded, so did the complexity of the computer models. By the late 1980's, it was feasible to simulate three-dimensional time-dependent flow, such as that occurring in a heart. Recently, a full Navier-Stokes equations was solved for flow in a rigid artificial heart with a relatively low Reynolds number.[4]

Yet, very few researchers have simulated the interaction of fluid with cardiovascular tissue (i.e., moving boundaries) in solving the Navier-Stokes equation. The Immersed Boundary Method[5] allows for the interaction of fluid with moving boundaries, which permits the solution of the blood-tissue system used in this research. They improved this method by extending this method to a three-dimensional model of the heart.[6,7] The goal of this study was to develop a model utilizing the immersed boundary approach that simulates physiologic phenomena of the left ventricle. The validity of the model was tested by comparison to clinically obtained normal left ventricular data. This validation was necessary so that further studies could be performed into the area of altered left ventricular function.

2 Methods

2.1 Computational methods: blood-tissue interaction

The blood-tissue interaction in the three-dimensional, computational model was accomplished by using Peskin's Immersed Boundary method.[5-7] The immersed boundary approach treats the fluid-structure system as two separate subsystems, the fluid subsystem and the structure subsystem. Each subsystem is solved separately at each time step. The results of the previous solution of one subsystem are used to solve the other one. In Peskin's approach, the fluid velocity field is used to deform the moving structure (cardiac tissue). Once deformed, the structural forces associated with the new configuration are computed, ignoring the fluid flow field. In reality, the boundary force would be in equilibrium with the fluid force; a weakly coupled method only approximates equilibrium. These forces are then applied to the fluid field, and a new fluid velocity field is computed. The solution method is a five step procedure at each time step and consists of the following:

1.Given at time step n the blood's velocity field ($\vec{\mathbf{u}}$) and pressure field (p), and heart's structural position ($\vec{\chi}$) calculate a structural velocity

$$\vec{U}_{lk}^{n} = \sum_{\vec{x}} \vec{u}^{n} \delta_h(\vec{x} - \vec{\chi}^{n}) h^3$$

where **(1)**

$$\delta_h(\vec{x} - \vec{\chi}^{n}) = \begin{cases} \frac{1}{4h}\left[1 + \cos\left(\frac{(\vec{x} - \vec{\chi}^{n})}{2h}\right)\right], & \left|(\vec{x} - \vec{\chi}^{n})\right| \le 2h \\ 0 & , \left|(\vec{x} - \vec{\chi}^{n}) > 2h\right| \end{cases}$$

2. Calculate the intermediate force in a fiber

$$\vec{\chi}_{lk}^{n+1,*} = \vec{\chi}_{lk}^{n} + \Delta t(\vec{U}_{lk}^{n}) + \lambda(\Delta t)^2 \vec{f}_{lk}^{n+1,*}$$

where **(2)**

$$\vec{f}_{lk}^{n+1,*} = g(\vec{\chi}_{lk}^{n+1,*}, \left|\vec{\chi}_{lk}^{n+1,*}\right|, T_m)$$

where $\vec{f}_{lk}^{n+1,*}$ is the force in a fiber on the structure at the intermediate step.

3. Calculate the force the structure has on the fluid

$$\vec{F}^{n+1,*}(\vec{x}) = \sum_{lk} \vec{f}_{lk}^{n+1,*} \delta_h(\vec{x} - \vec{\chi}_{lk}^{n}) \Delta a \Delta s \tag{3}$$

4. Solve the Navier-Stokes (incompressible fluids) Equations

$$\frac{\partial \vec{u}}{\partial t} + (\vec{u} \bullet \nabla \vec{u}) = -\frac{1}{\rho}\nabla p + \frac{\mu}{\rho}\nabla^2 \vec{u} + \frac{1}{\rho}\vec{F}^{n+1,*} \tag{4}$$

$$\nabla \bullet \vec{u} = 0$$

5. Update the position of the structure

$$\vec{\chi}_{lk}^{n+1} = \vec{\chi}_{lk}^{n} + \Delta t \sum_{\vec{x}} \vec{u}^{n+1} \delta_h(\vec{x} - \vec{\chi}_{lk}^{n}) h^3 \tag{5}$$

The three-dimensional Navier-Stokes equations are solved by an extension of Chorin's projection method,[8] which was extended to include a body force term.[9] While this type of solution provides a straightforward, efficient solution to the problem by use of Fast Fourier Transforms, it restricts the maximum cell Reynolds number to a value of two for stability.

2.2 Computational methods: left ventricle contractility

The cardiac tissue is represented as a set of totally immersed, neutrally buoyant, infinitely thin springs. The interaction between the fluid and the tissue is described by using local body forces on the fluid at the location of the contact point with the surface, rather than by prescribing boundary conditions at the contact point. The magnitude and direction of the body forces are determined by the tissue configuration (i.e. strain). Because the fluid is viscous, the velocity field can not have discontinuities, and since the structure is assumed to be infinitely thin and move at the velocity of the adjacent fluid, no slip can occur on either side of the tissue.

The 3-D computer ventricle floated in a 1 cm^3 fluid grid with 64 grid points along each direction. The computational length was scaled by a factor of 0.0725 (computational length equaled 0.0725 times physiologic length), as was the time scale. Thus, the velocity scale was unity. The computer model consisted of a thin-walled left ventricle and aorta. The aorta was functionally simulated by a capped cylinder with a flow sink located at its distal end. The computer left ventricle consisted of two layers, or sets of fibers, which ran orthogonal to one another. One set of fibers wrapped circumferentially around the left ventricle, while the other set of fibers ran in an apical-basal direction. This is in comparison to the actual heart which is thick-walled and has fibers running torroidally from the apex. Each computational left ventricular fiber consisted of a collection of points with each pair of adjacent points defined as a fiber link. The spatial distribution of the points on a fiber was such that the fiber link length never exceeded one half of a grid width. This minimum spacing between adjacent points was necessary to insure that the computer left ventricle did not rupture. The links on a fiber were either passive or contractile. Passive links represented passive tissue, such as chordae or mitral tissue; while contractile links represented muscle tissue. As in a native heart, muscle contraction was responsible for the systolic pressure rise. Also, the ventricular fibers were arranged so that the geometry of the computer left ventricle closely matched echocardiographic data of a normal human heart (Figure 1).

The two parameters which determined the biomechanical behavior of the passive cardiovascular tissue were also specified in the heart model. These two parameters were initial strain and the proportionality constant which related force to strain. The proportionality constant is calculated based upon an assumed Young's modulus for the material times the surface area associated with a structural point. The physiologic range of the physiologic Young's modulus is 10^7 dynes/cm^2 (elastin) to 10^{10} dynes/cm^2 (collagen), while strain of the mitral leaflet during early systole has been estimated as high as 15%. Any ventricular tissue not specified to be passive was specified to be contractile. The initial contractile fiber strain was set at 3%, while the proportionality constant was based upon a Young's modulus of 8×10^9 dynes/cm^2. The proportionality constant for the contractile tissue was not equal to that of passive tissue because the unstimulated muscle fiber is more

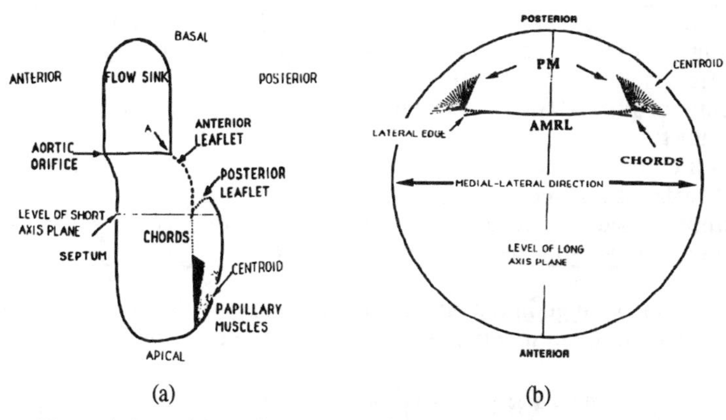

(a) (b)

Figure 1. Long (a) and Short (b) axis views of computational model.

elastic than mitral tissue. Contractile tissue was modeled by the same set of equations used for passive tissue, with one important difference. The resting length of a contractile tissue fiber was assumed to be a function of time; i. e. a contracting fiber was assumed to be functionally similar to a fiber with a decreasing resting length.

In the model, the left ventricular pressure could not be specified, but the degree of shortening for the individual fibers which compose the left ventricle could. The computational fibers produce the contraction force of the left ventricle by shortening which applies a force to the fluid, and thus produce a pressure field within the ventricle. The degree of shortening for a fiber link is the ratio of the link length at a time step n to its resting length, and is given by

$$C^n = \frac{\zeta^n}{\zeta^0} \equiv \text{degree of fiber shortening,}$$

$$\text{where } \zeta^0 \equiv \text{resting length of fiber link} \qquad (6)$$

$$\zeta^n \equiv \text{length of fiber link at time step n.}$$

In order to control the left ventricular pressure, it was necessary to formulate an equation to relate the degree of shortening for a fiber link to a desired left ventricular pressure. At the start of this procedure, the relationship between left ventricular pressure and degree of shortening is unknown. It was assumed that the change in ventricular pressure would be linearly related to the change in papillary muscle force and change in degree of shortening. A simple process control scheme was implemented to control the development of left ventricular pressure with degree of fiber shortening by a relationship between the change in ventricular pressure (p), change in papillary muscle

Table 1. List of Simulations and corresponding model dimensions (AML, anterior mitral leaflet;).

	Case 1	Case 2	Case 3	Case 4	Case 5	Case 6
Outflow Tract Area	5.2 cm^2	4.5 cm^2	4.5 cm^2	4.5 cm^2	3.2 cm^2	3.2 cm^2
Increase in Septal Wall Thickness	0	0	0	0	8 mm	8 mm
AML Chordal Slackness						
central	none	severe	moderate	moderate	none	moderate
lateral	none	severe	moderate	none	none	none

force (F), and change in degree of shortening (C). This relationship between parameters was determined by two adjustable parameters a_1 and a_2, and was given by

$$\left[\hat{p}^N - \hat{p}^{N-1}\right] = a_1\left[F^N - F^{N-1}\right] + a_2\left[C^N - C^{N-1}\right], \quad \text{for } N = n-2 \text{ to } n \tag{7}$$

$$\hat{p}^n = 16000 \sin\left[\frac{\pi}{280}(n+20)\right]^{.6425}, \quad \text{if } N = n \tag{8}$$

$$\hat{p}^{n-m} = \frac{1}{64}\left(\sum_{i=31}^{34}\sum_{j=31}^{34}\sum_{k=21}^{24} p_{i,j,k}^{n-m}\right), \quad \text{if } N \neq n \text{ for } m = 1,2 \tag{9}$$

note: superscripts in the above equations refer to the time step.

Equations 8 and 9 are used to calculate the needed pressures for equation 7. The parameters a_i were updated at each time step by a linear least squares fit by use of the previous time steps. Equation 8 defines the left ventricular sinusoidal waveform specified in the computer model. It was chosen because it simulates physiologic characteristics of left ventricular pressure during isovolumic contraction and early systole in that the left ventricular pressure equals the aortic pressure 50 ms after mitral valve closure and it peaks 140 ms after mitral valve closure.

3 Normal cardiac function simulation

3.1 Experimental Protocol

To examine the effectiveness of the thin-walled left ventricular model, simulations were performed during early systole, covering the isovolumic contraction (0 to 50 ms) and early systolic ejection (50 to 90 ms) phases of the cardiac cycle. All simulations start 20 ms after mitral valve closure with 0 ms denoting mitral valve closure and the beginning of isovolumic contraction. The time between 0 and 20 ms is used to pretreat the ventricle by increasing the left ventricular pressure to the desired level. The total simulation time is 70 ms (20-90ms). The dimensions used for the computational model are given in Table 1, with case 1 being the normal geometrical configuration of

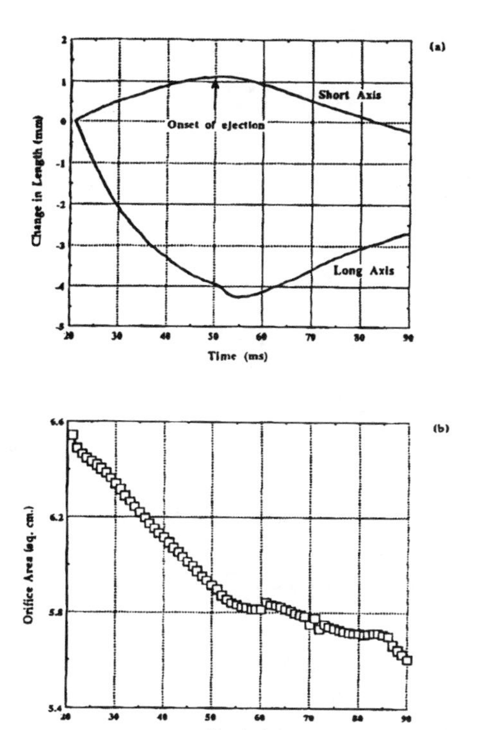

Figure 2. Change in length of (a) Left Ventricular Long and Short Axes and (b) Mitral Valve Orifice Area during early sytole.

the left ventricle and the other 5 cases being those of diseased geometries (described in the section 4).

3.2 Results

The validity of the model is demonstrated by the endocardial wall motion, and ventricular flow fields. It can be seen that the left ventricle undergoes dimensional changes during early systole (Figure 2). After the end of isovolumic contraction the long axis began to increase, while the short axis decreased. Figure 3 shows how the left ventricular structure points moved during both isovolumic contraction and systolic ejection. During isovolumic contraction the posterior wall moved basally and posteriorly, while the septal wall moved apically and anteriorly. In the short axis view (Figures 3c and 3d), it can be seen that during isovolumic contraction the increase in the short axis was not the result of a uniform wall motion away from the center of the heart. Rather, the septal wall was preferentially displaced away from the center of the ventricle. During ejection, the short axis of the left ventricle decreased and the long axis increased (Figure 3e). The long axis increased primarily because points on the posterior wall moved apically. The decrease in the short axis was the result of all of the ventricular walls moving inward toward the center of the left ventricle; however, the displacement of the anterior wall was approximately twice that of the posterior wall.

At the end of isovolumic contraction (t=50 ms), most of the flow field was moving from the apex toward the base. A recirculation region was observed in the small pocket formed by the concave anterior leaflet and disappeared during early ejection. It can be seen in Figures 4e and 4g that fluid in the anterior leaflet sinus was swept into the outflow tract at a velocity of 5 to 7 cm/s. The reverse flow region adjacent to the septal wall also began

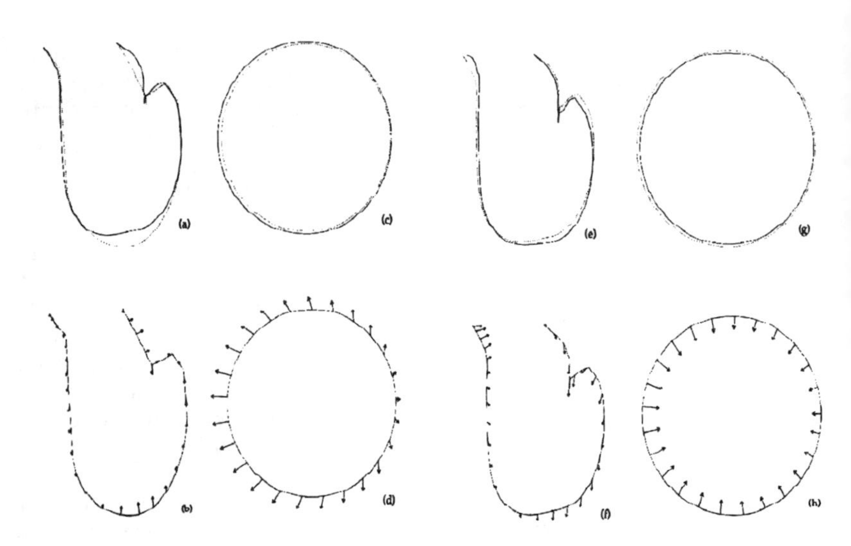

Figure. 3 Left Ventricular Wall Motion in Long and Short Axes between (a-d) t=20ms and t=50ms and (e-h) t=50 ms to t=90 ms.

to disappear during ejection. Apical to the mitral valve, a third reverse flow region was located adjacent to the left ventricular posterior wall and it was still present at t=90 ms. In this region, fluid moved apically from the vicinity of the posterior leaflet at a velocity of approximately 5 cm/s (Figures 4e and 4g). Prior to reaching the apex, some of the apically flowing fluid swirled toward the base and flowed into the outflow tract. During the first 40 ms of ejection, volumetric flow rate through the aortic orifice increased linearly from approximately zero to 200 cm^3/s. Therefore, one of the primary characteristics of the outflow tract flow field during ejection was temporal acceleration.

The other characteristic of the outflow tract flow field was spatial acceleration (Figures 4e and 4g). Fluid in the outflow tract spatially accelerated as it approached the aortic orifice. At t=70 ms, the average outflow tract velocities accelerated from 15 cm/s 20 mm below the aortic orifice to 30 cm/s in the aorta, and at t=90 ms the acceleration was from 35 to 70 cm/s. Spatial acceleration in the outflow tract was the result of decrease in the cross-sectional area perpendicular to the direction of flow, and the squeezing action of the left ventricle. The circumferential contractile force increased the volumetric flow rate along the outflow tract direction. The contribution of the ventricular contractile force to the spatial acceleration can be seen in Figures 4f and 4h. The flow pattern in the short axis plane during ejection can be characterized as one which converged on the outflow tract from all directions, with one exception. It can be seen in Figure 4f that the fluid adjacent to the anterior mitral residual leaflet (AMRL), and the AMRL

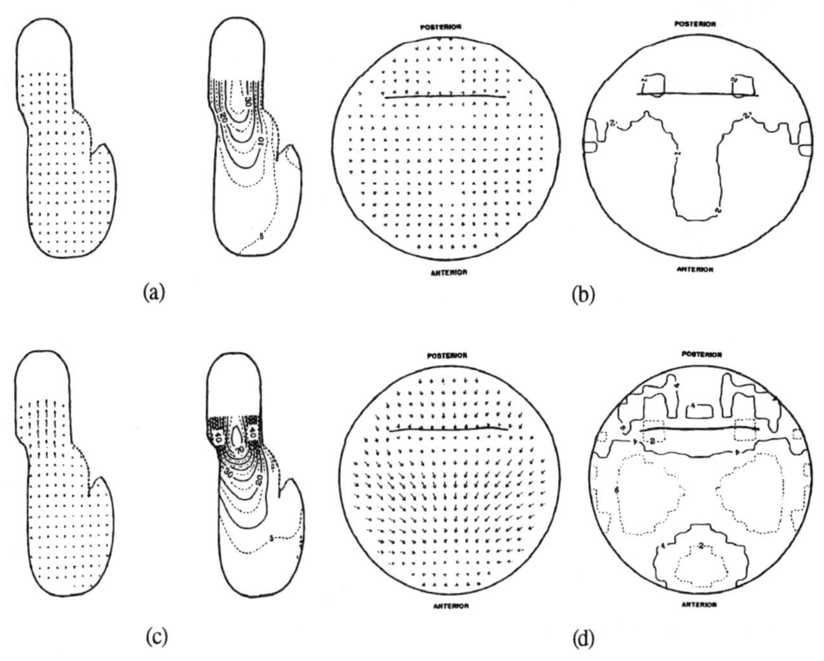

Figure 4. Vector and Contour plots of the Velocity Fields in the Long and Short Axes at (a-b) t=70ms and (c-d) t=90ms

itself, was moving posteriorly at a velocity of 1 to 2 cm/s during the first 30 ms of ejection. By t=90 ms, all of the fluid, including that between the AMRL and the posterior wall, was flowing toward the outflow tract (Figure 4h).

3.3 Discussion

During early systole, the human left ventricular long axis initially shortens 1 mm to 4 mm (isovolumic contraction), and then lengthens during early ejection.[10] It can be seen in Figure 2a that the computer left ventricle shortened approximately 4.2 mm during isovolumic contraction and then lengthened approximately 1.1 mm during early systolic ejection. An increase in the long axis implies that the apical portion of the ventricle was moving away from the base (Figures 3e-3f). The endocardial wall motion has been measured during ejection,[11] and it was found that the contraction of the native heart had three distinctive characteristics. First, virtually none of the contractile force is a pushing force, it is primarily a squeezing force. Second, the septal wall moved inward faster than the posterior wall. Finally, the basal

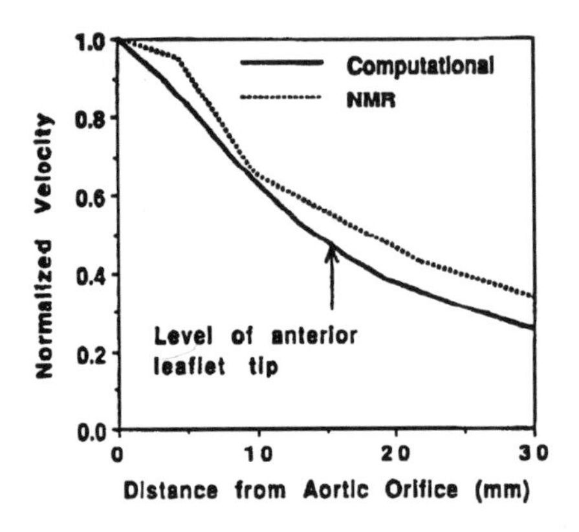

Figure 5. Comparison of Normalized Outflow Tract\ Velocities obtained by NMR and the Computational Model.

half of the left ventricle moves toward the apex during ejection. The present computer simulation of the left ventricle matched all three of these characteristics (Figures 3e-h). This agreement between physiologic and computational wall motion indicates that the contractile forces responsible for fluid flow were physiologic. This agreement validates the assumption about left ventricular function (a fixed ratio between squeezing and pushing forces).

The outflow tract flow field during systolic ejection has been measured clinically,[12] and it was concluded that the flow field has two dominant characteristics. First, the outflow tract velocity profile is relatively flat and skewed toward the septal wall. Second, there is rapid spatial acceleration in the outflow tract. Figures 4e and 4f show that the computational velocity profiles in the plane of the aortic orifice were also relatively flat and skewed toward the septal wall. To determine the variation between the computational and physiologic outflow tract field during ejection, computational velocities along the outflow tract centerline were compared to Nuclear Magnetic Resonance (NMR) imaging data using the phase velocity encoding technique. The computational and NMR velocities were normalized by their respective centerline velocities in the plane of the aortic orifice. The comparison established that the computational velocities along the outflow tract centerline were within 10% of those of the NMR data as shown in Figure 5.

Since the computer left ventricle was compliant (i.e. all the structures were free to move in response to fluid forces), agreement between the model and the native heart on structural motion would indicate that the forces altering the structure were physiologic. The simulation results were consistent with in vivo experimental and clinical data in the following key areas. In the heart, the mitral annulus contracts, which reduces the mitral orifice area. Agreement between native and computational annular contraction rates implies that the annular force exerted on the mitral valve was physiologic. The native mitral annulus begins to contract during late diastole and continues through late systole, and the rate of contraction is faster during late diastole and isovolumic contraction than during early systolic ejection.[13] It can be seen in Figure 2b that the mitral valve annular area decreased during the systole, with a faster rate of decrease during isovolumic contraction. The rate of mitral

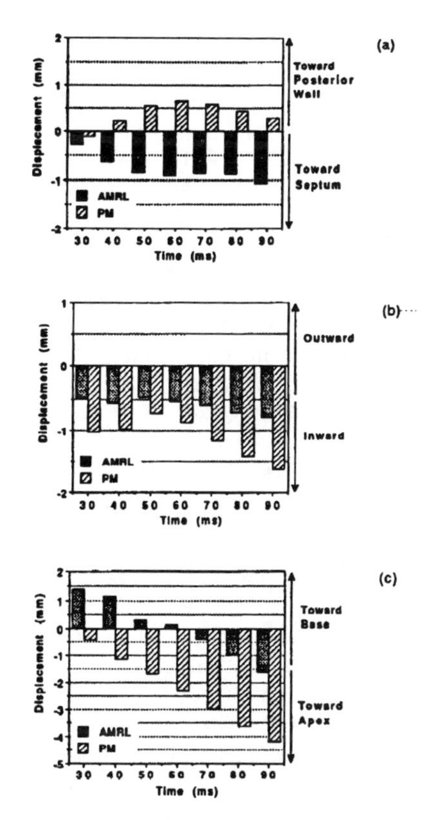

Figure 6. Displacement of AMRL and Papillary Muscles along (a) Anterior-Posterior, (b) Medial-Lateral, and (c) Apical-Basal directions

valve annular area decrease was about four times faster during isovolumic contraction than early ejection.

Also, the mitral valve-papillary muscle apparatus moved during the simulation. It is seen in Figure 6 that the tip of anterior leaflet was initially displaced 1.4 mm basally during isovolumic contraction, and then pulled apically during ejection. The displacement was in the direction of the strongest force acting on the mitral leaflet. The chordal forces pulled the leaflet tip apically, while the fluid pressure forces pushed it basally. The human mitral valve is also held in position against the transvalvular pressure forces during systole by the chordal forces. In a normal human heart, the chordal forces are not sufficient during isovolumic contraction to prevent a small basal displacement (2 mm) of the anterior leaflet tip, while during ejection, the chordal forces pull the leaflet tip apically.[14] Since the majority of forces acting on the mitral leaflet are along the apical-basal direction, this agreement between computer simulations and observations in the human left ventricle is very significant. It implies that the force balance on the computer mitral valve is physiologic. During systolic ejection (~ 280 ms), the papillary muscles also move in a normal human ventricle, and the papillary muscle tip separation in a human heart decreased 6 mm.[15] During the 40 ms of simulated systolic ejection, the papillary muscle tip separation decreased by 1.5 mm. Agreement in both magnitude and direction of the papillary muscle tip and AMRL displacement in the computer left ventricle indicates that the force balance on these structures was physiologic.

Chorin's algorithm is limited to simulating flows with a local cell Reynolds number of two or less. Thus, it was necessary to approximate the ventricular blood flow as a low Reynolds number flow. While this approximation is valid for early systole, it is possible that the local cell

Reynolds number limitation of the model could not accurately simulate peak or late systolic flow fields.

3.4 Conclusions

The computational results obtained from a thin-walled three dimensional left ventricular model during isovolumic contraction and early systolic ejection were consistent with the normal ventricular function of a human heart. The simulation produced velocity fields consistent with those occurring in a normally-contracting left ventricle, and endocardial wall motion which was consistent with human heart data obtained experimentally. The simulation results were also consistent with measurements of mitral valve-papillary muscle apparatus displacement, and the changes in the mitral valve orifice area. These observations support the conclusion that the use of the thin-walled, three-dimensional left ventricular model gives a realistic picture of left ventricular function.

4 Simulation of cardiac disease

4.1 Systolic anterior motion of the mitral valve leaflets

In systolic anterior motion of the mitral valve leaflets (SAM), the mitral valve leaflets move into the left ventricular outflow tract obstructing flow exiting the left ventricle. A symptom often associated with SAM is an abnormally narrowed outflow tract. In an effort to eliminate SAM, surgeons have widened the outflow tract.[16] The most widely accepted hypothesis for the initiation of SAM is the Venturi Mechanism.[16] The Venturi mechanism is as follows (Figure 7a): (i) A narrowed left ventricular outflow tract causes a high velocity stream during systolic ejection, which produces a low pressure region; (ii) the low pressure "sucks' the anterior mitral valve leaflet into the outflow tract; (iii) the mitral valve obstructs the exiting blood, increasing the outflow tract velocity. Reduction of early systolic ejection velocities, either by surgically widening the outflow tract, or by clinical treatment such as negative inotropes, will reduce the severity of SAM.

Unfortunately, the Venturi Hypothesis is too simplistic to explain clinical observations such as why SAM does not occur in all patients with hypertrophic cardiomyopathy (HCM),[15] and how SAM can occur in patients without a narrowed outflow tract . In addition to the above inconsistencies, other investigators have shown that sufficient papillary muscle restraint could prevent SAM, even with a severely narrowed outflow tract. Therefore, a reasonable hypothesis for SAM should include all of its characteristics and xplain all of the clinical observations. The altered papillary muscle-mitral valve geometry (PM-MV) hypothesis is an integrated mechanism which appears to accomplish this goal.[15] The hypothesis states that SAM is initiated in the following manner (Figure 7b):

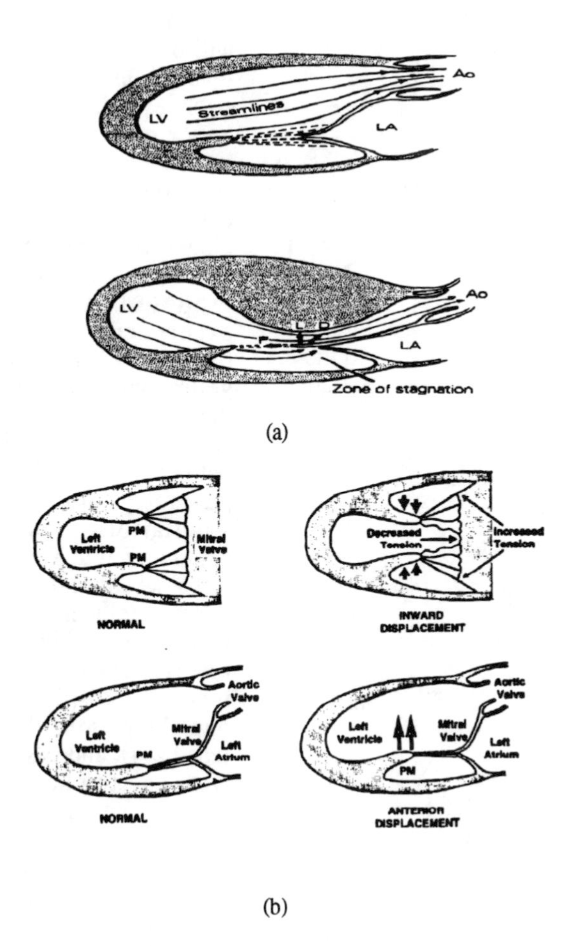

Figure 7. Schematics of (a) Venturi and (b) Altered PM-MV Geometry Hypotheses for the initiation of SAM.

(i) abnormal chordal tension distribution on the mitral valve causes chordal slackness; (ii) abnormal coaptation of the mitral valve will result, forcing it to coapt closer to the base of the anterior leaflet, which increases the fraction of the anterior leaflet exposed to left ventricular fluid forces and position the surplus leaflet closer to the outflow tract; (iii) the central-slack, lateral-restrained tension configuration will cause anterior buckling of the surplus leaflet when papillary muscle tension increases; (iv) the cowl faces directly into the path of the exiting blood and is drawn upward by drag forces; (vi) once initiated, SAM may be propagated by the drag forces and/or the Venturi Mechanism.

4.2 Experimental protocol

In the simulations of SAM, the extent of chordal slackness on the anterior mitral leaflet (AML) was varied. The degree of slackness is classified as either severe or moderate, depending on when force developed in the chords attached to the anterior residual leaflet. Chordal slackness that ispresent at

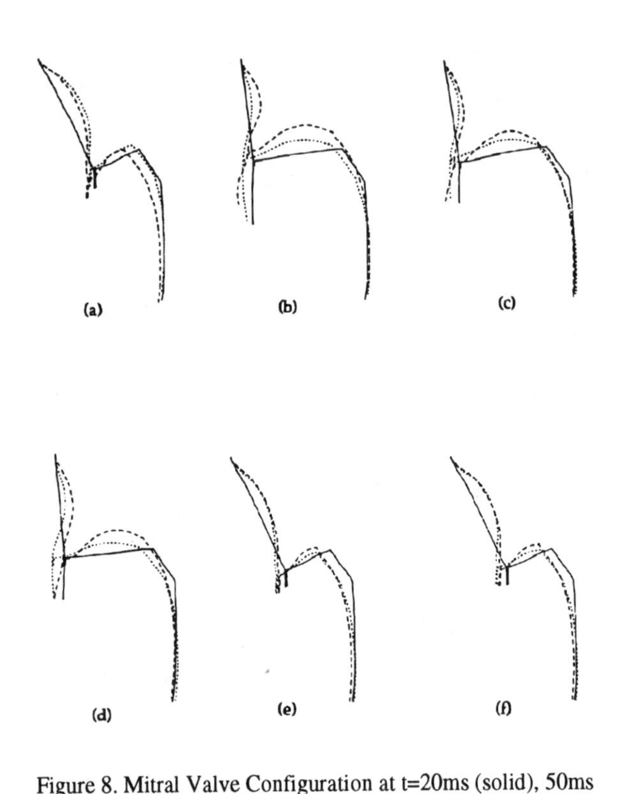

Figure 8. Mitral Valve Configuration at t=20ms (solid), 50ms (dotted) and 90ms (dashed) for (a) case 1, (b) case 2, (c) case 3, (d) case 4, (e) case 5 and (f) case 6. This shows the motion of the valve during early systole.

the onset of ejection will be referred to as severe, while chordal slackness which is only present during isovolumic contraction will be referred to as moderate. Since the intent of the proposed set of experiments is to only quantify geometric factors, the ejection flow rate will not be varied between cases. The geometry for the simulations are summarized in Table 1. The simulations conducted were the geometry involving anteriorly and inwardly displaced papillary muscles (case 2-4), and cases with a thickened septum (cases 5-6). Case 1 was presented in the previous section and is to be used as a comparison to normal left ventricular geometry / function.

4.3 Results

Examination of Figure 8 shows that SAM was initiated in a similar manner in cases 2, 3 and 4 during ejection, while cases 5 and 6 do not exhibit this pattern. The figure also shows that SAM could not be seen in the long axis view until after the onset of ejection (50 ms). During ejection, the coapted portion of the posterior and anterior leaflets moved toward the posterior wall, while the anterior leaflet tip moved toward the outflow tract. One exception to this pattern is case 4, where SAM occurred because the line between the coaptation point and the papillary muscle tips moved away from the anterior mitral residual leaflet (AMRL) tip.

The displacement of the AMRL was a function of the forces acting on the AMRL. In general, the magnitude of the AMRL displacement was inversely

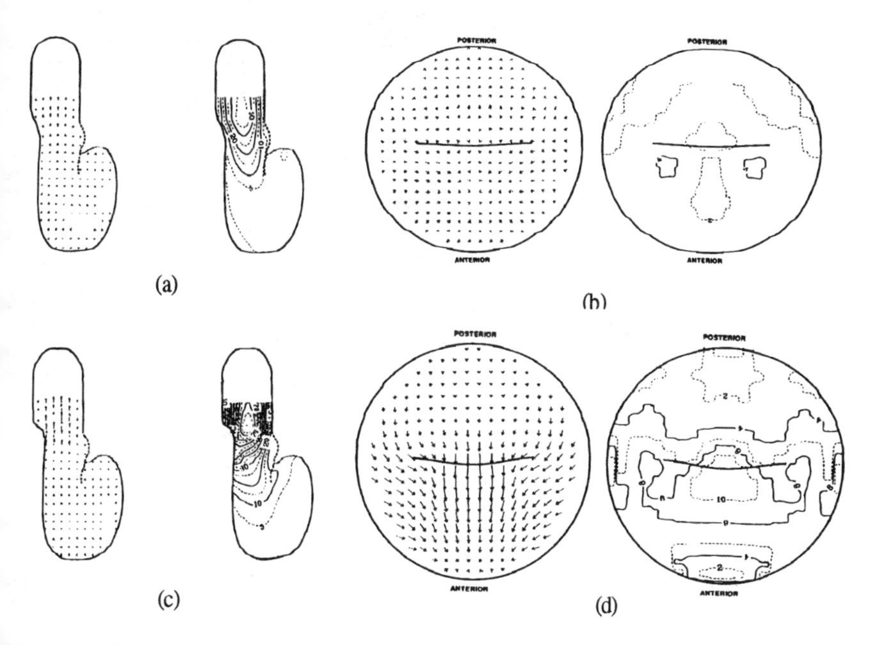

Figure 9. Velocitiy Fields in Long and Short Axes for Case 2 at (a-b) t=70ms and (c-d) t=90ms.

proportional to the chordal forces on the AMRL. In case 2 the AMRL moved 3 mm toward the base and 2.8 mm toward the septum. Conversely, case 4 had the largest amount of AMRL chordal restraint and the smallest AMRL displacement. The AMRL in case 4 moved only 1.5 mm toward the septum and its peak basal displacement was less than 1 mm. In cases 3 and 4, the papillary muscles were pulling the AMRL lateral edges inward during most of simulated systole, unlike case 2. In cases 3 and 4 the lateral edges of the AMRL was pulled 3 mm inward, versus 1 mm in case 2.

The velocities responsible for displacing the AMRL toward the septal wall in Case 2 can be seen in both the long and short axis planes. Fluid from a region posterior to the AMRL flowed toward the outflow tract in a direction which was approximately perpendicular to the AMRL surface, as shown in Figures 9 a-b. These velocities which pushed the AMRL toward the outflow tract increased during ejection. At t=90 ms, anteriorly-directed velocities at the AMRL tip were approximately 10 cm/s. In Figure 9b, it appears that some of the increase in the anteriorly-directed velocities was created by fluid which was pushed from the lateral edges of the AMRL toward the central portion. There were insignificant differences between the velocity fields of

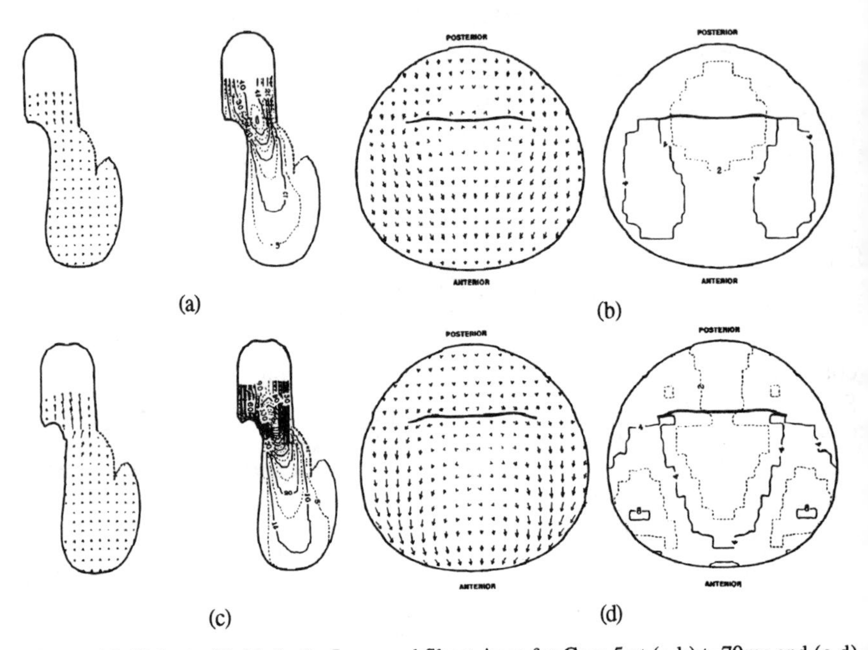

Figure 10. Velocity Fields in the Long and Short Axes for Case 5 at (a-b) t=70ms and (c-d) t=90ms.

cases 2 and 3.

In cases 5 and 6, the septum was thickened by positioning the septal wall 8 mm closer to the posterior wall than in the normal case. This decreased the distance between the central portion of the AMRL and the septum by 8 mm and narrowed the outflow tract area to 62% of the normal case outflow tract. In case 6, the papillary muscles were also positioned 8.4 mm inward of the normal case (and case 5) position. Despite the inward positioning of the papillary muscle, the AMRL remained relatively flat in both cases.

Since the chordal force distribution was not identical between the two cases, the AMRL motion varied between cases 5 and 6. During simulated systole, the papillary muscles and the lateral edges of the AMRL moved inward. In both cases the papillary muscles were displaced approximately 1.5 mm inward. In case 6, the inward displacement of the AMRL was 5 times that of case 5 (2.5 mm versus 0.5 mm). In addition, neither the papillary muscles nor the AMRL moved significantly along the apical-basal direction. In contrast, both the papillary muscles and AMRL were displaced apically approximately 0.8 mm and 1.1 mm respectively prior to the onset of ejection in case 5.

For case 5, the thickened septum narrowed the outflow tract, which increased the ejection velocities. In Figures 10a-b, it can be seen that the outflow tract velocities were skewed slightly toward the anterior wall. Also it can be seen in Figure 10b that there were posteriorly-directed velocities of 2 cm/s at the central portion of the AMRL, while fluid at the lateral edges flowed anteriorly at 4 cm/s. The velocity magnitudes in the region adjacent to the AMRL were 2 cm/s to 4 cm/s higher in case 6 than in case 5.

4.4 Discussion

The computer left ventricle was developed to gain insight into the initiation mechanism of SAM. Specifically, the model quantified how ventricular abnormalities associated with SAM affected the intraventricular flow field during the initiation of SAM. These flow field changes in a contractile left ventricle has been poorly defined previously since other researchers have concentrated on quantifying the outflow tract flow field after the initiation of SAM.[17] These researchers have attempted to ascertain the initiation mechanism of SAM based on the outflow tract velocities and pressures after SAM has initiated, rather than examining the entire intraventricular flow field during initiation. Abnormal outflow tract velocities after the initiation of SAM may be a symptom, rather than its cause. Regardless of why SAM initiates, abnormally high outflow tract velocities will always occur during the propagation of SAM Therefore, the initiation flow field must be characterized in order to gain insight into the forces responsible for the initiation of SAM.

<u>Role of Abnormal Chordal Force Distribution</u>

It has been shown that SAM is prevented by tension in the chords attached to the central portion of the AMRL. This tension caused fluid to flow parallel around the AMRL. Therefore, in the absence of central chordal tension on the AMRL fluid should flow in a direction perpendicular to the anterior leaflet tip. In case 2 (Figure 9a), the fluid flowed perpendicular to the leaflet tip because the central portion of the AMRL was free to move anteriorly. The combined results of cases 1 and 2 establish that the amount of chordal force applied to the central portion of the AMRL determined the flow direction in the near AMRL region.

<u>Relationship Between Central Chordal Force and the Degree of SAM</u>

The relationship between the degree of SAM and the amount of central chordal force has been unknown. The computational model made the measure of the chordal force possible. Case 2 had the least central chordal force, while case 4 had the largest. Correspondingly, the degree of SAM was largest in case 2 (2.5 mm), smaller in case 3 (2.0 mm), and smallest in case 4 (1.8 mm). This indicates that the degree of SAM is inversely related to central chordal force, which supports the PM-MV hypothesis.

<u>Role of Inwardly Positioned Papillary Muscles</u>

Positioning the papillary muscles inward introduces slackness in the centrally attached chords while maintaining chordal tension in the laterally attached chords. When the lateral edges are pulled inward, as in case 4, fluid is funneled from the lateral edges of the AMRL toward the centerline. The fluid pushed toward the centerline of the left ventricle was forced out of the region posterior to the AMRL. Fluid exiting this region pushed the AMRL tip toward the septum. Therefore, the inward pulling action of the papillary muscles enhanced the fluid forces responsible for SAM by funneling fluid from the lateral edges of the AMRL toward the central portion of the leaflet.

The effect of funneling fluid toward the central portion of the AMRL was not as pronounced in case 6. This was primarily a result of differences in the AMRL area. In case 6, the area of the AMRL was half that of case 4. Therefore, the volume of fluid affected by the inward pulling action of the AMRL was half of that of case 4. This indicates that the degree of SAM would be influenced by the residual leaflet length. These results also support the belief of other investigators that the severity of SAM is influenced by residual leaflet length.[18]

Effect of Anteriorly Positioned Papillary Muscles

Several researchers have observed that the mitral coaptation zone is displaced anteriorly in SAM patients.[15] If the papillary muscles and chords were unchanged, the result of either factor would be to increase slack in the chords attached to the posterior leaflet. In the simulations, virtually all of the papillary muscle force applied to the posterior leaflet opposed prolapse, or basal displacement. If chordal restraint on the posterior leaflet were decreased, the posterior leaflet should be displaced into the left atrium, not toward the outflow tract. Therefore, neither excessive posterior leaflet tissue nor abnormal annular contraction could shift the coaptation zone closer to the outflow tract. The present results indicate that only anterior positioning of the papillary muscles could create a force which would pull the coaptation zone closer to the outflow tract, which was predicted by in the altered PM-MV geometry theory.[15]

Effect of Septal Geometry

The outflow tract is bounded by the anterior leaflet, the septal wall and the lateral walls. Thickening the septum decreased the ability of the computational septal wall to generate a contractile force. Flattening of an hypertrophic septum also occurs in the human heart for the same reason, decreased contractility. It should be noted that septal flattening during ejection does not occur normally, either in the human heart or the simulations. A change in the ventricular contraction pattern changed how fluid entered the outflow tract. In the normal simulation, fluid entered the outflow tract by flowing in a direction perpendicular to the ventricular centerline. In cases 5 and 6, blood entered the outflow tract primarily by flowing in a direction parallel to the ventricular centerline (Figure 10). This implies that if the ejection flow rate had been allowed to vary between cases, the early ejection rate would have been higher in cases 5 and 6 than in the normal case. It has been shown that one of the characteristics of HCM is an abnormally high early ejection flow rate.[17] The results of the cases 5 and 6 suggest that contractile force is greater during early ejection because of a change in ventricular wall motion. Therefore, the increased early ejection flow rate observed in hypertrophic cardiomyopathic patients could be related to fundamental changes in how the ventricle contracts. Since there was less chordal restraint on the central portion of the AMRL in case 6, the rate a which the AMRL was pushed posterior was approximately twice that of case 5.

4.5 Conclusions

It was found that the intraventricular flow field was substantially altered by changes in the chordal force distribution. In the absence of chordal force on the central portion of the AMRL, fluid flowed from a region posterior of the AMRL toward the outflow tract in a direction perpendicular to the AMRL. This fluid pushed the central portion of the AMRL into the outflow tract. Once the AMRL was positioned in the outflow tract, it altered the outflow tract flow field. The high velocity outflow tract flow field was shifted adjacent to the septal wall.

Also, it was demonstrated that a thickened septum altered how the ventricle contracts. Because the ventricular contraction pattern was altered, the rate that fluid was forced toward the aorta increased. Since the ejection flow rate was not allowed to vary between cases, the ejection flow was choked in the thickened septal simulations. This created a flow pattern which pushed the AMRL toward the posterior wall.

First, the simulation results indicate that fluid forces necessary for the initiation of SAM were always present in the normal human ventricle. Second, it was shown that SAM was initiated as the result of a fluid lift forces which are unopposed by chordal tension. Third, the inverse relationship between chordal forces on the central portion of the AMRL and degree of SAM was established. Finally, the importance of inwardly and anteriorly positioned papillary muscles was shown.

5 Conclusions

A method has been presented and validated for the computational study of cardiac function. An anatomically-correct, thin-walled left ventricle model was constructed to examine early systolic function for both normal and diseased cases. Simulations of normal cardiac function provided a validation of using a thin-walled approximation of the thick-walled cardiac muscle, and showed that the results agreed well with those seen clinically. Next, an investigation into the initiation of SAM of the mitral valve leaflets was performed using this model. These studies showed that the mechanism by which the disease was believed to propagate is incomplete, and a newer methodology should be used when examining patients with this disease so that the true cause of the disease is treated and not just a symptom that is associated with it.

Acknowledgments

This research was made possible by a grant from Cray Research Inc. for computer time on the CRAY YMP located at the Pittsburgh Supercomputing

Center. The financial support of Medtronic Inc. and the Department of Energy for Graduate Student Fellowships is also acknowledged.

References

1. Greenfield, J. & Kolff, W. (1972) The prosthetic heart valve and computer graphics, JAMA 219, 69-74.

2. Underwood, F. N. & Mueller, T. J. (1979) Numerical study of the steady axisymmetric flow through a disk-type prosthetic heart valve in an aortic shape chamber, J. Biomechan. Eng, 101, 198-204.

3. Hung, T. K. & Schuessler, G. B. (1977) An analysis of the hemodynamics of the opening of aortic heart valves, J. Biomech., 10, 597-606.

4. Rogers, S., Kutler, P., Kwak, D. & Kiris, C. (1989) Numerical simulation of flow through an artificial heart, Proceedings of the Fourth International Conference on Supercomputing, L Karatashev and S Karatashev, eds. International Supercomputing Institute Inc., St. Peterburgs, pp.1-20.

5. Peskin CS (1972) Flow Patterns Around Heart Valves: A Digital Computer Method for Solving the Equations of Motion, PhD Thesis, Albert Einstein College of Medicine.

6. Peskin CS, & McQueen DM (1989a) A three-dimensional computational method for blood flow in the heart: I. Immersed. elastic fibers in a viscous incompressible fluid, J Comp Physics, 81:372-405.

7. Peskin CS, & McQueen DM (1989b) A three-dimensional computational method for blood flow in the heart: II. Contractile fibers, J Comp Physics, 82:289-97.

8. Chorin, A. J. (1967) A numerical method for solving incompressible viscous flow problems, J. Comp. Physics, 2:12-26.

9. Peskin CS (1977) Numerical analysis of blood flow in the heart. method, J Comp Physics, 25:220-252.

10. Hawthorne, E. W. (1961) Instantaneous dimensional changes of the left ventricle in dogs, Circ. Res., 9:110-119.

11. Slager, C. J., et al. (1986) Quantitative assessment of regional left ventricular wall motion using endocardial landmarks, J. Am. Coll. Cardiol., 7:317-326.

12. Goldberg, S. H., et al. (1988) Doppler Echocardiography, 2nd edition, Lea and Febiger, Philadelphia, pp.45-65.

13. Tsakiaris, A. G., et al. (1971) Size and motion of the mitral valve annulus in anesthetized intact dogs, J. Appl. Physiol., 30:611-618.

14.Karas, S. & Elkins, R. C. (1970) Mechanism of function of the mitral valve leaflets, chordae tendineae and left ventricular papillary muscles in dogs, Circ. Res., 26:689-696.

15. Jiang L, Levine RA, King ME, & Weyman AE (1987) An integrated mechanism for systolic anterior motion of the mitral valve in hypertrophic cardiomyopathy based on echocardiographic findings, Am Heart J, 113:633-44.

16. Wigle ED, Adelman AG, & Silver MD (1981) Pathophysiological Considerations in Muscular Subaortic Stenosis, Wolstenholme GEW, O'Connor M, eds. Hypertrophic obstructive cardiomyopathy CIBA Foundation Study Group No. 37, London: JA Churchill , Ltd, 63-76.

17. Jenni R, Ruffmann K, Vieli A, & Anliker M (1985) Dynamics of aortic flow in hypertrophic cardiomyopathy, European Heart J, 6:391-98.

18. Shah PM, Taylor RM, & Wong M (1981) Abnormal mitral valve coaptation in hypertrophic obstructive cardiomyopathy: Proposed role in systolic anterior motion of mitral valve, Am J Cardiol, 48:258-62.

Chapter 3

Numerical modelling of blood flow in compliant arteries and arterial bifurcations

X.Y. Xu & M.W. Collins
Thermo-Fluids Engineering Research Centre, City University, London, UK

Abstract

The study of blood flow in the larger arteries presents interesting and challenging problems. Such flows are pulsatile, and pass though vessels which are complex both in geometry and their compliant nature. The complete cardiovascular system is far too complicated to be amenable to a comprehensive analytical treatment. However, recent advances in Computational Fluid Dynamics (CFD) and computer hardware have brought the numerical modelling power closer to a stage where real physiological fluid flow problems can be addressed. In this chapter, we describe a novel numerical modelling approach for the coupled solution of the equations of motion for the flow and wall in the larger arteries. The numerical model is able to solve for three-dimensional, unsteady flows with complex, moving boundaries, as well as the full, time-dependent displacement and stress fields. A thorough treatment of code validation aspects is presented. This is followed by analysis of flow in a compliant artery with various EDRF-related interventions, and flow in a distensible bifurcation model.

Nomenclature

a Internal radius

b	External radius
D	Diameter
d_n	Displacement (n=1,2,3)
E	Young's modulus
f	Frequency
P	Pressure
Q	Volume flow rate
R	Radius
r	Radial coordinate
Re	Reynolds number ($Re = \frac{\rho \bar{u} D}{\mu}$)
t	Time
t_p	Pulse time
u_n	Velocity (n=1,2,3)
x_n	Coordinates (n=1,2,3)
α	Womersley number ($\alpha = \frac{D}{2}\sqrt{\frac{\rho\omega}{\mu}}$)
Δt	Time step
μ	Viscosity
ν	Poisson's ratio
ρ	Density
σ_{ij}	Stress
τ	Shear stress
ω	Angular frequency

1 Introduction

1.1 Arterial fluid dynamics

There are many factors which complicate the fluid dynamics of the larger arteries. One of the complexities is that the vessels in which blood flows are often curved and branched, producing such phenomena as secondary flows, entrance effects and flow separation. The distensibility of the blood vessels, which deform in a viscoelastic fashion, and the non-Newtonian behaviour of blood, also add further complications. However, the major factor is the unsteady or pulsatile nature of the flow. At any given point in the arterial system, the blood pressure and velocity change periodically according to the rhythmic pumping action of the heart. Typical velocity and pressure waveforms for the femoral artery in a dog are shown in Fig.1.

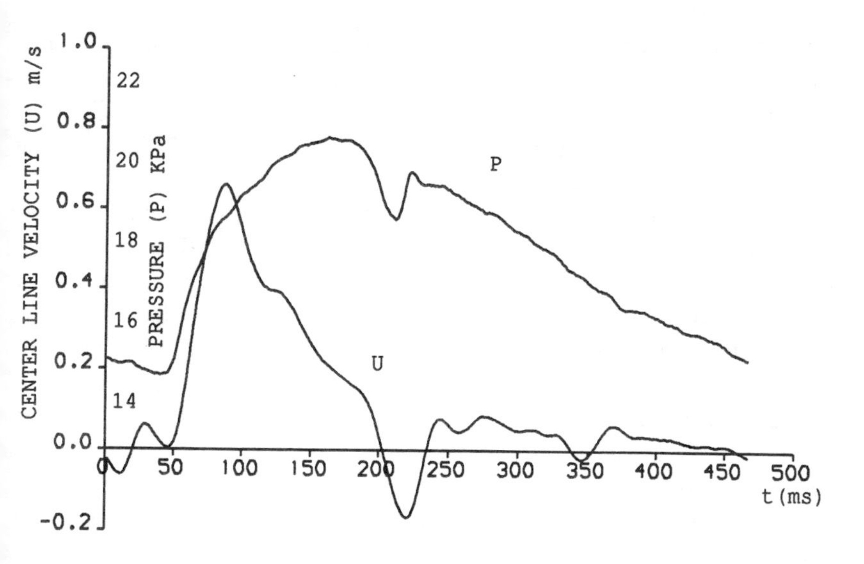

Figure 1: Typical velocity and pressure waveforms for a canine femoral artery.

Studies devoted to understanding the dynamics of blood flow date back centuries. However, it is only in the last several decades, motivated by the possible role of the haemodynamics in early development of atherosclerosis, that attention has been focused on the detailed characteristics of blood flow. Although such flows are still beyond fully detailed understanding at present, much has been learned through extensive experimental and numerical studies. The fluid dynamics features which are found to be related to flow in the arterial system can be summarised as follows.

(a) The nature of the flow is by and large laminar.[1] For normal arteries, i.e., without any disfigurement due to disease, the only exception is the aorta where turbulence could occur at certain circumstances. Once blood leaves the aorta and travels down to the next level of vessel, e.g., the carotid, coronary and femoral arteries, the Reynolds number becomes an order of magnitude less and the flow is laminar throughout the cycle.

(b) The flow is substantially three-dimensional with secondary flows in curved tubes and branches.[2] These are fluid motions which occur in a plane perpendicular to that of the main axial flow and thus are secondary to the primary direction of motion. They are characterised by the spiral flow pattern in the bifurcation plane and counter-rotating vortices in the transverse plane. Secondary flows are induced by vessel branching and curvature where flow is forced to change direction.

(c) The flow in branches has the tendency to separate from the wall, giving rise to a recirculating, 'dead-water' region.[3] This is because of the adverse pressure gradient associated with the deceleration of the fluid as it passes from the parent vessel into branches whose combined cross-sectional area exceeds that of the parent vessel. An example of separated flow in a model T-junction is illustrated in Fig.2. For a upstream Reynolds number of 245 and a flow division of $Q_1/Q_2 = 0.25$, two flow separation zones are clearly present: a large separation zone in the main tube at the entrance to the daughter tube; and a small secondary corner separation zone at the entrance to the side branch. It should be noted that due to the pulsatile nature of the flow, the occurrence of flow separation in the arterial system is an unsteady phenomenon. The recirculation region may change location and size or even disappear and then reappear as the flow pulses.

(d) Wall shear stress not only varies with time, but also with space.[4] The spatial variation is associated with the asymmetric velocity patterns produced by the asymmetric geometry, including vessel branching and curvature. For example, at a bifurcation the flow in the upstream parent

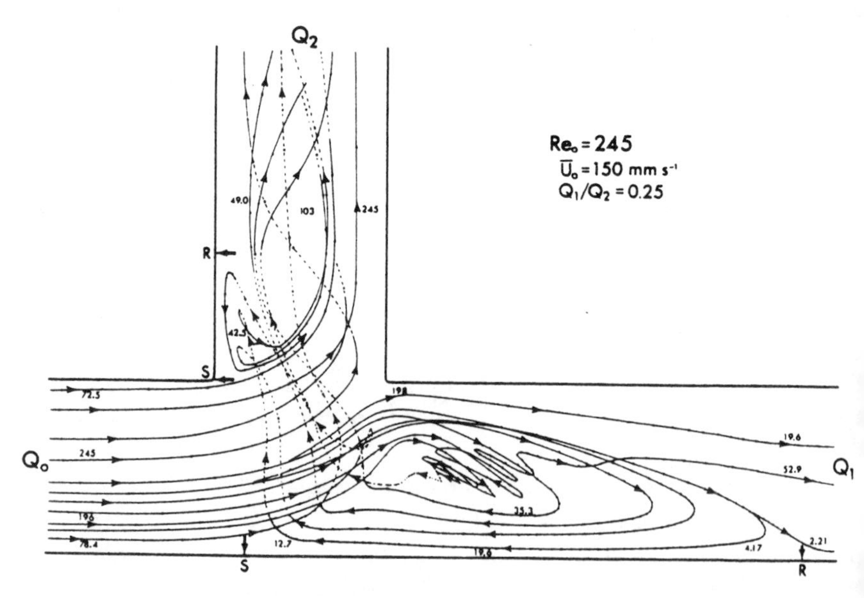

Figure 2: Particle paths in a model T-junction (from Karino *et al*[11]).

vessel divides into the two daughter vessels so as to bring relatively high velocity blood at the centre of the parent vessel in close proximity to the wall of the flow divider. The result is that the velocity profiles are skewed towards the flow divider, producing higher shear stress on the divider wall and lower shear stress on the outer wall.

In concluding this section, it should be emphasised that our knowledge of the fluid dynamics characteristics in the larger arteries, especially at bifurcations, has mainly been derived from theoretical and experimental studies which often involve various degree of simplification. A better understanding should be achieved by taking into account such physiological phenomena as wall compliance, nonlinear wave travel and flow-dependent dilatation.

1.2 Vessel wall mechanics

Blood vessel walls are viscoelastic inhomogeneous multi-layered tissues. They are composed mainly of endothelial lining, elastin, collagen and smooth muscle arranged in a complex way. The endothelial lining is the innermost layer of the vessel wall that comes in contact with the flowing blood. It provides a smooth surface and offers a selective permeability to various substances carried in the blood stream but is too soft to contribute to the elastic properties of the wall. Elastin is a rubber-like material which is very extensible. In stretching of up to about 60% its unstressed length, elastin follows Hooke's law. However, for extensions beyond this point it will become highly nonlinear. Collagen is much stiffer than elastin and may be stretched by only 10-30% above its original length. The behaviour of collagen is non-Hookean, and exhibits some plastic deformation. Owing to the complicated structure, the stress and strain relation for arterial walls is generally nonlinear and viscoelastic.[5] An extensive literature exists on the arterial wall mechanics (see bibliographies in Bergel[6] and Fung[5]).

A large variety of mathematical models has been developed with various assumptions of arterial wall behaviour, such as, membrane, thick shell, elastic, viscoelastic, linear, nonlinear, small strain, and finite deformation. It has been found that in the actual circulation the radial dilatation of large arteries is up to 2-15%, while the longitudinal movement is very small because of the tethering of vessels to the surrounding tissues. For this reason, a uniaxial model is commonly used, which describes the wall motion in the radial direction only. In this case, the mechanical properties of arteries can be defined in terms of a distensibility or compliance, which is the fractional change of cross sectional area divided by the distending pressure.

The interactions between the flowing blood and arterial wall are complex. On one hand, there is the direct effect of the force of the flow on the wall. The total load imposed on the arterial wall by the flow of blood is manifested through both a normal component essentially equal to the fluid pressure and a tangential component, which is the wall shear stress. The macroscopic effect of these two haemodynamic forces is to distend and distort the wall and to produce stresses and strains within the wall itself. In this aspect it is the pressure, which is much the larger of the two forces, that plays the principal role. Microscopically, however, both the pressure and wall shear stress are of interest. This is associated with the increasingly recognised control mechanisms within the cardiovascular system. These mechanisms include: (a) the pressure dependence of wall elasticity, and (b) the phenomenon of flow-dependent dilatation, which is mediated by endothelium-derived relaxing factor (EDRF), a muscle relaxant released in response to changes in fluid shear stress.[7,8] The result of vasodilatation is to modulate the elastic properties of the artery.[9] The other side of the blood/arterial wall interactions is that changes in shape and size of the arterial wall will change the geometry which bounds the flow, resulting in altered velocity and stress distributions in the artery.

1.3 Earlier work and modelling approaches

There has been an enormous number of studies designed to understand the detailed characteristics of blood flow in larger arteries, particularly at arterial bifurcations. Experimental techniques including flow visualisation, velocity measurement and shear stress measurement have played an important role in obtaining both qualitative and quantitative information about the flow field of interest. Flow visualisation provides the best overall but the least quantitative picture. The methods normally used in the study of arterial flow are: hydrogen bubble,[10] particle tracer,[11] dye injection,[12] and streaming birefrigence.[13] Velocity measurement has been useful in offering detailed and quantitative data. At the present time, the most widely used velocity measuring instruments are Laser-Doppler anemometry[4] and Doppler ultrasound,[14] which are both non-invasive and of relatively high resolution. Another non-invasive method is magnetic resonance imaging (MRI). Three-dimensional *in vivo* measurements have now become possible through technological developments in the fields of Doppler ultrasound and magnetic resonance imaging.[15] However, measurement of shear stress has proved difficult. Although approaches are available, such as direct measurement using a hot-film probe[16] or an electrochemical technique,[17] and indirect estimation using an extrapolation method,[18] they do not give accurate information for pulsatile flows, let alone *in vivo* conditions.

Theoretically, the pioneering and most extensive work was carried out by Womersley[19] based on a linearisation of the basic equations. In this analysis, the effects of wall elasticity, fluid viscosity and the coupling between fluid and wall motion have been taken into account, and the oscillatory velocity profile across the vessel due to an imposed sinusoidal pressure gradient are predicted as a function of the parameter α (defined as $\alpha = \frac{D}{2}\sqrt{\frac{\rho\omega}{\mu}}$). Womersley's analysis has provided a basis for understanding many aspects of unsteady flow in elastic tubes, as well as a useful set of analytical solution for validating numerical models.

In addition, there is an increasing role for computer modelling in our efforts to understand the fluid dynamics phenomena in large arteries, particularly at bifurcations. Earlier numerical simulations were confined to either two-dimensions or steady flow (see Xu & Collins[20] for a summary), primarily due to a lack of computing power at that time. It is only very recently that three-dimensional calculations of unsteady flow in a bifurcation have become possible. Dinnar *et al*[21] was probably the first to present such an attempt. While the finite difference method they used was found to be extremely efficient when compared to a finite element approach, it was restricted to simple bifurcation geometries with rectangular cross section. Perktold *et al*[22–25] carried out a series of studies of three-dimensional, pulsatile flow in rigid models of the human carotid artery bifurcation. The numerical model is based on the pressure correction technique and uses the Galerkin finite element method. The finite element discretisation employs eight-node isoparametric elements with tri-linear velocity approximation and constant pressure. The model incorporates a physiological velocity waveform at inlet, and a non-Newtonian viscosity model[24,25] described by the Casson's relation has been included. These simulations have been able to predict the expected oscillating wall shear stress at the outer sinus wall, and comparison of Newtonian and non-Newtonian calculations have revealed little differences in the basic flow characteristics.

Xu *et al*[26,27] have predicted three-dimensional flows through canine femoral bifurcations with a rigid wall assumption. The calculations were performed using a hybrid code ASTEC,[28] in which the discrete versions of the Navier-Stokes equations are derived using a finite volume approach, but are solved on a finite element mesh. The unstructured mesh consists of arbitrarily shaped eight-node elements. The procedure for solving the discrete equations is iterative and based on the SIMPLE scheme of Patankar & Spalding.[29] A preconditioned conjugate gradient algorithm is used to calculate the pressure corrections. A vector upwinding scheme is employed to minimise the false diffusion errors associated with the advection term. The models' geometry and inlet velocity waveform are based on data from *in vivo* measurements,[15] and non-Newtonian effects (described by the power law relation) have been taken into account. Numerical results demonstrate the maximum wall shear stress near the flow divider on the divider wall, and the lowest shear stress on the outer wall of the larger daughter artery. It has also been found that for the two models studied, variations of the wall shear stress generally follow their flow waveform in shape, with no rapid oscillation being noticed on the outer wall as shown to be present at the outer sinus of a human carotid artery bifurcation.[22] Comparison of the results for Newtonian and non-Newtonian simulations appear to agree with the findings of Perktold *et al*[24] in that the general features of the flow are little affected by the choice of the non-Newtonian model, although they could have local significance. Comparison between predictions and *in vivo* measurements has been made; while there is a reasonable agreement in general, local discrepancies still exist. It is postulated that this disagreement may be attributed to several factors, such as, the rigid wall assumption in the prediction, and the errors and uncertainties involved in the measurements.

Three-dimensional modelling of pulsatile flow in a compliant bifurcation has rarely been performed. Up to now, only two such publications are known to the authors. In the study presented by Reuderink[30] the problem was decoupled by first finding the wall motion due to the pressure distribution obtained from a one-dimensional wave analysis, and then applying the resulting wall motion as boundary conditions to a moving wall flow model. The wall was modelled as a thick shell, ignoring any variation of stress over the wall thickness. The flow and wall motions were calculated using the finite element method. In the study presented by Rappitsch *et al*,[31] a weakly coupled approach was used. The wall was modelled as a linearly elastic, geometrically nonlinear thin shell. Calculations of the wall displacements and stresses used the finite element code ABAQUS (version 4,9). Calculations of flow were performed using the pressure correction technique and the finite element method.

The authors' group are currently working on a project to develop a fully coupled numerical modelling approach for the wall-fluid problems in the cardiovascular system. We have recently been able to simulate the flow in a compliant tube by solving the equations of fluid and wall motion simultaneously using two different approaches, namely the finite volume method,[32] and the finite element method.[33]

2 The numerical model

Numerical modelling of arterial flow including wall motion is complicated by the necessity of solving the equations of motion for the flow and wall simultaneously. The pioneering study of flow in elastic tubes by Womersley[19] is an example of this approach. However, the various assumptions and approximations made in this study preclude it from being extended to complex geometries, such as arterial bifurcations, where atherosclerotic lesions are likely to occur. A general approach which can solve the coupled system of equations governing the flow and wall motion in arbitrary geometries is needed more than ever. We have investigated the possibility of developing such a comprehensive modelling approach using two different methods, namely the finite volume (FV) and finite element (FE) methods. Details of these are outlined below.

2.1 Finite volume method

In the FV method, the equations governing the unsteady, axisymmetric, flow of a Newtonian fluid are solved together with the equations governing the motion of a linearly elastic, incompressible tube undergoing axisymmetric infinitesimal strains. Mathematical details of the coupling procedure have been described by Henry & Collins,[32] and only a brief outline will be given here. Basically, the solution procedure used to solve the wall equations mirrors that for the fluid equations. The fluid equations are solved using CFDS, FLOW3D (release 2.4),[34] which is a general purpose code for the numerical solution of laminar and turbulent flows and heat transfer. FLOW3D uses the FV method to solve the governing equations on general, three-dimensional, non-orthogonal (body-fitted) grids. It allows both the grid and solution domain boundaries to move in some prescribed manner. The code solves the discrete equations using the velocity-pressure coupling schemes known as SIMPLEC,[35] a variation of the original SIMPLE scheme.[29] The Rhie-Chow algorithm[36] is incorporated into the solution procedure to allow the use of non-staggered grids. However, the standard version of FLOW3D requires the position of each moving surface to be a known function of time, while in the current problem the position of the interface is part of the solution. Hence, it was necessary to modify FLOW3D so that the routine in which the new wall position is defined could be called repeatedly, until some predetermined convergence condition was met. This was achieved by using a predictor/corrector scheme. For further details of the coupled procedure, readers are referred to Henry & Collins.[32,37]

2.2 Finite element method

In the FE method, three sets of equations are solved: (1) the full time-dependent Navier-Stokes equations governing the flow of a Newtonian fluid, (2) the linear-elastic small-displacement stress equations for the wall, and (3) the mesh displacement equations. Mathematical details of the coupled solution procedure are outlined below.

2.2.1 Fluid motion

The flow simulation is based on the time-dependent, three-dimensional Navier-Stokes equations for incompressible Newtonian flows with constant properties. Here the blood is assumed to be a Newtonian fluid because recent studies have found that the non-Newtonian effects on the basic features of the flow in large arteries are little. In a fixed grid system with Cartesian coordinates, the momentum and the continuity equations can be written as

$$\rho_f \frac{\partial u_n}{\partial t} + \rho_f u_m \frac{\partial u_n}{\partial x_m} - \frac{\partial}{\partial x_m}\left[\mu_f\left(\frac{\partial u_n}{\partial x_m} + \frac{\partial u_m}{\partial x_n}\right)\right] + \frac{\partial P}{\partial x_n} - F_n = 0 \quad (1)$$

$$\frac{\partial u_m}{\partial x_m} = 0 \quad (2)$$

where ρ_f is the fluid density, μ_f the molecular viscosity, u_n the velocity component in the x_n direction (n=1,2,3), and P is the pressure. The computational region at $t = t_o$ is denoted as Ω^o and that at time $t > t_o$ as Ω. Suppose any point in Ω^o with coordinates $\mathbf{X}^o$ moves to point $\mathbf{X}$ at time t, then $\mathbf{X}$ is related to $\mathbf{X}^o$ by

$$\mathbf{X} = \mathbf{X}(\mathbf{X}^o, t) \quad (3)$$

Defining the grid velocity $\mathbf{u^g}$ as

$$\mathbf{u^g}(\mathbf{X}^o, t) = \left(\frac{\partial \mathbf{X}(\mathbf{X}^o, t)}{\partial t}\right)_{\mathbf{X}^o} \tag{4}$$

means the derivatives of a variable ϕ can be modified as follows

$$\left(\frac{\partial \phi}{\partial x_n}\right)_t = \left(\frac{\partial \phi}{\partial x_m^o}\right)_t \Gamma_{mn}^{-1}(\mathbf{X}^o, t) \tag{5}$$

$$\left(\frac{\partial \phi}{\partial t}\right)_{\mathbf{X}} = -\left(\frac{\partial \phi}{\partial x_m^o}\right)_t \Gamma_{mq}^{-1} u_q^g + \left(\frac{\partial \phi}{\partial t}\right)_{\mathbf{X}^o} \tag{6}$$

where u_q^g is the grid velocity, Γ_{mn}^{-1} is the transformation matrix defined as

$$\Gamma_{mn}(\mathbf{X}^o, t) = \left(\frac{\partial x_m}{\partial x_n^o}\right)_t \tag{7}$$

Thus the usual Navier-Stokes equations defined on Ω can be transformed to Ω^o by modifying the derivative terms

$$\begin{aligned} & \rho_f \left(\frac{\partial u_n}{\partial t}\right)_{\mathbf{X}^o} - \rho_f \left(\frac{\partial u_n}{\partial x_m^o}\right)_t \Gamma_{mq}^{-1}(u_q^g - u_q) \\ - & \left\{\frac{\partial}{\partial x_q^o}\mu_f \left[\frac{\partial u_n}{\partial x_p^o}\Gamma_{pm}^{-1} + \frac{\partial u_m}{\partial x_p^o}\Gamma_{pn}^{-1}\right]\right\} \Gamma_{qm}^{-1} \\ + & \frac{\partial P}{\partial x_p^o}\Gamma_{pn}^{-1} - F_n = 0 \end{aligned} \tag{8}$$

$$\frac{\partial u_m}{\partial x_p^o}\Gamma_{pm}^{-1} = 0 \tag{9}$$

These equations are solved in the region Ω^o subject to appropriate boundary conditions.

The grid velocity u_q^g is determined by solving the mesh displacement equations. By viewing the original mesh as an elastic solid material having a zero Poisson's ratio (a very squashy sponge), the volume equations governing the displacements of the mesh are

$$\frac{\partial}{\partial x_m^o}\left(\frac{\partial u_m^g}{\partial x_n^o} + \frac{\partial u_n^g}{\partial x_m^o}\right) = 0 \tag{10}$$

Suitable boundary conditions for the above equations are

$\mathbf{u}^g = \mathbf{u}$	for $\mathbf{x}^o$ on $\partial\Omega_A{}^o$ (fluid/wall interface)
$\mathbf{u}^g \cdot \mathbf{n} = 0$	for $\mathbf{x}^o$ on $\partial\Omega_I{}^o$, $\partial\Omega_O{}^o$ (inlet, outlet)

2.2.2 Wall motion

It is known that the arterial wall undergoes finite deformation, i.e., the equations of classical (infinitesimal strain) elasticity cannot accurately describe the motion of the wall. Also, the stress and strain relationship for the arterial wall is generally nonlinear and viscoelastic. However, as a starting point, it is instructive to consider a simpler model before moving on to the more difficult problem of finite strain and nonlinear elasticity. The equations governing the motion of a linearly elastic, incompressible, isotropic solid undergoing small deformations are, in Cartesian coordinates

$$\rho_w \frac{\partial^2 d_n}{\partial^2 t} - \frac{\partial}{\partial x^o_m} \lambda \delta_{nm} \frac{\partial d_p}{\partial x_p} - \frac{\partial}{\partial x^o_m} G \left(\frac{\partial d_n}{\partial x^o_m} + \frac{\partial d_m}{\partial x^o_n} \right) = F_n \tag{11}$$

where d_n is the displacement in the x_n (n=1,2,3) direction, ρ_w the wall density, and λ and G are elastic constants which are related to the Young's modulus E and the Poisson's ratio ν by

$$G = \frac{E}{2(1+\nu)} \tag{12}$$

$$\lambda = \frac{\nu E}{(1+\nu)(1-2\nu)} \tag{13}$$

For a linear elastic material, the velocity of material particle is defined as

$$u_n = \frac{\partial d_n}{\partial t} \tag{14}$$

Substituting eqn(14) into eqn(11) yields the wall momentum equations which are compatible with those for the fluid

$$\rho_w \frac{\partial u_n}{\partial t} - \frac{\partial}{\partial x^o_m} \lambda \delta_{nm} \frac{\partial d_p}{\partial x_p} - \frac{\partial}{\partial x^o_m} G \left(\frac{\partial d_n}{\partial x^o_m} + \frac{\partial d_m}{\partial x^o_n} \right) = F_n \tag{15}$$

The displacement d_n at time t_{r+1} can be approximated by

$$d_n^{r+1}(\mathbf{x}^o, t) = d_n^r + \frac{(u_n^r + u_n^{r+1})}{2}\Delta t \tag{16}$$

Equations (8), (9) and (15) are solved simultaneously together with appropriate boundary conditions. Details of the numerical solution algorithm of these equations are discussed in the next section.

2.2.3 The coupled solution algorithm

The momentum equations for both the fluid and the wall are discretised using the Galerkin finite element method, with the continuity equations being incorporated by a penalty-augmented Lagrangian-multiplier (PALM) method.[38] It should be noted that no upwind weighting of the convective terms is employed. Twenty-noded brick elements with quadratic velocity and linear pressure variations are used in the whole region.

An implicit predictor/corrector method is used to solve the fully-coupled time-dependent equations. The predictor is the second order explicit Adams-Bashforth method and the corrector is the implicit Crank-Nicholson scheme, which is non-dissipative, completely stable and second order accurate. The non-linear system of algebraic equations is solved using the Newton-Raphson iteration, using a direct frontal solution method to handle the linear system at each iteration. The mesh displacement equations are solved separately at each time step to obtain updated grid velocities which will then be used to calculate the transformation matrix introduced in the modified fluid equations. By doing so, the transformed fluid equations are always solved on the original computational mesh, regardless of the fact that the physical grid is moving.

The above coupled solution procedure has been implemented in a special version of a FE-based computational fluid dynamics code FEAT.[39] Currently the wall model is limited to linear elastic and small strains, but work is under way to extend it to allow nonlinear elastic and large strains.

3 Code validation

In order to validate the modelling approaches described above, a careful validation exercise has been carried out using well-established theoretical results. This has been done in three stages: (1) decoupled solution for wall displacement, (2) decoupled solution for flow field with moving boundary, and (3) coupled solution for wall motion and fluid flow. For each case, numerical predictions have been attempted using both the FV and FE codes.

3.1 Decoupled model testing - wall movement of an incompressible elastic tube

For the decoupled solid model, the problem considered is that of a tube with an internal diameter of $10mm$, and a wall thickness of $1.0mm$. Young's modulus and density for the wall were taken to be $5 \times 10^5 Pa$ and $1.0 \times 10^3 kg/m^3$ respectively. The internal and external pressures were set to $2668Pa$ and zero, respectively.

It was first assumed that the cylinder was infinitely long. Hence, no axial boundary conditions were required, and the problem reduces to that of finding the distribution of displacement and stress in the radial direction. The equilibrium distributions of radial displacement and stress for a thick-walled cylinder having an internal radius a and external radius b, under uniform pressure are given by[40]

Radial displacement

$$d_r = \frac{a^2 P}{(b^2 - a^2)E}[(1+\nu)b^2 + (1-\nu)r^2]\frac{1}{r} \tag{17}$$

Radial stress

$$\sigma_r = \frac{a^2 P}{b^2 - a^2}\left(1 - \frac{b^2}{r^2}\right) \tag{18}$$

Azimuthal stress

$$\sigma_\theta = \frac{a^2 P}{b^2 - a^2}\left(1 + \frac{b^2}{r^2}\right) \tag{19}$$

Figure 3 gives a comparison of radial displacement calculated using the analytical solution of eqn(17) and a prediction using the FE code. Predicted and analytical distributions of radial and azimuthal stresses are given in Fig.4. In each case the agreement between the prediction and analytical solution is very good. Good agreement has also been demonstrated by Henry & Collins[37] by using the FV code.

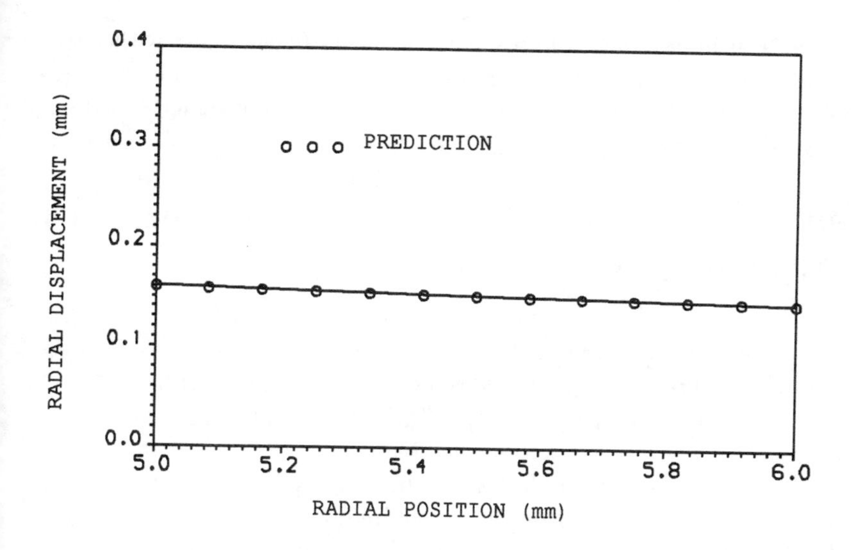

Figure 3: Analytical and predicted radial displacements in a cylinder under uniform pressure with free ends.

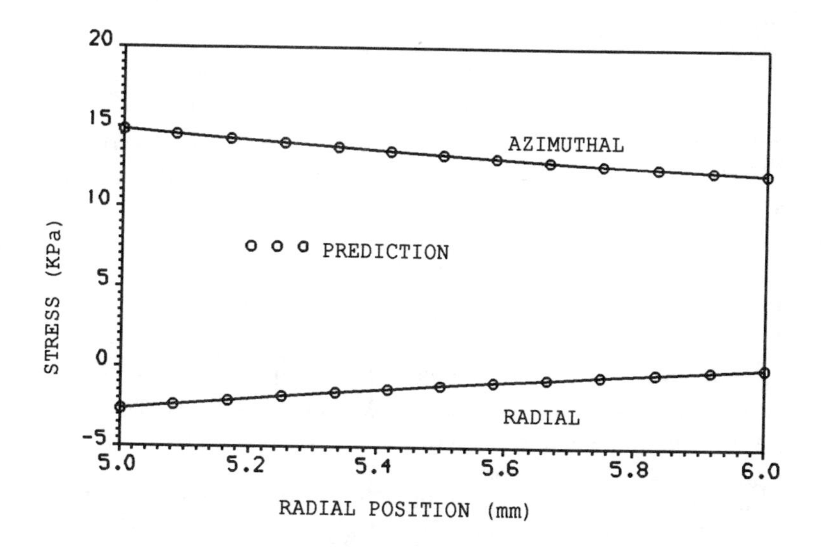

Figure 4: Analytical and predicted stresses in a cylinder under uniform pressure with free ends.

The cylinder was then assumed to be of finite length with rigidly fixed ends. The tube length was taken to be equal to two diameters, i.e., $20mm$. All other conditions were as described above. Analytical solutions for the radial and axial displacements of the mid-surface of a thin cylindrical tube with fixed ends and under a uniform pressure are given by[40]

Radial displacement

$$\begin{aligned} d_r &= a_1 \frac{\nu a}{1-\nu^2} - \frac{\nu^2 a^2}{Eh} P + \frac{1-\nu^2}{Eh} a^2 P \\ &+ e^{\frac{qx}{a}} \left(A_1 \cos \frac{qx}{a} + B_1 \sin \frac{qx}{a} \right) \\ &+ e^{-\frac{qx}{a}} \left(A_2 \cos \frac{qx}{a} + B_2 \sin \frac{qx}{a} \right) \end{aligned} \tag{20}$$

Axial displacement

$$\begin{aligned} d_x &= a_2 + a_1 \frac{x}{1-\nu^2} - \frac{\nu a x}{Eh} P \\ &+ \frac{\nu}{2q} \left[e^{\frac{qx}{a}} \left((A_1 - B_1) \cos \frac{qx}{a} + (A_1 + B_1) \sin \frac{qx}{a} \right) \right. \\ &- \left. e^{-\frac{qx}{a}} \left((A_2 + B_2) \cos \frac{qx}{a} - (A_2 - B_2) \sin \frac{qx}{a} \right) \right] \end{aligned} \tag{21}$$

where a is the radius of the mid-surface, $a_1, a_2, A_1, A_2, B_1, B_2$ are constants to be determined by the boundary conditions, and q is a constant defined by:

$$q = \sqrt{\frac{a}{h}} \left\{ 3(1-\nu^2) \right\}^{\frac{1}{4}} \tag{22}$$

Predicted displacements using the FE code are compared with the analytical solutions in Fig.5. Good agreement between the predictions and theory can be seen. However, a similar exercise carried out by Henry & Collins[37] using the FV code did not achieve such a good agreement. In addition, the FV solution took approximately five times longer to reach a converged solution than did the FE solution, demonstrating the advantage of the FE method over the FV method in solving solid mechanics problems.

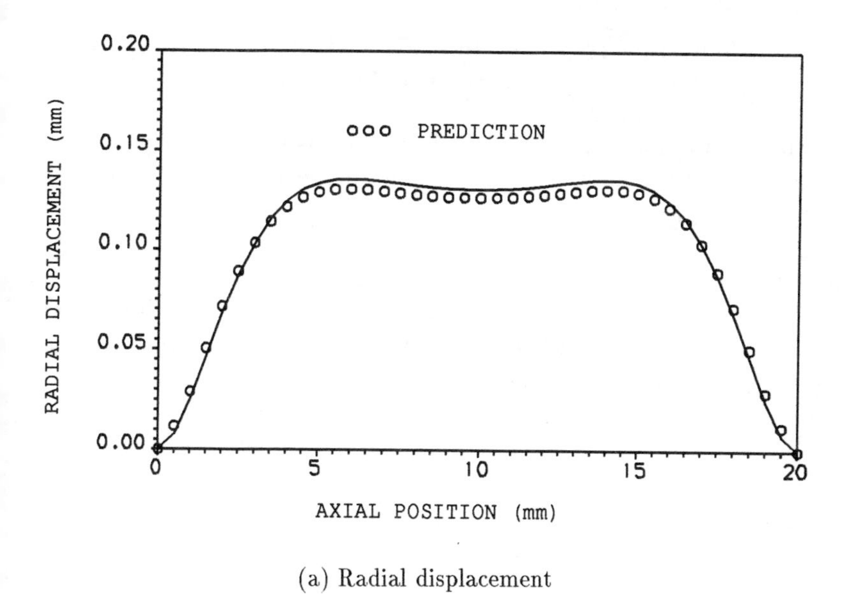

(a) Radial displacement

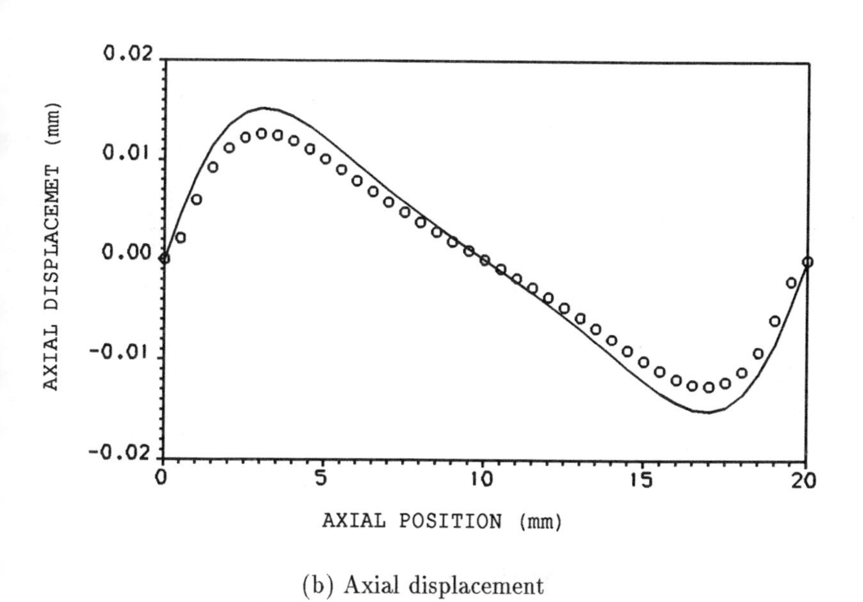

(b) Axial displacement

Figure 5: Analytical and predicted displacements in a cylinder under uniform pressure with fixed ends.

3.2 Decoupled model testing - unsteady flow in an expanding/contracting tube

To validate the moving boundary algorithm used in our coupled codes, the problem of unsteady flow in a semi-infinite expanding or contracting circular tube with one closed end was considered. In this problem, the flow is driven by a single expansion or contraction of the wall, and a full solution of the Navier-Stokes equations similar in both space and time has been derived by Uchida & Aoki.[41] By introducing a non-dimensional parameter $\alpha(t) = \dot{a}a/\nu$, where a is the radius of the tube ($a = a(t)$ and $\dot{a}$ is the velocity of the wall ($\dot{a} = da/dt$)), and further assuming that α is a constant, a similar solution for axial and radial velocity distributions can be expressed as follows:

Axial velocity

$$\frac{u_x}{u_m} = -\frac{F'/\eta}{2\alpha} \tag{23}$$

Radial velocity

$$\frac{u_r}{\dot{a}} = -\frac{F/\eta}{\alpha} \tag{24}$$

where $\eta = r/a$, u_m and $\dot{a}$ are given by

$$\frac{u_m}{\nu/a_o} = \frac{-2\alpha}{1 + 2\alpha(\nu t/a_o{}^2)}\frac{x}{a_o} \tag{25}$$

$$\dot{a} = \frac{\alpha\nu}{a_o}[1 + 2\alpha(\nu t/a_o{}^2)]^{-\frac{1}{2}} \tag{26}$$

with a_o being the initial radius. $F(\eta)$ is calculated by numerical intergration of the following equation

$$\left(\frac{F'}{\eta}\right)'' + \left(\frac{1}{\eta} + \frac{F}{\eta} + \alpha\eta\right)\left(\frac{F'}{\eta}\right)' - \left(\frac{F'}{\eta} - 2\alpha\right)\frac{F'}{\eta} = K \tag{27}$$

Since eqn(27) is singular at the origin, for the region $0 \leq \eta \leq 0.01$ $F(\eta)$ is calculated from the power series:

$$F(\eta) = b_2\eta^2 + b_4\eta^4 + b_6\eta^6 + b_8\eta^8 + \ldots \tag{28}$$

where

$$b_4 = \frac{1}{16}[K + 4(b_2 - \alpha)b_2] \tag{29}$$

$$b_6 = \frac{1}{12}(b_2 - 2\alpha)b_4 \tag{30}$$

$$b_8 = \frac{1}{72}(9\alpha b_6 - 2{b_4}^2) \tag{31}$$

and for $n \geq 2$

$$b_{2n+4} = -\frac{1}{(2n+2)^2(2n+4)}\{[(2n-4)b_2 + (2n+2)\alpha](an+2)b_{2n+2} + \sum_{m=1}^{n-1}(2n-2-4m)(2n+2-2m)b_{2m+2}b_{2n+2-2m}\} \tag{32}$$

Values of b_2 and K are determined to satisfy the boundary conditions.

In the analytical solution, the tube is closed at one end by a compliant membrane which prevents only axial motion of the fluid, leaving radial motion unrestricted. In the numerical solution, this membrane was approximated by a plane of symmetry. The initial tube diameter was set to $10mm$, and its length to 15 times the mean diameter. The tube outlet plane was set to a surface of constant pressure. While this is an obvious approximation, it was found not to affect the resulting velocity profiles away from the outlet.

Predicted axial and radial velocities using the FE code for cases of an expansion ($\alpha = 1.0$), and a contraction ($\alpha = -1.0$) are given in Figs 6 and 7, respectively. Included in these figures are theoretical curves obtained from eqns(23) and (24). It is demonstrated that the predictions are in excellent agreement with the analytical solutions. However, it is noted that it was not possible to achieve a correct solution using the FE code for the case of a more rapid expansion ($\alpha = 1.67$) in which the flow reverses near the wall. This case was predicted correctly with the FV code (see Iudicello *et al*[42]). Also, the time required to reach a converged solution for the FE code was almost an order of magnitude larger than that used for the FV code, indicating the relative efficiency of the FV method when applied to flow modelling.

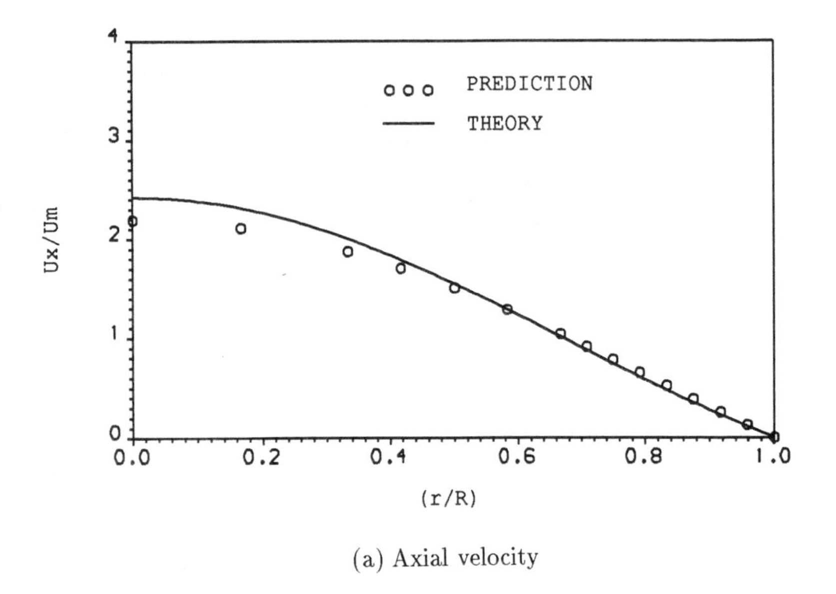

(a) Axial velocity

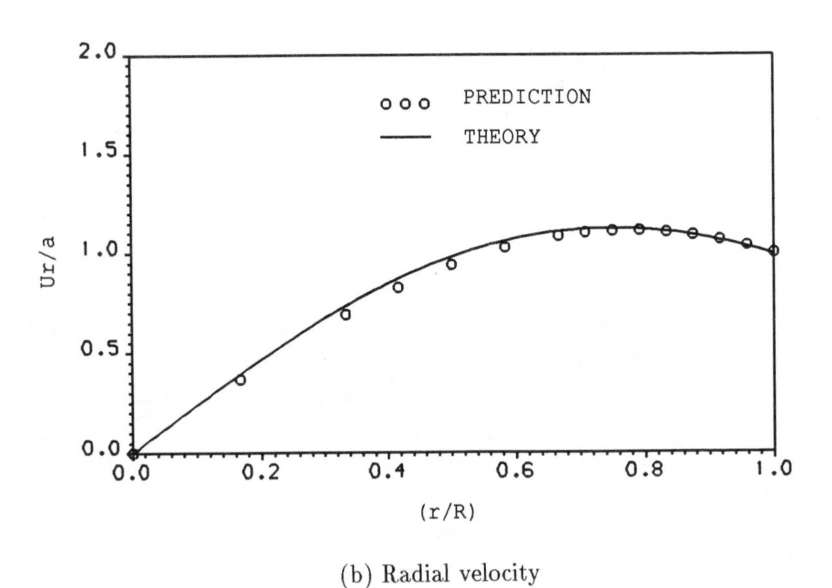

(b) Radial velocity

Figure 6: Analytical and predicted velocities in an expanding tube.

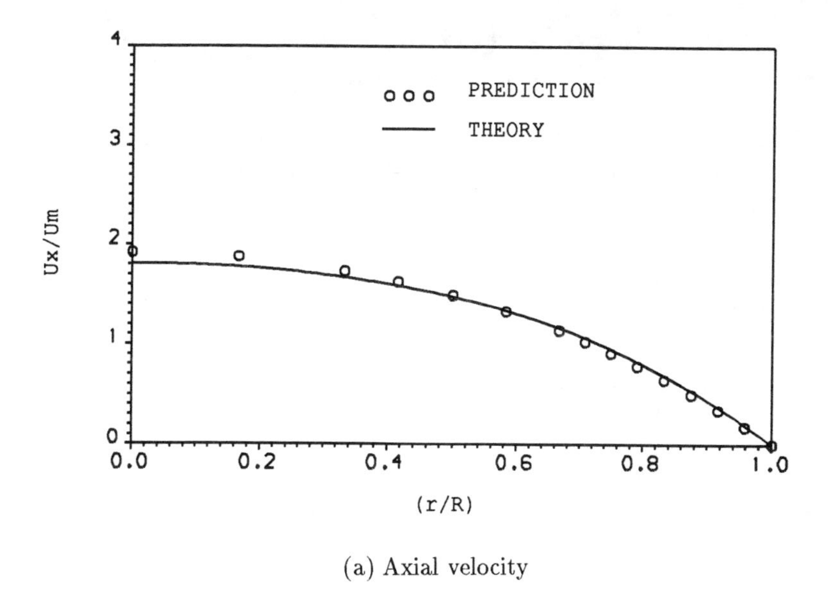

(a) Axial velocity

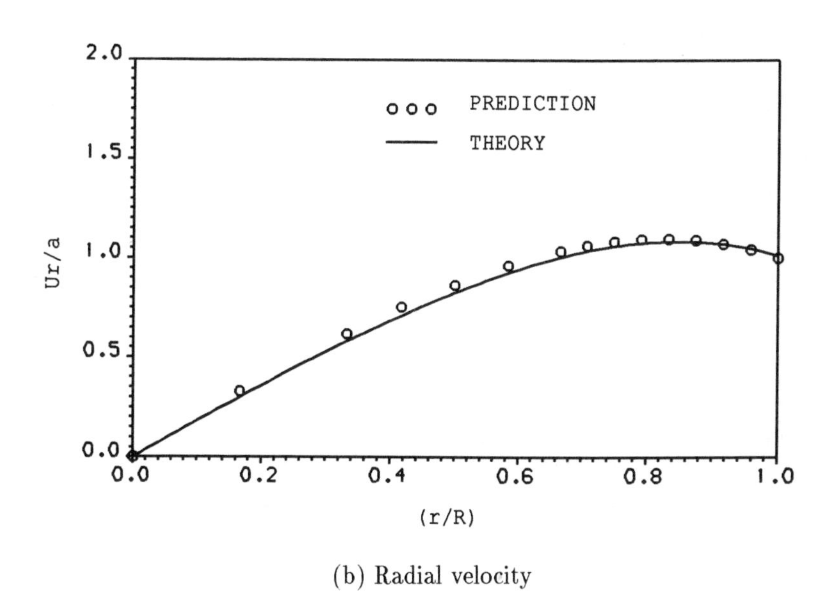

(b) Radial velocity

Figure 7: Analytical and predicted velocities in a contracting tube.

3.3 Coupled model testing - unsteady flow in an incompressible, linearly elastic tube

Finally, the codes were applied to coupled fluid/solid cases. The problem considered here is the flow in a segment of elastic tube which has a $10mm$ internal diameter, a $1.0mm$ thick wall, and is $50mm$ long. The tube is incompressible, linearly elastic, and is fixed at both ends. The whole geometry of the tube was calculated, thus avoiding the use of symmetry plane boundary conditions. The tube is subdivided into 8 fluid elements and 4 solid elements in the radial direction, and 20 elements in the axial direction. A steady flow calculation was performed first. However, as the final wall position was unknown, the calculation had to be solved as a unsteady problem even though the inlet conditions were fixed in time. Fig.8 presents the predicted distribution of radial displacement in the wall midway along the length of the tube and its comparison with the analytical solution. It can be seen that the prediction matches well the analytical values.

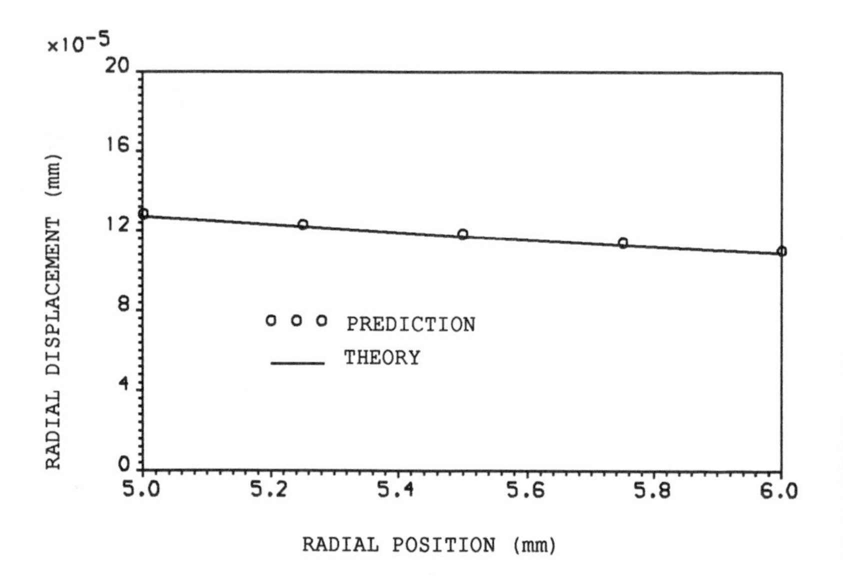

Figure 8: Analytical and predicted radial displacement for steady flow in a compliant tube.

The unsteady flow driven by a sinusoidally varying inlet pressure was then considered. An analytical solution for oscillatory flow in a freely-moving elastic tube has been established by Womersley.[19] If the flow is driven by an oscillatory pressure gradient expressed as $-\partial P/\partial x = Ae^{int}$, by assuming the fluid velocity is small compared to the wave speed, an approximate solution for axial and radial velocities can be obtained:

Axial velocity

$$u_x = \frac{A}{\rho}\frac{1}{in}\left\{1 + \eta\frac{J_o(\alpha y i^{3/2})}{J_o(\alpha i^{3/2})}\right\} e^{int} \tag{33}$$

Radial velocity

$$u_r = \frac{RA}{2\rho c}\left\{\eta\frac{2J_1(\alpha y i^{3/2})}{\alpha i^{3/2} J_o(\alpha i^{3/2})} + y\right\} e^{int} \tag{34}$$

where $y = r/R$, J_o and J_1 are Bessel functions of the first kind and order zero and order one respectively, c is the wave speed and η is a constant which can be determined by the boundary conditions.

For the unsteady flow calculation, the pressure variation was given by

$$P = \begin{cases} P_{max}\cos(\omega t) & \text{inlet} \\ 0 & \text{outlet} \end{cases}$$

where $P_{max} = 100$ pa, $\omega = 2\pi/t_p$, and t_p is the cycle period which was set to 0.75s. Each cycle consists of 100 equally spaced time steps. The calculation was carried out for three cycles, and each of these took approximately 25 hours on a Sun SPARC10 workstation (with one processor).

Shown in Fig.9 are predicted axial flow velocity profiles at the mid-point along the length of the tube at four different times during a cycle, and their corresponding theoretical profiles generated from eqn(33). In general, there is a good agreement between prediction and theory. Differences can be noticed at the wall where the analytical axial wall velocity are considerably larger than the predicted values. Results of a similar calculation using the FV method can be found in Henry & Collins.[32] However, the fully coupled FV code is currently restricted to axisymmetric geometries.

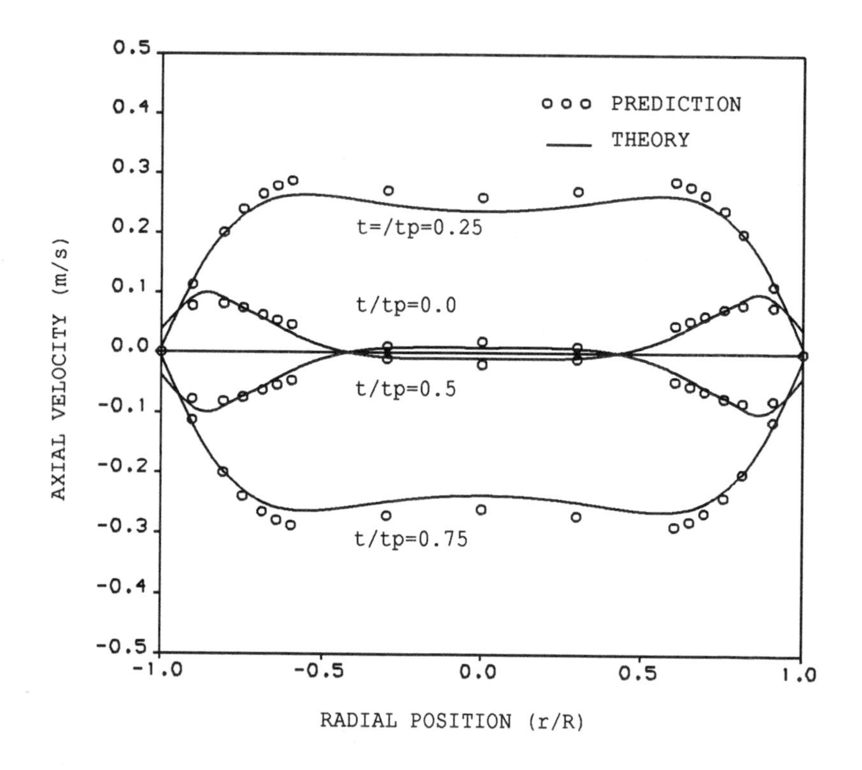

Figure 9: Analytical and predicted axial velocity profiles for unsteady flow in a compliant tube.

4 Modelling of flow-dependent dilatation with EDRF-related interventions

Flow-induced dilatation appears to be a universal phenomenon, demonstratable in almost all classes of animals, in microvessels as well as conduit arteries[43,44] and in veins as well as arteries.[45] It contributes to the hydrodynamic efficiency of the circulation by dynamically coupling metabolic requirement with tissue perfusion. Recent experimental studies have suggested that the phenomenon of flow-dependent dilatation in large arteries is caused by the release of a locally active vasodilator agent from the endothelial lining of the arterial wall in response to changes in local shear stress. Since the shear stress varies considerably both in time and space, flow-dependent dilatation also will vary regionally. Indeed, the frequency and amplitude of the pulsatile flow, superimposed on the mean flow rate, have been shown to determine the release of endothelium-derived relaxation factor (EDRF), the substance which most often causes flow-dependent dilatation.[46] Although flow-dependent dilatation increases the diameter of a large artery by only around 5%, this may increase wall distensibility by as much as 50%,[9] a change which will in turn significantly lower wave speed and thus the pressure and flow distributions within the vessel. Therefore, the relationships between the flow and EDRF release and the diameter and the elastic behaviour of the arterial wall are extremely complex in the intact circulation. Advanced high-precision ultrasonic techniques can determine overall flow parameters and changes in arterial geometry, as performed by Griffith *et al*[47] in their very recent *in vivo* studies. Nevertheless, direct measurements of shear stress and wall mechanical properties, which are the immediate cause and effect of EDRF release are lacking. It is these parameters that prohibit further understanding of the mechanisms involved in the flow-dependent dilatation.

Here the wall shear stress history, which is unobtainable through *in vivo* measurement, was predicted using the FV code described previously. Using ultra high-precision ultrasound *in vivo* data[47] as input, flows in porcine common carotid arteries were calculated for three different interventions: (1) ipsilateral infusion of $0.1\mu M$ bradykinin, (2) contralateral infusion of $1.0\mu M$ bradykinin, and (3) ipsilateral infusion of $0.1\mu M$ bradykinin after endothelial denudation. The artery was modelled as a distensible tube of circular cross-sections. The segment considered was 10 diameters long. The *in vivo* ultrasound data for flow rate were applied at the inlet together with the assumption of fully developed flow. The measured pressures were used as the downstream boundary condition. Wall movement was prescribed based on the ultrasound data for diameter change.

Since the experiments were carried out over long periods (300-700s), a large time step was desirable, for reasons of computational efficiency. To justify whether predictions using a large time step are valid, calculations were performed using three different time steps, i.e., Δt=1, 5 and 10 cycles respectively. Results for Δt=1 and 10 cycles are presented in Figs 10(a) and 10(b), which gives the *in vivo* data for pressure, velocity, and diameter change for Case No. 1 (ipsilateral infusion of 0.1μM bradykinin), and the predicted shear stress corresponding to the *in vivo*

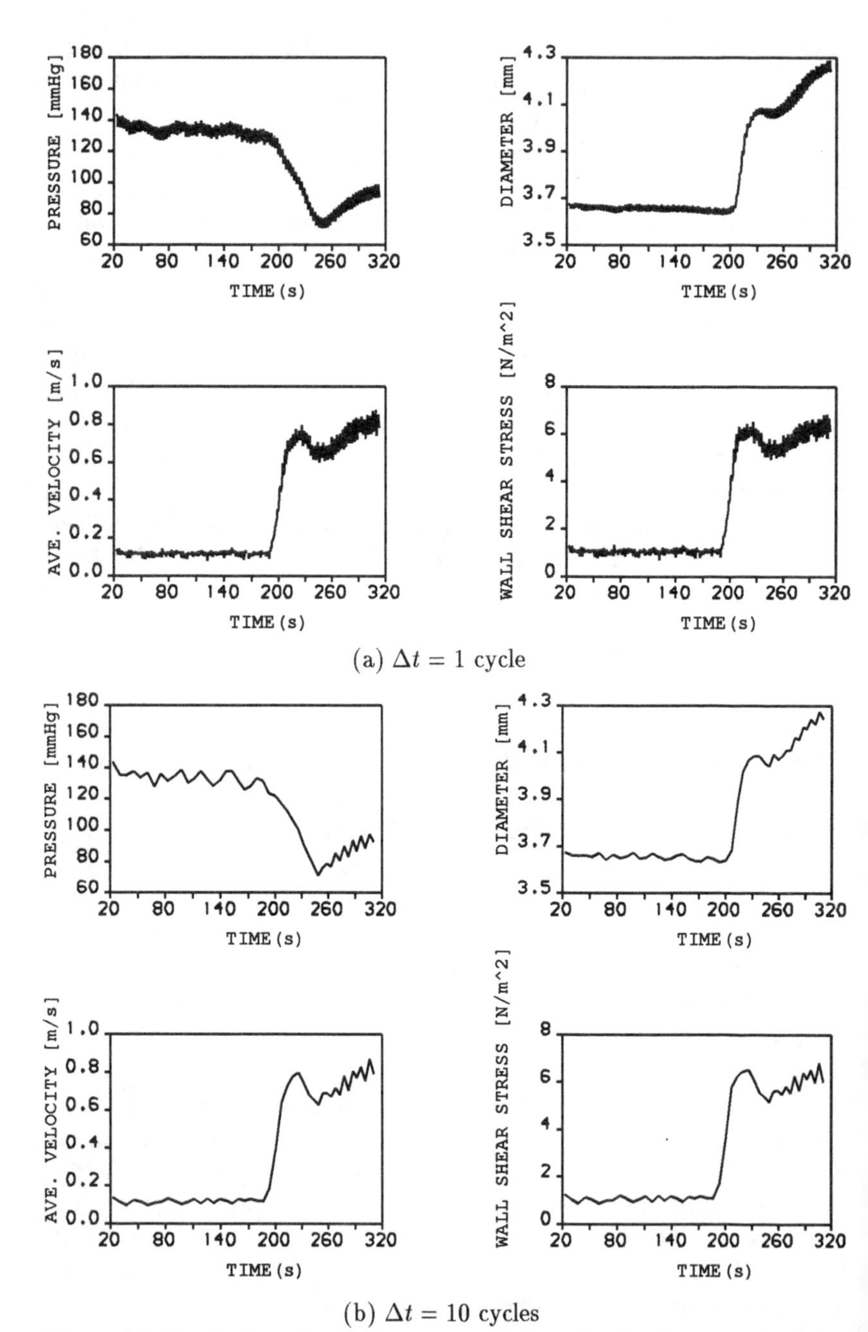

(a) $\Delta t = 1$ cycle

(b) $\Delta t = 10$ cycles

Figure 10: Prediction of wall shear stress history for flow in a porcine left common carotid artery with ipsilateral infusion of $0.1\mu M$ bradykinin using time steps of (a) $\Delta t = 1$ cycle, and (b) $\Delta t = 10$ cycles.

data. It was found that a time step of 10 cycles could be used without a marked loss in shear stress information, indicating that for the numerical problem treated here, smaller cycle-to-cycle effects can be stepped-over without subsequent noticeable differences in the final solution. From the combined numerical and experimental data it can be observed that the infusion of bradykinin into the ipsilateral side induced a synchronous increase in flow, increase in heart rate, increase in wall shear stress and decrease in local intravascular pressure. These haemodynamic changes enhanced EDRF release and resulted in an increase in diameter.

Results for cases No.2 (contralateral infusion of 1.0μM bradykinin) and No.3 (ipsilateral infusion of 0.1μM bradykinin after endothelial denudation) using a time step of 10 cycles are presented in Figs.11 and 12 respectively. Fig.11 demonstrates that the infusion of bradykinin into the contralateral side induced a fall in pressure, increase in heart rate and decrease in mean wall shear stress with no change in the mean flow rate. There was a transient initial fall in diameter but this was followed by a slow and sustained increase.

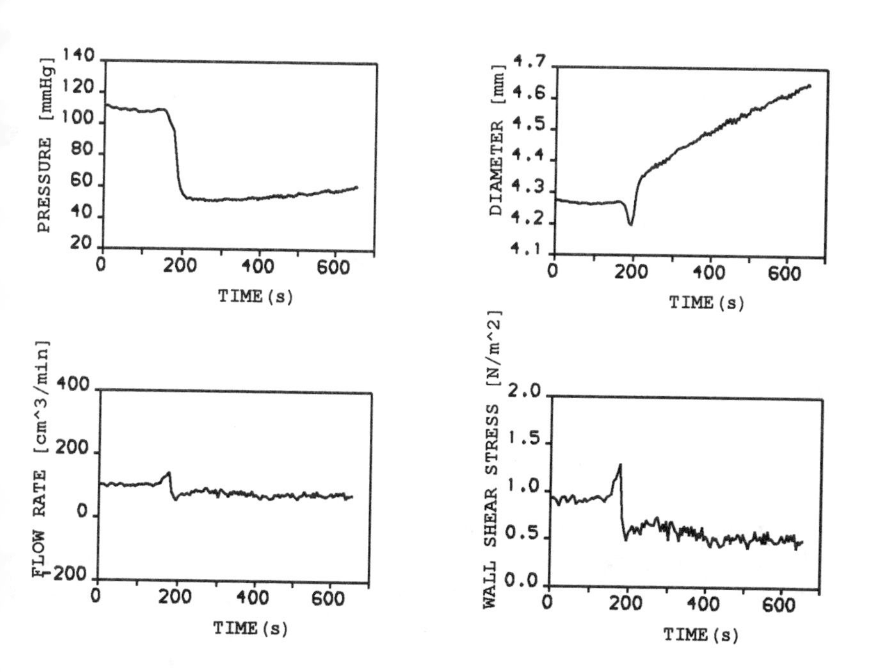

Figure 11: Predictions of wall shear stress history for flow in a porcine left common carotid artery with contralateral infusion of 1.0μM bradykinin.

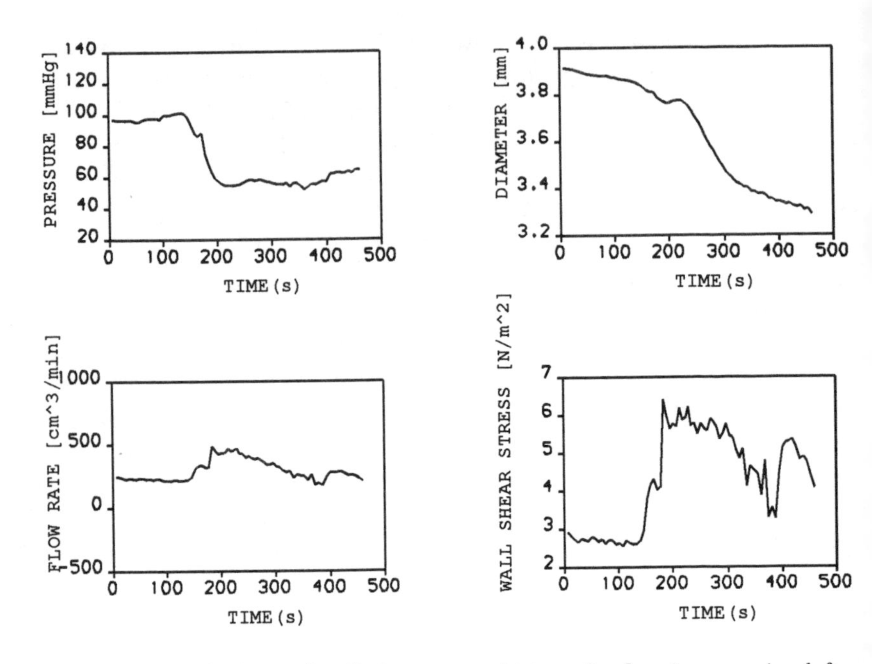

Figure 12: Predictions of wall shear stress history for flow in a porcine left common carotid artery with ipsilateral infusion of $0.1\mu M$ bradykinin after endothelial denudation.

Fig.12 indicates that following the local endothelial denudation the ipsilateral infusion of bradykinin induced increases in flow and heart rate, increase in wall shear stress and a fall in pressure. This was followed by a decrease, rather than an increase in diameter, probably as a consequence of passive collapse of the denuded carotid artery segment.

As an initial attempt to investigate the effects of EDRF release on wall mechanical properties, the instantaneous wall compliance for case No.1 defined by $\Delta D/\Delta P$ was calculated from the *in vivo* ultrasound data for intravascular pressure and diameter changes. This is presented in Fig.13 as a function of time, demonstrating that the ipsilateral bradykinin infusion has increased the instantaneous wall compliance. To predict the wall movement using the definition of wall compliance, an iterative predictor/corrector scheme was implemented in the FV code. The prediction was performed at a uniform time step of 10 cycles. The comparison between the calculated arterial diameters and the measurements is presented in Fig.14, and as can be observed, good agreement was achieved.

This provided a valuable code validation test for predicting flow in a distensible tube when its distensibility or compliance is known.

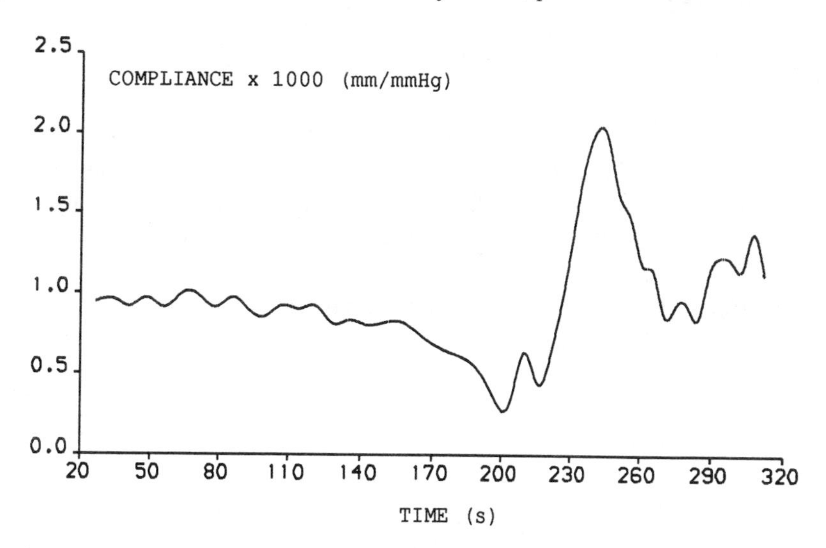

Figure 13: Wall compliance ($\Delta D/\Delta P$) calculated from ultrasound data.

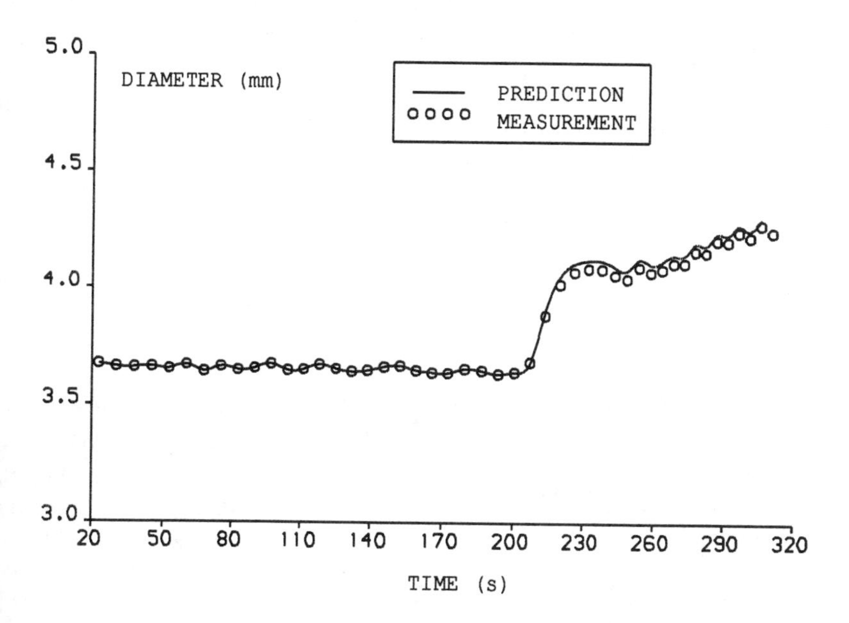

Figure 14: Comparison between predicted diameters and measurements.

5 Modelling of flow in a compliant arterial bifurcation

Arterial branchings are of special interest with regard to atherosclerosis because of their frequent involvement with deposits. A thorough understanding of the haemodynamics associated with flow at vessel ramifications could help unravel some of the mysteries of atherogenesis. We have previously investigated flow in right angle T-junction models and physiologically realistic canine femoral arterial bifurcations by means of state-of-the-art computational fluid dynamics techniques. Due to the high mortality associated with extracranial carotid artery disease, it is also important to study the detailed fluid dynamics in a carotid bifurcation. Presented here are our preliminary results of this study.

The carotid artery bifurcation consists of a main branch, namely the common carotid artery, which asymmetrically divides into two branches, the internal and external carotid arteries. The internal carotid artery is characterised by a widening in its most proximal part, the sinus or the bulb. Bharadvaj *et al*[10] developed an "average" geometry of the human adult carotid artery bifurcation. Two assumptions were made in this model : (a) the arteries and the carotid sinus are circular in cross-section; (b) the parent and the daughter vessels are in the same plane at the bifurcation. Based on these assumptions and their geometrical data, the 3D numerical grid for a carotid bifurcation was generated as shown in Fig.15. The essential geometrical parameters are common, internal and external carotid artery diameters ($D_{com} = 8.0mm, D_{int} = 8.3mm, D_{ext} = 5.6mm$), the maximum carotid sinus diameter ($D_{sin} = 8.9mm$) as well as the common internal carotid and common external carotid angles (25° respectively).

Numerical predictions were carried out for pulsatile flow in both rigid and compliant models of the carotid artery bifurcation using the FV code. *In vitro* flow waveforms obtained by Ku *et al*[48] were applied as the time-dependent boundary conditions. As shown in Fig.16, the mean flow rate in the common carotid artery calculated from the flow waveform is $5.1ml/s$. The mean inflow velocity is $0.1015m/s$ and the corresponding mean Reynolds number is $Re = 213$. The pulse frequency was assumed to be 80 strokes/min. The corresponding Womersley parameter is 5.93.

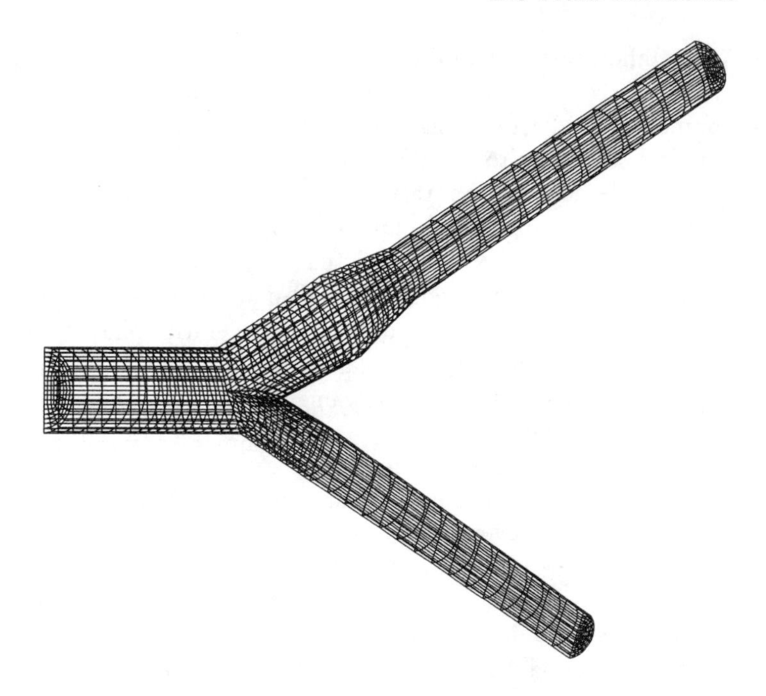

Figure 15: Grid configuration of a human carotid bifurcation model.

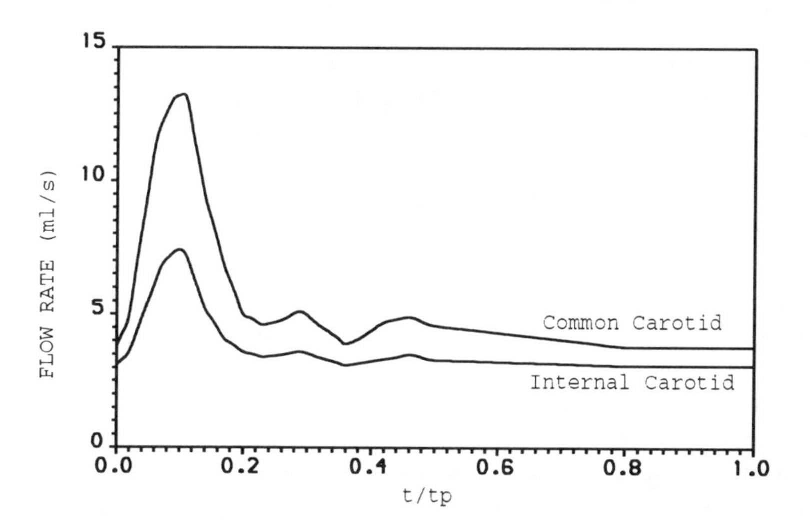

Figure 16: Flow waveforms in the common and internal carotid arteries according to Ku et al.[48]

At the inflow boundary, fully developed time-dependent velocity profiles were prescribed. The profiles correspond to the pulse waveform in the common carotid artery (Fig.16) and were calculated as long straight tube profiles according to the Womersley solution.[19] At outflow boundaries, a variable flow division ratio during the pulse cycle was specified together with the assumption of fully developed flow.

Given in Figs.17 and 18 are predicted velocity vector fields in the bifurcation plane of the rigid model at specified phases during the pulse cycle (systolic peak flow t/tp=0.1 and systolic deceleration phase t/tp=0.14). In the common carotid artery upstream of the branching area, the symmetric inflow profiles remain relatively unaffected. In the entrance region to the branches, at high flow rates (systolic phase) a high axial velocity can be found near the flow divider, and steep velocity gradients occur at the divider wall. In the carotid sinus which is a region of special interest, stagnated and reversed flow occur extending from the outer wall. At peak systolic flow (t/tp=0.1) proximal to the sinus the development of flow separation can be observed. In the systolic deceleration phase (t/tp=0.14), significant flow separation and reversed flow in the sinus occur. In the external carotid artery away from the outlet, the axial velocity profiles are shifted towards the inner wall.

In order to investigate the effect of wall distensibility on the basic flow characteristics in a carotid artery bifurcation, numerical predictions were carried out for a simplified compliant model with the same boundary conditions. In the calculation it was assumed that the arterial wall expands in the radial direction only, the variation of radius being dependent upon pressure. It was also assumed that the variation of radius does not exceed 10% of its original value. Given in Figs.19 and 20 are predicted velocity vector fields in the bifurcation plane of the compliant model at systolic peak flow (t/tp=0.1) and systolic deceleration phase (t/tp=0.14) respectively. Comparison with results corresponding to a rigid wall assumption does not show significant difference. The size and strength of flow separation in the carotid sinus are slightly smaller in the compliant model.

Here the wall distensibility was treated using a simple pressure-diameter relationship, and predictions were performed using the FV code. Due to the complexity of the geometry, further investigations of wall distensibility effects will be carried out using the coupled FE code, which does not have any restriction in geometry.

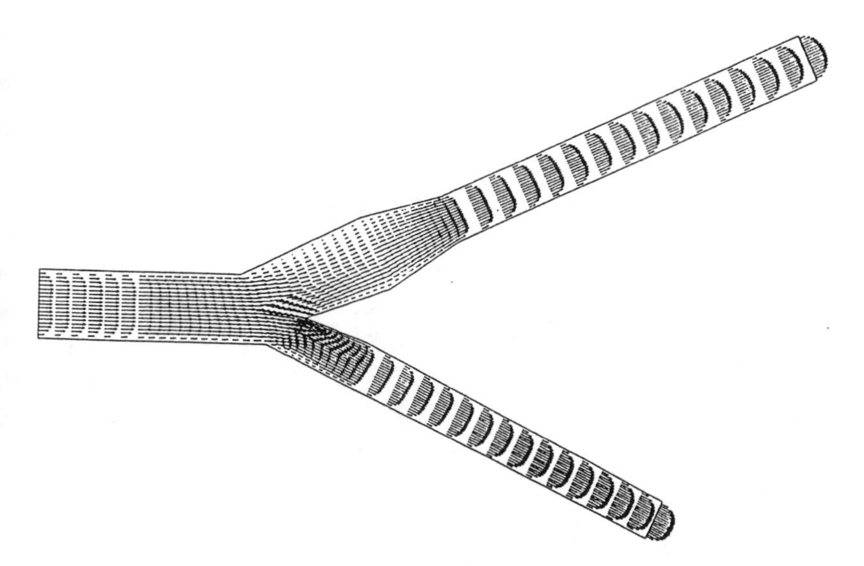

Figure 17: Velocity vector field in the bifurcation plane of a rigid model (systolic peak flow t/tp=0.1).

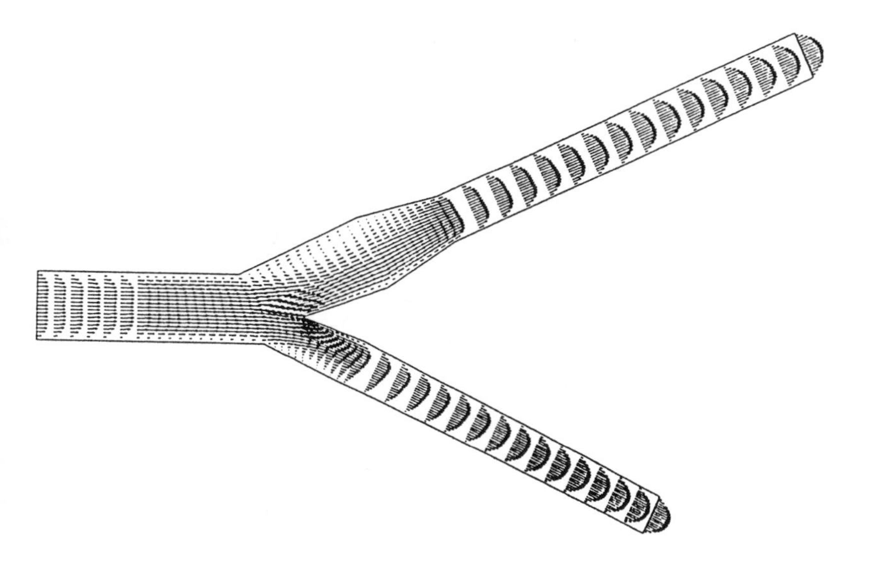

Figure 18: Velocity vector field in the bifurcation plane of a rigid model (systolic deceleration phase t/tp=0.14).

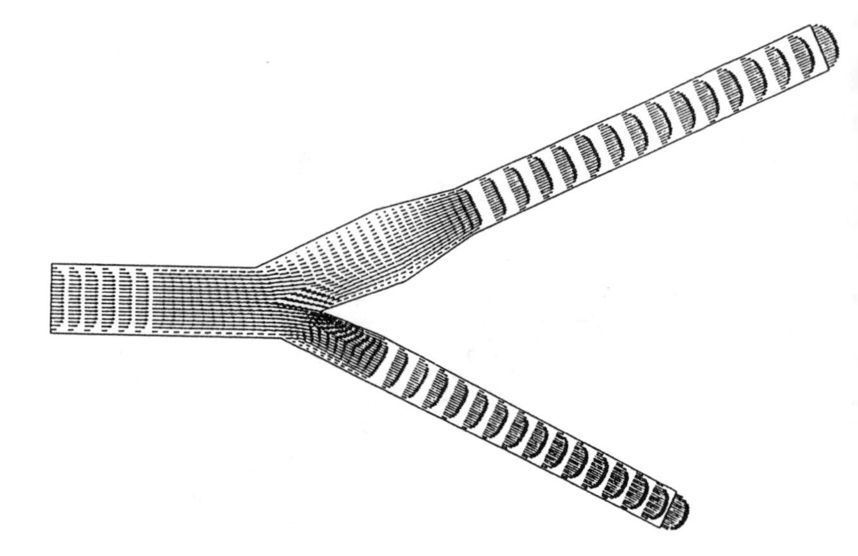

Figure 19: Velocity vector field in the bifurcation plane of a compliant model (systolic peak flow t/tp=0.1).

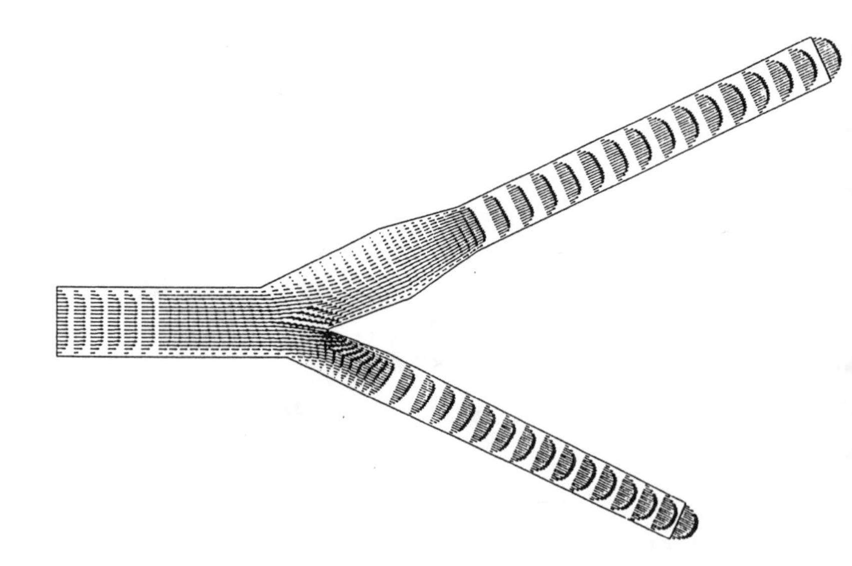

Figure 20: Velocity vector field in the bifurcation plane of a compliant model (systolic deceleration phase t/tp=0.14).

6 Conclusion

Recent progress in developing a comprehensive numerical modelling approach for the coupled solution of fluid/wall interaction problems in the arterial system has been presented. Two distinctly different numerical approaches, namely the finite volume method and finite element method, have been investigated. Validations of numerical predictions against established analytical solutions have demonstrated that either approach can give reliable quantitative information.

Predictions have been presented for flow in a compliant artery with various EDRF-related interventions. It is demonstrated that numerical predictions can play an important role in enhancing *in vivo* data. A combined *in vivo* and numerical study is able to provide information, such as wall shear stress, not easily obtained by experimental methods alone. The comparison between predicted arterial diameters and *in vivo* measurements has demonstrated that the numerical modelling approach can predict accurately the change in diameter induced by an EDRF intervention, using the wall compliance derived from the experimental data. Preliminary investigation of flow in a compliant carotid artery bifurcation appears to suggest that the effect of arterial wall compliance on the basic flow features is not significant.

Our future work will include modifying the wall model used in the current codes to allow more complexity in the stress-strain relationship, and, in the long term, developing a unique coupled code which uses the finite volume method for the fluid equations and the finite element method for the solid.

Acknowledgement

The authors are very grateful to Drs. A.G.Hutton and T.H.Lan of the Berkeley Technology Centre, Nuclear Electric Plc, for their help and advise. This project is funded by a British Heart Foundation grant.

References

1. Liepsch, D.W. Flow in tubes and arteries: a comparison, *Biorheology*, 1986, **23**, 395-433.

2. Fukushima, T. & Azuma, T. The horseshoe vertex: a secondary flow generated in arteries with stenosis, bifurcation, and branchings, *Biorheology*, 1982, **19**, 143-154.

3. Karino, T. & Motomiyza, M. Flow visualisation in isolated transparent natural blood vessel, *Biorheology*, 1983, **20**, 119-127.

4. Ku, D.N. & Giddens, D.P. Laser Doppler anemometer measurements of pulsatile flow in a model carotid bifurcation, *J.Biomechanics*, 1987, **20**, 407-421.

5. Fung, Y.C. *Biomechanics: Mechanical Properties of Living Tissues*, Springer-Verlag, New York, 1981.

6. Bergel, D.H. The properties of blood vessels, *Biomechanics: Its Foundations and Objectives*, ed. Y.C.Fung, N.Perrone & M.Anliker, Pretice-Hall, Englewood Cliffs, N.J. 1972, pp.105-140.

7. Rubanyi, G.M., Romero, J.C. & Vanhoutte, P.M. Flow-induced release of endothelium-derived relaxing factor, *Am.J.Physiol.*, 1986, **250**, H1145-H1149.

8. Hutcheson, I.R. & Griffith, T.M. Release of endothelium-derived relaxing factor is modulated both by frequency and amplitude of pulsatile flow, *Am.J.Physiology*, 1991, **260**, H257-262.

9. Tardy, Y., Meister, J-J, Perret, F., Brunner, H.R. & Ardite, M. Non-invasive estimate of the mechanical properties of peripheral arteries from ultrasonic and photoplethysmographic measurements, *Am.Phys.Physiol. Meas.*, 1991, **12**, 39-50.

10. Bharadvaj, B.K., Madon, R.F. & Giddens, D.P. Steady flow in a model of human carotid bifurcation. I. Flow visualisation, *J.Biomechanics*, 1982, **15**, 349-362.

11. Karino, T., Kwong, H.H.M. & Goldsmith, H.L. Particle flow behaviour in models of branch vessels, *Biorheology*, 1979, **16**, 231-248.

12. Walburn, F.J., Sabbah, H.N. & Stein, P.D. Flow visualisation in a model of an atherosclerotic human abdominal aorta, *J.Biomech.Eng.*, 1981, **103**, 168.

13. Liepsch, D., Poll, A., Strigberger, J. Stabbah, H.N. & Stein, P.D. Flow visualisation studies in a mold of the normal human aorta and renal arteries. *J.Biomech.Eng.*, 1989, **111**, 222-227.

14. Ku, D.N., Giddens, D.P., Phillips, D.J. & Strandness, D.E. Hemodynamics of the normal human carotid bifurcation: *in vitro* and *in vivo* studies, *Ultrasound in Medicine and Biology*, 1985, **11**(1), 13-26.

15. Jones, C.J.H., Lever, M.J., Ogasawara, Y., Parker, K.H., Hiramatsu, O.; Mito, K., Tsujioka, K. & Kajiya, F. Blood velocity distributions within intact canine arterial bifurcations, *Am.J.Physiol.*, 1992, **32**(5), H1592-1599.

16. Brech, R, & Bellhouse, B.J. Flow in branching vessels, *Cardiovascular Res.*, 1973, **7**, 593.

17. Lutz, R.J., Cannon, J.N., Bischoff, K.B. & Dedrick, R.L., Stiles, R.K. & Fry, D.L. Wall shear stress distribution in a model canine artery during steady flow, *Circ.Res.*, 1977, **41**, 391-399.

18. Bharadvaj, B.K., Mabon, R.F. & Giddens, D.P. Steady flow in a model of the human carotid bifurcation. II. Laser-Doppler anemometer measurements, *J.Biomechanics*, 1982, **15**, 363-378.

19. Womersley, J.R. The mathematical analysis of the arterial circulation in a state of oscillatory motion, WADC-TR56-614, Wright Air Development Centre, 1957.

20. Xu, X.Y. & Collins, M.W. A review of the numerical analysis of blood flow in arterial bifurcation, *Proc.Instn.Mech.Engrs. Part H: J.Eng. in Medicine*, 1990, **204** , 205-216.

21. Dinnar, U., Enden, G. & Israeli, M. A numerical study of flow in a three-dimensional bifurcation, Cardiovascular Systems Dynamics Society Meeting, Canada.

22. Perktold, K. & Resch, M. Numerical flow studies in human carotid artery bifurcations: basic discussion of the geometric factor in atherogenesis, *J.Biomed.Eng.*, 1990, **22**, 111-123.

23. Perktold, K., Resch, M. & Peter, R.O. Three-dimensional numerical analysis of pulsatile flow and shear stress in the carotid artery bifurcation, *J.Biomechanics*, 1991, **24**, 409-420.

24. Perktold, K., Resch, M. & Florian, H. Pulsatile non-Newtonian flow characteristics in a three-dimensional human carotid bifurcation model, *ASME J.Biomech.Eng.*, 1991, **113**, 464-475.

25. Perktold, K., Peter, P.O., Resch, M. & Langs, G. Pulsatile non-Newtonian blood flow in three-dimensional carotid bifurcation models: a numerical study of flow phenomena under different bifurcation angles, *J.Biomed.Eng.*, 1991, **13**, 507-515.

26. Xu, X.Y., Collins, M.W. & Jones, C.J.H. Flow studies in canine artery bifurcations using a numerical simulation method, *ASME J.Biomech.Eng.*, 1992, **114**, 504-511.

27. Xu, X.Y. Numerical analysis of blood flow in 3-D arterial bifurcations. Ph.D Thesis, City University, UK, 1992.

28. Lonsdale, R.D. An algorithm for solving thermalhydraulic equations in complex geometries: The ASTEC code, UKAEA Report, 1988.

29. Patankar, S.V. & Spalding, D.B. A calculation procedure for heat, mass and momentum transfer in three dimensional parabolic flows, *Int.J. Heat Mass Transfer*, 1972, **15**, 1787-1806.

30. Reuderink, P.J. Analysis of the flow in a 3D distensible model of the carotid artery bifurcation, Ph.D Thesis, Eindhoven University of Technology, The Netherlands, 1991.

31. Rappitsch, K., Perktold, K. & Guggenberger, W. Numerical analysis of intramural stresses and blood flow in arterial bifurcation models, *Computational Biomechanics*, ed, K.D.Held, C.A.Brebbia *et al*, Computational Mechanics Publications, Southampton Boston, 1993, pp149-156.

32. Henry, F.S. & Collins, M.W. A novel predictive model with compliance for arterial flows, *1993 Advances in Bioengineering*, ASME 1993, **BED-Vol.26**, pp131-135.

33. Xu, X.Y., Griffith, T.M., Collins, M.W., Jones, C.J.H., & Tardy, Y. Coupled modelling of blood flow and arterial wall interactions by the finite element method, *Computers in Cardiology 1993*, IEEE Computer Society Press, Washington, Brussels, Tokyo, 1993, pp687-690.

34. Anonymous, FLOW3D Release 2.4: User manual, AEA Industrial Technology, Harwell Laboratory, UK, 1991.

35. Van Doormal, J.P. & Raithby, G.D. Enhancements of the SIMPLE method for predicting incompressible fluid flows, *Numer. Heat Transfer*, 1984, **7**, 147-163.

36. Rhie, C.M. & Chow, W.L.A. A numerical study of the turbulent flow past an airfoil with trailing edge separation, *AIAA J.*, 1983, **21**, 1525-1532.

37. Henry, F.S. & Collins, M.W. Prediction of transient wall movement of an incompressible elastic tube using a finite volume procedure, *Computational Biomechanics*, ed, K.D.Held, C.A.Brebbia *et al*, Computational Mechanics Publications, Southampton Boston, 1993, pp156-172.

38. Gresho, P.M., Lee, R.L. & Sani, R.L. On the time dependent solution of the incompressible Navier-Stokes equations in two and three dimensions, *Recent Advances in Numerical Methods in Fluids*, 1990, Vol.1, Pineridge Press, Swansea.

39. Anonymous. User guide to FEAT. Engineering Analysis Centre, Nuclear Electric Plc, UK, 1991.

40. Love, A.E.H. A Treatise on the Mathematical Theory of Elasticity, Cambridge University Press, 1952.

41. Uchida, S. & Aoki, H. Unsteady flow in a semi-infinite contracting or expanding pipe, *J.Fluid Mechanics*, 1977, **82**(2), 371-387.

42. Iudicello, F., Henry, F.S., Collins, M.W., Shortland, A., Jarvis, J.C. & Salmons, S. Numerical simulation of the flow in model skeletal muscle ventricles, *Computers in Cardiology 1993*, IEEE Computer Society Press, Washington, Brussels, Tokyo, 1993, pp377-380.

43. De Mey, J.G. & Gray, S.D. Endothelium-dependent reactivity in resistance vessels. *Prog.Appl.Microcir.*, 1985, **8**, 181-187.

44. Owen, M.P. & Bevan, J.A. Acetylcholine induced endothelium-dependent vasodilatation increases as artery diameter decreases in the rabbit ear. *Experentia*, 1985, **41**, 1057-1058.

45. De Mey, J.G. & Vanhoutte, P.M. Heterogeneous behaviour of the canine arterial and venous wall. *Circ.Res.*, 1982, **51**, 439-447.

46. Henderson, A.H. ST Cyres lecture: endothelium in control, *Br.Heart Journal*, 1991, **65**, 116-125.

47. Griffith, T.M. *et al* Internal communication.

48. Ku, D.N., Giddens, D.P., Zarins, C.K. and Glagov, S. Pulsatile Flow and Atherosclerosis in the Human Carotid Bifurcation, *Arteriosclerosis*, 1985, **5**(3), 293-301.

Chapter 4

Numerical simulation of arterial hemodynamics

S. Cavalcanti & G. Gnudi

Department of Electronics, Computer Science and Systems, University of Bologna, Viale Risorgimento 2, I-40136 Bologna, Italy

Abstract

Modelling of blood flow in the arteries is a challenging application of fluid mechanics, while simultaneously contributing to a better understanding of the pathogenesis of arterial disease and design of prosthetic devices, such as arterial grafts. Numerical simulation of the arterial hemodynamics is aimed at gaining a deeper insight into the effects caused by the mechanical and geometrical properties of arteries, as well as pressure and flow conditions on various hemodynamic variables such as velocity profiles, parietal shear stress, etc.. This chapter is intended to present a method to mathematically describe and numerically simulate arterial hemodynamics as well as a review of several results obtained by simulation, indicating problems of relevance to researchers and clinicians. The simulator describes the motion of a Newtonian viscous fluid in a non-linearly deformable tapered tube. The results refer to the canine femoral artery. The effects on axial velocity profiles, shear-stress and their time course during a cardiac cycle produced by changing the values of physical and geometrical parameters of the simulated artery are presented and discussed. In particular, the role of the pressure-radius curve in combination with the unstressed radius is investigated as well as the effects of a mild stenosis. The influence of natural taper angle on pressure wave propagation is analysed and discussed. The attention is focused also on the influence of nonlinearities on the longitudinal impedance when flow rate undergoes large variations, as may occur in physiological conditions.

Nomenclature

t	time
r	radial coordinate
z	axial coordinate
$R(z,t)$	inner radius
$\eta = r/R$	adimensional radial coordinate
$P(z,t)$	transmural pressure
$u(\eta, z, t)$	radial velocity component
$w(\eta, z, t)$	axial velocity component
$\alpha_k(z,t)$	coefficient of the series expansion (9)
$Q(z,t)$	blood flow (18)
$rm(z)$	unstressed midwall radius
$h(z)$	unstressed wall thickness
$\delta(z,t)$	radial extension ratio
σ_0, β	stress-strain parameters (20)
ρ	blood density
μ	blood viscosity
ν	blood kinematic viscosity
μ_w	wall viscosity
τ_w	wall shear stress
C_p	compliance of proximal load (Fig. 1)
Z_{cp}	characteristic resistance of proximal load
Z_{pp}	peripheral resistance of proximal load
C_d	compliance of distal load (Fig. 1)
Z_{cd}	characteristic resistance of distal load
Z_{pd}	peripheral resistance of distal load
Z_l	arterial longitudinal impedance
f	frequency

1 Introduction

The study of pulsatile blood flow in arteries is a complex problem for which a completely satisfactory solution is not yet available. This is due to arterial geometry (e.g. tapering, bifurcations, curvature) and physical properties (e.g. viscoelasticity, anisotropy, and non-homogeneity of the wall, suspension- like rheology of blood) as well as to intrinsically complex equations governing the motion of blood. All that makes it difficult to determine a mathematical model for an arterial segment, which gives a good description of the biophysical phenomena and, in the same time, is sufficiently simple.

The availability of such a model is of basic importance not only to evaluate how the local motion of blood depends on the various physical and geometrical parameters, but also to build, by summation, a comprehensive simulator of the arterial tree which allows to evaluate global properties as: wave propagation, cardiac loads, blood distribution in the various districts, etc..

Besides, it is remarked that modelling of blood flow in the arteries is of primary importance for a better understanding of the pathogenesis of arterial disease and design of prosthetic devices, such as arterial grafts.

Arterial models of the systemic and of the pulmonary arterial tree have a fairly long history. The oldest models were of the windkessel type.[1] Models that included the travelling waves of pressure and flow emerged later, when studies on elastic tubes were carried out[2,3] and the arterial system was then approached from elastic-tube theory and the relations were described in the frequency domain.[4] The first complete treatment of this subject is due to Womersley,[2] whose theory has been the basis for several applications[5,6] during the Sixties. This theory yields a linear model of arterial hemodynamics, in fact it was deduced assuming as negligible both the nonlinear terms of blood motion equation, i.e. convective accelerations, and vessel wall deformations. Integrating the motion equation in these hypotheses , the relationship between pressure gradient and flow becomes an impedance not dependent on the wall elastic characteristic. Furthermore, Womersley's theory does not allow to take account of the vessel taper whose effects are mainly revealed through the above mentioned nonlinear terms.

Subsequent studies have taken into consideration the nonlinear properties of the arterial wall, including hysteresis (viscoelasticity), the large radius variations, the vessel taper, the axial curvature etc.[1,7] Afterwards, Streeter proposed a theory[8] extending to arteries the methodology of the technical fluid-dynamics. It is founded on the simplification of the motion equation, which is directly written in terms of mean velocity within the cross-section without the preliminary determination of velocity profiles. Of course, the problem remains, and it is again encountered when evaluating the viscous losses. In this theory the problem is coped with by assuming pre-defined velocity profiles which however are not always plausible.

More recently, contributions along this line have been published, however the Streeter theory does not seem to be promising with regard to arterial modelling, as proved by the results obtained so far, which correspond very scarcely to the experimental data.

In 1972 Ling & Atabek[7] confirmed the need of taking into account the wall deformations in the large arteries and emphasised the importance of the convective acceleration, particularly in the evaluation of the mean flow value. The model of Ling & Atabek is partial derivative and highly nonlinear for both the nonlinear elastic characteristic of the wall and, above all, the presence of convective acceleration terms in the equation of motion. The basic idea in this theory is the assumption of a small shape variation along the axis of the axial velocity profile, which allows the approximate calculation of the velocity components in a section of artery without explicit dependence on the axial coordinate. Even though Ling & Atabek do not consider the viscoelastic behaviour of the arteries, it could be easily included in their model. A comparison with the fit to experimental data given by the linear theory, clearly points out the limits of the Womersley model with particular reference to the mean flow and the axial velocity profiles, but also to the pulsatile flow waveform. In conclusion, the approximate Ling & Atabek's theory allows to set up a mathematical model which fits very well the experimental data.[1,7,9,10]. Nevertheless it presents the disadvantage of being a partial differential model, thus requiring the use of finite difference solving algorithms that are conceptually simple, but relatively expensive from the point of view of the requested computer time and memory. As a consequence this model is in practice applicable only to study local hemodynamics when pressure, radius and their longitudinal gradients are known in a vessel section.

In view of simulating hemodynamics in a segment of artery or an arterial network, it is necessary to build mathematical models computationally less expensive and moreover allowing an easier interpretation of the complex phenomena which characterise the blood motion in an arterial vessel. The problem has been studied and solved in Refs 10 and 11, where, starting from the same assumptions at the basis of the Ling & Atabek model, a mathematical model of the blood motion in a segment of artery was deduced. This theory is presented in the first part of this chapter and then it is utilised to simulate the hemodynamics of the canine femoral artery. Local flow fields and related parameters (e.g. the shear stress) are analysed in basal conditions, after alterations of the elastic properties of the wall, and after a large variation of flow rate.

2 Statement of the mathematical problem

The equations governing arterial blood flow are deduced from the Navier-Stokes and continuity equations, taking into account the nonlinear pulsatile behaviour of the arterial wall as well as the nonlinear inertial terms. The artery is modelled as a rectilinear, deformable shell of isotropic, incompressible, viscoelastic material, with a circular cross-section and without longitudinal movements. Lumen radius, wall thickness and wall elasticity are dependent on the axial position in order to simulate a different cross-section and wall stiffness along the vessel. Having assumed the arterial vessel axisymmetric and the axis straight, the analysis is of course restricted to a short part of an artery not too near a bifurcation. On the basis of this assumption, the mathematical problem can be greatly simplified, since it is realistic to assume a two-dimensional blood flow, i.e. with only the axial and the radial velocity components. Moreover, the lumen radius, R, is sufficiently smaller than the wavelength of the pressure wave, therefore the radial Navier-Stokes equation simply reduces to $\partial p/\partial r = 0$.[12] The results of Deshpande *et al.*[13] on the pressure distribution along a rigid tube with a severe obstruction well support this condition. It is then reasonable and convenient to assume the pressure as independent of the radial coordinate r. The Newtonian rheology of the blood is also stated, in agreement with several previous works, [14,15] proving that no significant difference exists between the flow patterns evaluated by Newtonian and non-Newtonian fluid models.

2.1 Equations governing blood motion

In the above assumptions, continuity and axial Navier-Stokes equations, can be written in the following way

$$\frac{\partial \mathrm{u}}{\partial r} + \frac{\mathrm{u}}{r} + \frac{\partial \mathrm{w}}{\partial z} = 0 \quad , \tag{1}$$

$$\frac{\partial \mathrm{w}}{\partial t} + \mathrm{u}\frac{\partial \mathrm{w}}{\partial r} + \mathrm{w}\frac{\partial \mathrm{w}}{\partial z} = -\frac{1}{\rho}\frac{\partial p}{\partial z} + \nu\left(\frac{\partial^2 \mathrm{w}}{\partial r^2} + \frac{1}{r}\frac{\partial \mathrm{w}}{\partial r} + \frac{\partial^2 \mathrm{w}}{\partial z^2}\right) \quad , \tag{2}$$

where t is time, ρ the blood density, ν the blood kinematic viscosity; $\mathrm{w} = \mathrm{w}(r,z,t)$ and $\mathrm{u} = \mathrm{u}(r,z,t)$ are the components of velocity in axial (z) and radial (r) directions respectively, and $p = p(z,t)$ is the pressure. Equations (1) and (2), can be regarded as the fundamental equations of hemodynamics (for details, see Ref. 1).

In order to integrate eqns (1), and (2) in a region of space (r,z,t) it is necessary to specify the appropriate boundary conditions. In particular, on the basis of the above assumptions boundary conditions with respect to r at the wall are

$$\mathrm{w}(r,z,t)\Big|_{r=R} = 0 \quad , \qquad \mathrm{u}(r,z,t)\Big|_{r=R} = \frac{\partial R}{\partial t} \quad , \tag{3.a}$$

while the conditions on the vessel axis are

$$\frac{\partial \mathrm{w}}{\partial r}\Big|_{r=0} = 0 \quad , \qquad \mathrm{u}(r,z,t)\Big|_{r=0} = 0 \quad . \tag{3.b}$$

Conditions (3.*a*) reflect the hypothesis that the longitudinal motion of the arterial wall is negligible because of perivascular tethering, as fully justified in

Refs 16-18, whereas conditions (3.*b*) are a consequence of the axial symmetry of the velocity profile. Since $R = R(z,t)$ is a function of t, due to vessel deformability, boundary conditions on the wall (3.*a*) are prescribed for a moving line. These kinematic boundary conditions make complex the problem solution in the frame (r,z,t). To overcome the mathematical difficulties stemming from the kinematic conditions it is convenient to introduce the adimensional radial coordinate η

$$\eta = \frac{r}{R(z,t)} \quad . \tag{7}$$

Boundary conditions on the wall (3.*a*) may be expressed in the reference frame (η,z,t) with respect to a fixed line

$$w(\eta,z,t)\Big|_{\eta=1} = 0 \quad , \qquad (8.a) \qquad\qquad u(\eta,z,t)\Big|_{\eta=1} = \frac{\partial R}{\partial t} \quad , \qquad (8.b)$$

where $w = w(\eta,z,t)$ and $u = u(\eta,z,t)$ are respectively axial and radial velocity components in the new reference frame. In the new coordinates, eqns (1) and (2) can be rewritten in the form

$$\frac{1}{R}\frac{\partial u}{\partial \eta} + \frac{u}{\eta R} + \frac{\partial w}{\partial z} - \frac{\eta}{R}\frac{\partial R}{\partial z}\frac{\partial w}{\partial \eta} = 0 \quad , \tag{9}$$

$$\frac{\partial w}{\partial t} = \frac{1}{R}\left(\eta\left(\frac{\partial R}{\partial t} + w\frac{\partial R}{\partial z}\right) - u\right)\frac{\partial w}{\partial \eta} - w\frac{\partial w}{\partial z} - \frac{1}{\rho}\frac{\partial P}{\partial z} + \frac{\nu}{R^2}\left(\frac{\partial^2 w}{\partial \eta^2} + \frac{1}{\eta}\frac{\partial w}{\partial \eta}\right) \quad . \tag{10}$$

Axial Navier-Stokes eqn (10) is written by dropping the term $\partial^2 w/\partial z^2$ from the last right-hand term, since in a previous study[19] it was verified that for the problem under study it is negligible in comparison with the radial derivatives.

2.2 Mathematical model of blood motion

Multiplying the continuity eqn (9) by $\eta\, R$ and integrating it with respect to the radial coordinate between zero and η one finds

$$u(\eta,z,t) = \eta\, w\frac{\partial R}{\partial z} - \frac{2}{\eta}\int_0^\eta \eta\,\frac{\partial R}{\partial z}\, w\, d\eta - \frac{R}{\eta}\int_0^\eta \eta\,\frac{\partial w}{\partial z}\, d\eta \quad . \tag{11}$$

For $\eta = 1$ and imposing the conditions at the wall, (8.*a*), then (11) becomes

$$-\int_0^1 \eta\,\frac{\partial w}{\partial z}\, d\eta = \int_0^1 \eta\left(\frac{2}{R}\frac{\partial R}{\partial t} + \frac{2}{R}\frac{\partial R}{\partial z}\, w\right) d\eta \quad . \tag{12}$$

This result suggests a way to obtain an expression of the term $\partial w/\partial z$: extending the equality between the integrals to the integrands it is possible to write[19]

$$-\frac{\partial w}{\partial z} = \frac{2}{R}\frac{\partial R}{\partial t} + \frac{2}{R}\frac{\partial R}{\partial z}\, w \quad . \tag{13}$$

By inserting (13) into (11) one finds the following expression for the radial velocity

$$u(\eta, z, t) = \eta\left(\frac{\partial R}{\partial t} + w\,\frac{\partial R}{\partial z}\right) \quad . \tag{14}$$

Expressions (13) and (14) are a particular solution of the continuity equation (9). The expression (13) is very favourable since axial Navier-Stokes equation (10) can be put in the form

$$\frac{\partial w}{\partial t} = \frac{2}{R}\frac{\partial R}{\partial z}w^2 + \frac{2}{R}\frac{\partial R}{\partial t}w - \frac{1}{\rho}\frac{\partial P}{\partial z} + \frac{\nu}{R^2}\left(\frac{\partial^2 w}{\partial \eta^2} + \frac{1}{\eta}\frac{\partial w}{\partial \eta}\right) \quad , \tag{15}$$

which clearly shows the presence of two nonlinear terms directly depending on radius derivatives (with respect to both axial and time coordinates) and axial velocity, and inversely depending on the radius. In the case of a cylindrical tube with rigid walls – i.e. when $\frac{\partial R}{\partial t} = \frac{\partial R}{\partial z} = 0$ – eqns (13), and (14) express the laminar flow conditions and (15) reduces to the motion equation of Womersley's theory.[2]

To solve eqn (15) with respect to η the axial velocity component is expressed by the series expansion[10]

$$w(\eta, z, t) = \sum_{k=1}^{N} \alpha_k(z,t)\left(\eta^{2k} - 1\right) \quad . \tag{16}$$

This form becomes zero for $\eta = 1$ as required by (8.a) and is axially symmetric as required by the axial symmetry of flow (3.b). Expressing $w(\eta, z, t)$ in such a way the radial velocity component (14) becomes

$$u(\eta, z, t) = \eta\left(\frac{\partial R}{\partial t} + \frac{\partial R}{\partial z}\sum_{k=1}^{N} \alpha_k(z,t)\left(\eta^{2k} - 1\right)\right) \quad . \tag{17}$$

The N functions $\alpha_k(z,t)$ must satisfy[19] the following system of N nonlinear, partial differential equations

$$\begin{cases} \dfrac{\partial \alpha_k}{\partial t} = \dfrac{2}{R}\dfrac{\partial R}{\partial z}A_k + \dfrac{2}{R}\dfrac{\partial R}{\partial t}\alpha_k + \dfrac{4\nu}{R^2}(k+1)^2\alpha_{k+1} \qquad (k{=}1,..,N\text{-}1) \quad , \\[2ex] \dfrac{\partial \alpha_N}{\partial t} = -\dfrac{2}{R}\dfrac{\partial R}{\partial z}\left(T^2 + \displaystyle\sum_{k=1}^{N-1} A_k\right) + \dfrac{2}{R}\dfrac{\partial R}{\partial t}\alpha_N - \dfrac{4\nu}{R^2}\displaystyle\sum_{k=1}^{N} k^2\,\alpha_k + \dfrac{1}{\rho}\dfrac{\partial P}{\partial z} \quad . \end{cases} \tag{18}$$

The coefficients that appear in (18) are defined in the following way:

$$T \equiv \sum_{k=1}^{N} \alpha_k \quad ,$$

$$A_1 \equiv -2\,\alpha_1 T \quad , \qquad A_k \equiv -2\alpha_k T + \sum_{j=1}^{k-1} \alpha_j\,\alpha_{k-j} \qquad (k = 2, ..., N) \quad .$$

Equation set (18) governs the blood motion and to complete the statement of the mathematical model it is then necessary to prescribe the motion of the arterial wall and assign the constitutive relation between the transmural pressure and the artery lumen radius.

2.3 Mathematical model of wall motion

The wall motion may be evaluated by employing the continuity equation written in the integral form

$$\frac{\partial R}{\partial t} = -\frac{1}{2\pi R}\frac{\partial Q}{\partial z} \quad , \tag{19}$$

where Q indicates the flow that, taking account of eqn (16), can be expressed as

$$Q(z,t) = -\pi R^2 \sum_{k=1}^{N} \frac{k}{k+1}\alpha_k \quad . \tag{20}$$

The constitutive equation is deduced by assuming the transverse stress in the wall and the inertial effect from the wall mass as negligible. In such a hypothesis, taking into account the elastic response to the wall strain and the viscous force caused by the wall motion, the average Eulerian wall stress, acting in the circumferential direction, may be calculated as follows

$$\sigma_\theta(z, R(z,t)) = \sigma_e(z, R) + \mu_w \frac{\partial R}{\partial t} \quad . \tag{21}$$

The mechanical equivalent of eqn (21) is the classic Kelvin-Voigt model (e.g. see Ref. 20) which has a linear dash pot with viscosity coefficient μ, coupled in parallel with a nonlinear spring. For the incompressibility of the wall material, the radial extension ratio, δ, can be expressed as a function of the inner radius $R(z,t)$ in the following way

$$\delta(z, R(z,t)) \equiv \frac{R_m}{r_m} = \frac{R + \sqrt{R^2 + 2\, r_m\, h}}{2\, r_m} \quad , \tag{22}$$

where $R_m(z,t)$ indicates the midwall radius of the vessel in the deformed state, whereas $r_m(z)$ and $h(z)$ respectively denote the midwall radius and the wall thickness in the steady unstressed state ($P = \partial P/\partial t = 0$).

Assuming that the elastic behaviour of the wall is governed by a single smooth muscle fibre located in the middle of the wall, the static strain-stress relationship of the nonlinear spring is[11]

$$\sigma_e(z, R(z,t)) = \sigma_o\left(e^{\beta(\delta-1)} - 1\right) \quad . \tag{23}$$

For a relatively thin walled blood vessel the circumferential Eulerian stress is

$$\sigma_\theta(z, R(z,t)) = \left(\frac{r_m}{h}\delta^2 - \frac{1}{2}\right)P \quad . \tag{24}$$

Then, from eqns (21)-(24) we have the following constitutive equation

$$P(z,t) = \frac{\mu_w \frac{\partial R}{\partial t} + \sigma_o \left(e^{\beta(\delta-1)} - 1\right)}{\frac{r_m}{h}\delta^2 - \frac{1}{2}} \quad . \tag{25}$$

The parameters σ_0 and β are independent of the axial coordinate and of time, on the assumption that the smooth muscle fibres are uniformly active along the whole vessel. On the contrary, the unstressed thickness $h(z)$, and the midwall radius depend on the axial position according to the vessel geometry which is to be simulated.

The problem of solving (9) and (10) with respect to time, radial and axial coordinates, is reduced, by means of (16), to integrate (18) and (19) with respect to t and z only. The expression of axial (16) and radial (17) velocity components, the expression of blood flow (20), and the pressure-radius relationships (22) and (25), complete the mathematical model specification.

2.4 Numerical integration scheme

The set of differential equations (18)-(19) is nonlinear and time dependent. It is then necessary to resort to numerical integration to find the solution by choosing uniformly-spaced mesh points at $z_i = i\,\Delta z$ with $i = 0, 1, 2, \ldots.., M$ where $M\Delta z = L$ and L is the length of the artery. The whole vessel is then broken up into M short elements of length Δz. At each section z_i the following definitions hold:

$$\begin{aligned}
\alpha_{k,i}(t) &\equiv \alpha_k(i\Delta z, t) \quad , \\
Q_i(t) &\equiv Q(i\Delta z, t) \quad , \\
R_i(t) &\equiv R(i\Delta z, t) \quad , \\
P_i(t) &\equiv P(i\Delta z, t) \quad , \\
\delta_i(t) &\equiv \delta(R_i, i\Delta z) \quad , \\
r_{mi} &\equiv r_m(i\Delta z) \quad , \\
h_i &\equiv h(i\Delta z) \quad .
\end{aligned} \tag{26}$$

If the length of the arterial element Δz is sufficiently small, then it is possible to approximate - in each section z_i - the derivatives with respect to the axial coordinate z with the following finite difference centred scheme

$$\begin{aligned}
\frac{\partial Q_i}{\partial z} &= \frac{Q_{i+1} - Q_{i-1}}{2\Delta z} \quad , \\
\frac{\partial R_i}{\partial z} &= \frac{R_{i+1} - R_{i-1}}{2\Delta z} \quad , \\
\frac{\partial P_i}{\partial z} &= \frac{P_{i+1} - P_{i-1}}{2\Delta z} \quad .
\end{aligned} \tag{27}$$

Writing eqns (18) and (19) for each $i = 1, ..., (M-1)$ and inserting (26) and (27), one obtains[11] a set of $(N+1)(M-1)$ ordinary nonlinear differential equations that must be integrated with respect to time only, starting from an assigned initial value for the state variables $\alpha_{k,i}(t)$ and $R_i(t)$. Assuming, for the sake of simplicity, that the fluid is initially at rest and the vessel is in an unstressed condition, $(p = 0)$, the initial conditions are

$$\begin{aligned} \alpha_{k,i}(t)\Big|_{t=0} &= 0 \qquad for \quad k = 1, ..., N; \quad i = 1, ..., M-1 \\ R_i(t)\Big|_{t=0} &= r_{m\,i} - \frac{h_i}{2} \qquad for \quad i = 0, ..., M \end{aligned} \tag{28}$$

The midwall $r_{m\,i}$ and the wall thickness h_i change along the vessel according to the geometry of the simulated artery.

To complete the statement of the numerical problem and in particular to calculate eqns (27) for $i = 1$ and $i = (M-1)$, the conditions for the proximal $(i = 0)$ and distal $(i = M)$ vessel extremities must be assigned besides the initial conditions. In order to simulate the circulatory system downstream and upstream of the vessel, its proximal and distal ends are each closed on a load described by a classic three-element windkessel (Fig. 1). In this way, the distributed effects caused by the remaining vessels of the arterial tree are taken into account by lumping them at the proximal and distal vessel extremities by means of suitable load impedances.[21]

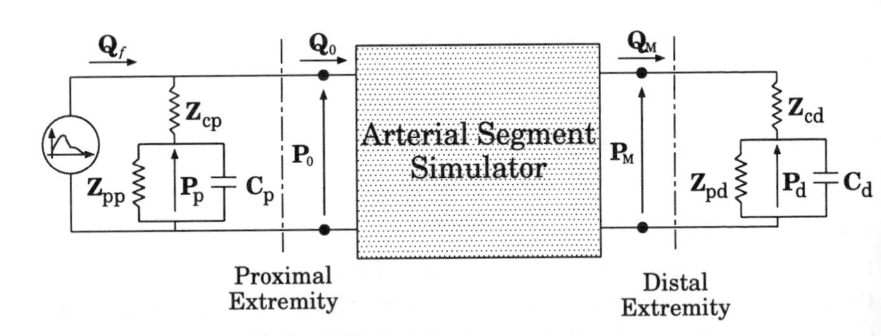

Figure 1: Electrical analogue of the mathematical model used for the simulations.

A pulsatile flow, $Q_f(t)$ is applied at the proximal extremity to simulate the artery input blood flow (Fig. 1). With reference to Fig. 1, it is possible to write for the proximal section ($i = 0$) the following equations

$$\frac{\partial P_p}{\partial t} = \frac{1}{C_p}\left(Q_f - Q_0 - \frac{P_p}{Z_{pp}}\right) \quad ,$$
$$P_0 = P_p + Z_{cp}(Q_f - Q_0) \quad , \tag{29}$$

and for the distal one ($i = M$) the equations

$$\frac{\partial P_d}{\partial t} = \frac{1}{C_d}\left(Q_M - \frac{P_d}{Z_{pd}}\right) \quad ,$$
$$P_M = P_d + Z_{cd} Q_M \quad . \tag{30}$$

The Q_0 and Q_M flows are calculated by writing the equation of mass conservation in the following way

$$Q_0 = Q_1 + 2\pi\, R_0 \frac{\partial R_0}{\partial t} \Delta z \quad ,$$
$$Q_M = Q_{M-1} - 2\pi\, R_M \frac{\partial R_M}{\partial t} \Delta z \quad . \tag{31}$$

Finally, to calculate wall motion at the vessel extremities, it is convenient to rearrange eqn (25) in the following way

$$\mu_w \frac{\partial R}{\partial t} = \left(\frac{r_m}{h}\delta^2 - \frac{1}{2}\right) P - \sigma_o \left(e^{\beta(\delta-1)} - 1\right) \quad . \tag{32}$$

The Gear predictor-corrector method[22] was used to integrate differential equations because they are highly nonlinear, very stiff, and characterised by widely varying time constants.

2.5 Simulation parameters

The numerical simulator has been set up in order to reproduce the pulsatile flow and pressure curves measured on the exposed femoral artery of an anaesthetised dog. Geometrical parameters were surveyed in an unstressed state of the artery, i.e. null transmural pressure. Wall viscosity was set equal to 100 dyn s/cm^3, blood density equal to 1.06 g/cm^3, and kinematic viscosity equal to 3.3 10^{-2} cm^2/s. The windkessel load parameters were estimated[19] in order to fit the experimental flow and pressure curves, obtaining characteristic resistance (Z_{cd}) equal to $1.1\,10^4$ dyn s/cm^5, peripheral resistance (Z_{pd}) equal to $1.2\,10^5$ dyn s/cm^5, and compliance (C_d) equal to $1.1\,10^{-5}$ cm^5/dyn .

The parameters of the proximal windkessel load were assumed equal to the distal ones. To validate the numerical simulator several data found out from literature[1,23,24] for canine femoral arteries were employed. The static strain-pressure curve of the arterial wall is of the type shown in Fig. 2. Flow and pressure curves as well as the physical and geometrical vessel parameters were found close to the ones measured by Milnor & Bertram.[23]

3 Effects of changes in the pressure-radius curve

In this section, the effects on axial velocity profiles, shear-stress and their time course during a cardiac cycle, as produced by changing the values of physical and geometrical parameters of the simulated artery are presented and discussed. Particular emphasis is on the role of the pressure-radius curve in combination with the unstressed radius. The parameters σ_o and β of the elastic characteristic (25) as well as the unstressed radius r_m determine the arterial vessel geometry and stiffness in a deformed state. To investigate how these properties affect hemodynamics, three different sets of parameters were considered, as reported in Table 1

Table 1. Arterial vessel data

	Artery-1	
r_m	2.13	[mm]
σ_o	1 10^5	[dyn/cm^2]
β	21.6	
	Artery-2	
r_m	1.65	[mm]
σ_o	10 10^5	[dyn/cm^2]
β	3.6	
	Artery-3	
r_m	1.82	[mm]
σ_o	1 10^5	[dyn/cm^2]
β	21.6	

Geometrical and physical characteristics of the three simulated arteries were chosen within a physiological range for a canine femoral artery, whose total length is 30 cm and whose unstressed wall thickness is 0.1 mm. The static radius-pressure curves of the three simulated arteries are shown in Fig. 2. Due to the pulsatile blood pressure, the inner radius of arteries as well as the wall thickness change during the cardiac cycle; the parts of curves covered during a cycle are indicated in Fig. 2.

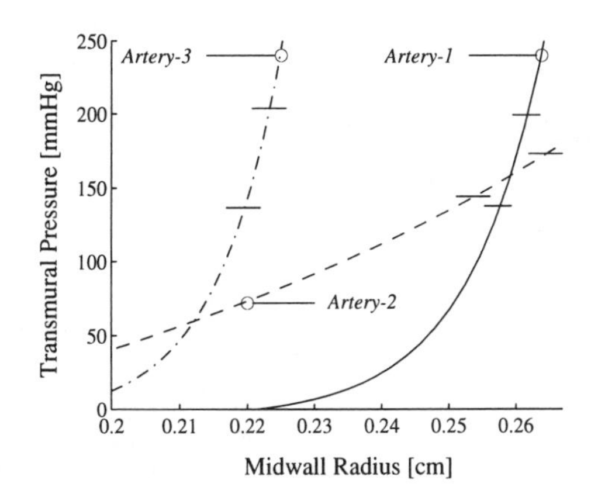

Figure 2 : Static curves relating midwall radius to transmural pressure for the three simulated arterial vessels. The parts of curves covered during a cardiac cycle are indicated.

For each artery the incremental elastic modulus was calculated (see Ref. 1) by considering the maximum systolic and minimum diastolic values marked on each radius-pressure curve of Fig. 2. In particular, *Artery-1* and *Artery-3* exhibit the same elastic modulus ($Ep = 5.4\ 10^{-6}$ dyn/cm^2) and a different inner radius, while *Artery-1* and *Artery-2* have the same time average inner radius and a different elastic module ($Ep = 1\ 10^{-6}$ dyn/cm^2 for *Artery-2*). Then *Artery-2* is significantly more elastic than *Artery-1*; on the contrary, *Artery-3* is as stiff as *Artery-1* but has a significantly smaller lumen. Comparing the simulation results obtained for these three arterial vessels, it is possible to separate the effects of wall elasticity from the effects of artery lumen.

In order to avoid the effects induced by the proximal and distal connections, the results presented are referred to a middle region of the arteries and in particular, to a cross-section 15 cm far from the distal and proximal ends.

In all the three simulated arteries there are no significant differences in the mean values over a cardiac cycle both for the pressure (Fig. 3) and flow (Fig. 4) curves.

Although wall elasticity does not affect mean values, it influences the pulsatile component inducing significant changes in pressure and flow waveforms. In particular, because of its greater elasticity, *Artery-2* undergoes the largest deformation (Fig. 2), provides a maximum (systolic) pressure about 15% lower (Fig. 3) and a systolic flow peak about 26% larger (Fig. 4).

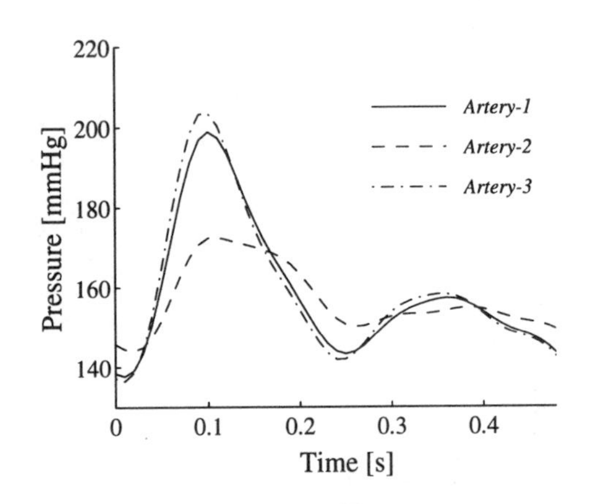

Figure 3: Pressure waveforms within a cardiac cycle as obtained for the three simulated arterial vessels.

On the contrary, arterial inner radius in the stressed state has very little influence on both pressure and flow curves as clearly appears by comparing *Artery-1* and *Artery-3*.

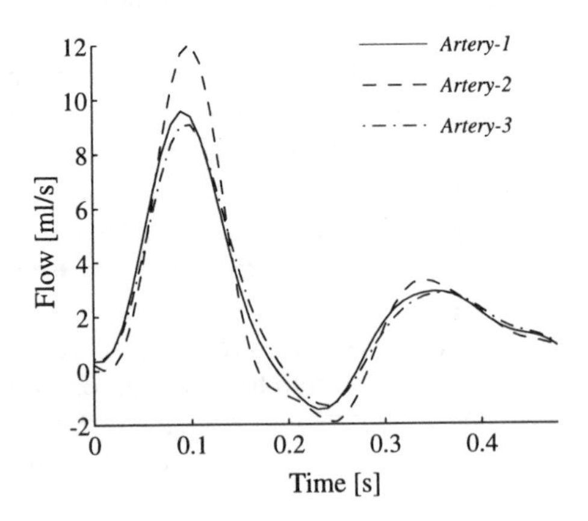

Figure 4: Flow waveforms within a cardiac cycle as obtained for the three simulated arterial vessels.

Time evolution of the circumferential stress in the three simulated arterial vessels is shown in Fig. 5. The wall elasticity as well as inner radius influence the mean values of the circumferential stress in general, while they have limited effects on the pulsatile component.

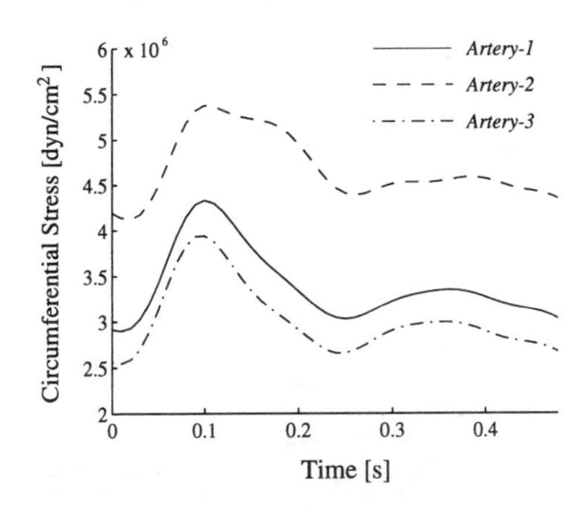

Figure 5: Time course of the circumferential stress in the three simulated arterial vessels.

In the unstressed condition, all the arteries have the same thickness but a different midwall radius (Table 1). In deformed state, *Artery-1* and *Artery-2* have the same time-averaged radius, therefore the wall of *Artery-2* becomes thinner than that of *Artery-1*. As a consequence, circumferential stress is higher (about 25%) in *Artery-2* than in *Artery-1*. Having the same stiffness as *Artery-2* but a smaller lumen, *Artery-3* has the lowest circumferential stress.

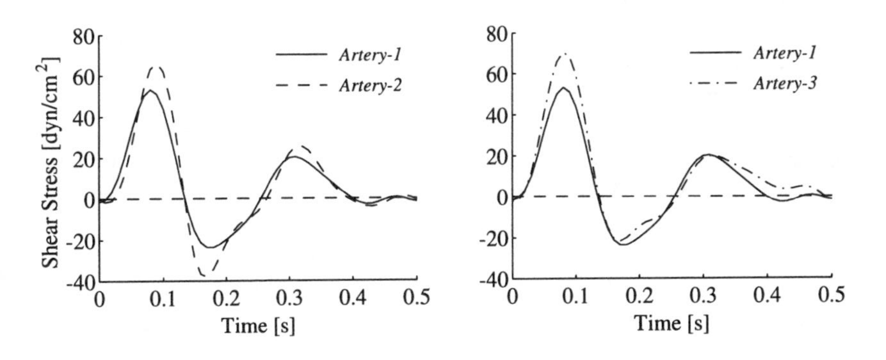

Figure 6: Shear stress at the wall as a function of time.

Parietal shear stress is sensitive to wall elasticity (Fig. 6, left panel) and to arterial lumen (Fig. 6, right panel). In *Artery-2*, because of the greater elasticity, the maximum shear stress increases of about 25% and the minimum decreases of about 62% with respect to *Artery-1*. The large rate of stress causes a greater shearing action on the inner surface of the wall of *Artery-2*. The significant increase of the negative peak of shear stress is caused by a layer of reversal flow occurring near the wall during the diastolic phase. This region of backward flow is well evident in Fig. 7, where time evolution during a cardiac cycle of the axial velocity profile is shown.

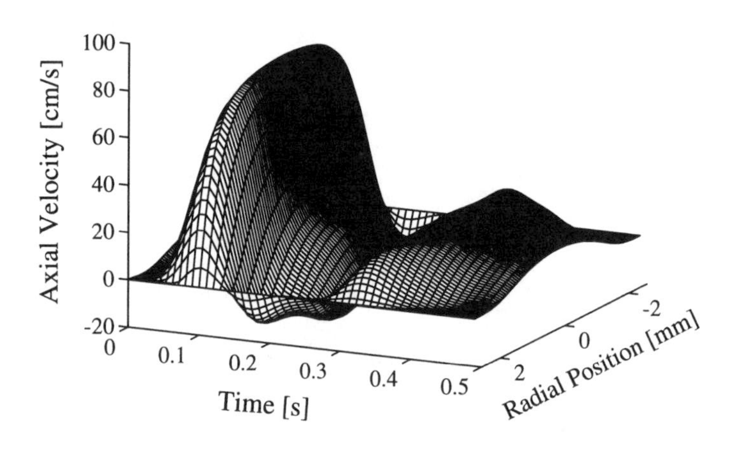

Figure 7: Three-dimensional view of the axial velocity profile versus time, during a cardiac cycle (*Artery-2*).

Wall elasticity significantly affects the flow separation region and as a consequence the shear rate near the wall. The lumen radius reduction in *Artery-3* causes an increase only of the positive peak of the wall shear stress and therefore, a significant increase of the mean value (about 63%).

A detailed view of axial velocity profiles at some characteristic instants of the cardiac cycle is shown in Fig. 8. It is well evident that over the whole cardiac cycle the profile has a shape very different from a parabolic one. Also during systole, when all the blood elements move forward and in particular when the flow peak occurs, the profile is more flat than the parabolic one. As a consequence, the velocity gradient at the wall (also referred to as wall shear rate) is always larger than the one calculated by assuming a steady laminar flow condition.

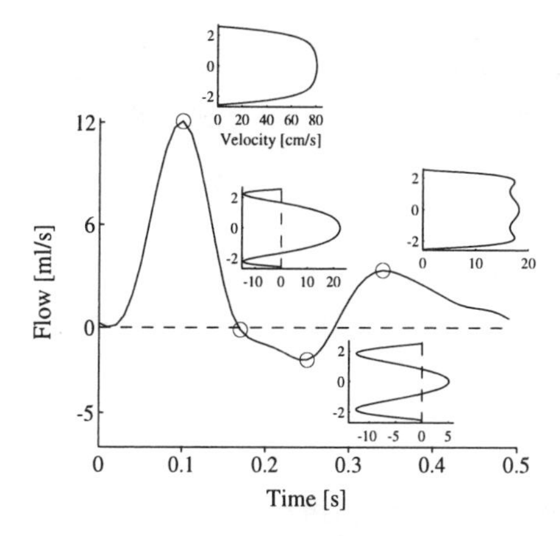

Figure 8: Axial velocity profiles at some characteristic instants of the cardiac cycle (*Artery-2*).

In fact, in the last assumption the shear stress at the wall can be expressed[1] as a function of the blood flow rate, Q, by the following relationship:

$$\tilde{\tau}_w = 4\,\mu\,\frac{Q}{\pi R^3} \quad , \tag{33}$$

where μ is the blood viscosity. In the case of *Artery-2*, a comparison between the wall shear stress calculated by eqn (33) and that obtained with the simulation is shown in Fig. 9.

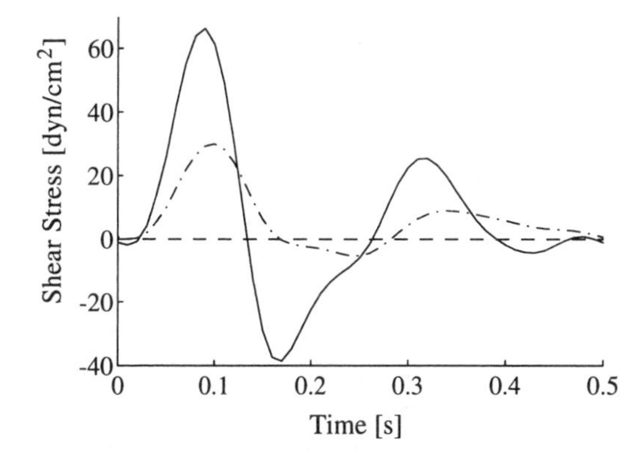

Figure 9: Wall shear stress versus time calculated by eqn (33) as compared with that provided by the simulation in the case of *Artery-2*.

Surprisingly, the observed differences are significant and the error that one commits using eqn (33) to estimate wall shear stress is dramatically large, particularly in the negative phase of the shear stress.

4 Effects of a mild stenosis

Hemodynamics in a stenotic condition have been extensively investigated in the last few decades and significant contributions have been obtained in these studies by means of the numerical simulation of the arterial hemodynamics. One of the main reasons stimulating these investigations is that the flow disturbances occurring in an atherosclerotic vessel are believed to have diagnostic value in recognising the pathological sites. In fact, the atherogenic process induces a restricted thickening of the artery intimal layer that locally reduces wall elasticity and artery lumen with consequent influence on vessel geometry. Alterations in vessel shape give rise to significant local modification of the time and spatial distribution of blood velocity. By means of non-invasive diagnostic techniques it is possible to discover when an altered velocity field is occurring in an artery and then to guess that an atherosclerotic process is probably taking place.[25] To this aim it is clinically significant to know the flow patterns occurring in a stenotic vessel to usefully construe measurements carried out by means of high resolution Doppler and MRI velocimeters[26] or digital cineangiography[27] and, in particular, to detect pathological sites in the early stage of disease.

There is also another important reason that has urged theoretical studies of hemodynamics in a stenotic vessel. The interaction between the pulsating arterial wall and the blood flow plays a pivotal role in the atherogenic process and several hemodynamic factors influenced by wall motion are believed to participate in the aetiology of atherosclerosis (e.g. see Ref. 28). Such factors as, for instance, wall shear stress, are not easily *in vivo* measurable because of the difficulties that arise when the neighbourhood of the moving wall must be investigated. Unfortunately, the more the experimental conditions tend to mimic the real system behaviour, the less the resolution of the experimental method. The *in vivo* methods so far proposed do not have sufficient resolution to show many of the complex phenomena occurring in a stenotic vessel: it is not possible to distinguish between turbulence and vorticity nor it is possible to determine the wall shear stress, whose role in damaging the endothelial layer that lines the artery lumen has still to be coherently explained. Moreover, the complexity of hemodynamics laws and, in particular, their marked nonlinear nature, causes the system behaviour to be strongly dependent on the experimental conditions. For instance, time-average pressure drop through a stenotic segment is not only dependent on the mean flow component but also on the pulsatile flow component superimposed, so that the time-average pressure drop is different from the drop caused only by the mean flow. For the same reason it is possible to make an appropriate remark on the large investigative effort made to establish the critical Reynolds numbers for which the flow separation or the transition from a laminar to a turbulent flow occurs downstream of the stenosis. Also these numbers are strongly dependent on the experimental conditions: there is a difference between steady and pulsatile flow conditions[29] and the critical numbers are influenced by the frequency of the pulsatile component,[30] by wall elasticity[31] and by local vessel geometry.[32] Thus these numbers are less meaningful than is generally believed. To overcome such experimental limitations it is then right to resort to the numerical simulation that allows a precise, quick evaluation of hemodynamic effects produced by changing the physical and the geometrical properties of an arterial vessel.

Almost all theoretical studies reported in literature analyse the case of a severe stenotic condition in a rigid tube.[13,33,34,35] Therefore, the numerical simulator here presented, which includes pulsatile wall motion, was used to examine the hemodynamics in a mild stenosis (for more details see Ref. 19). The early phase of atherogenic process when a neointimal hyperplasia or an intimal fibrocellular hypertrophy occurs was investigated simulating, by means of the model, a local, slight increase in the wall thickness of a canine femoral artery. An axisymmetric thickening of the intimal layer placed in the middle of the vessel was simulated in such a way to induce in the stenotic region a cosine shaped axial profile of the unstressed inner radius. The length of the thickened segment was equal to 1 cm and the maximum reduction, in the unstressed condition, of the inner radius was 1.1% corresponding to a 2.2% of area reduction, that represents a very mild stenotic condition. The wall elastic behaviour changes in the thicker region according to the constitutive expression (23).

Simulation results show that a stenosis has very little influence on pressure and flow waveforms or on their mean values as long as the area reduction is slight, in agreement with the experimental observation[36] that the pressure drop through a mild stenosis is very small compared with the pressure itself. For this reason a mild stenosis does not cause any clinical disorder. Just during the systolic phase -i.e. when a large flow rate occurs and the blood has its highest acceleration- the pressure drop across the stenosis slightly increases. On the contrary, also when the stenosis is mild the local flow is significantly perturbed: spatial velocity distribution is disturbed by stenosis even when the local growth of wall thickness is small and the flow rate is low. Moreover, disturbances are clearly evident whatever instant of the cardiac cycle is considered. To make the disturbances more apparent it is useful to show distribution along the vessel of the axial velocity profiles at fixed instants of the cardiac cycle.[37]

In the systolic phase, when the flow peak occurs, all blood elements are moving forward and in the stenotic region the blood is accelerated due to the vessel narrowing. During this phase of the cardiac cycle the most evident disturbance induced by the stenosis is that the mid-vessel velocity is higher -about 14%- in the post-stenotic region than in the rest of the vessel.

After the systolic peak, the speed of blood begins to reduce and the sign of pressure gradient reverses. The blood particles close to the wall having the lower momentum are the first to reverse their direction moving backward. Since the lower velocities occur in the post-stenotic region, the profile inversion takes place first here (Fig. 10). As an effect of the blood deceleration, a flow separation occurs close to the wall and a little vortex takes place in the post-stenotic region. In the short area where the vortex occurs there is a very large wall shear stress (about 50 dyn/cm^2), whereas in the neighbouring zones the shear has a negative value (about -20 dyn/cm^2). The vortex circulation then induces a very high gradient of instantaneous wall shear stress in the post-stenotic region.

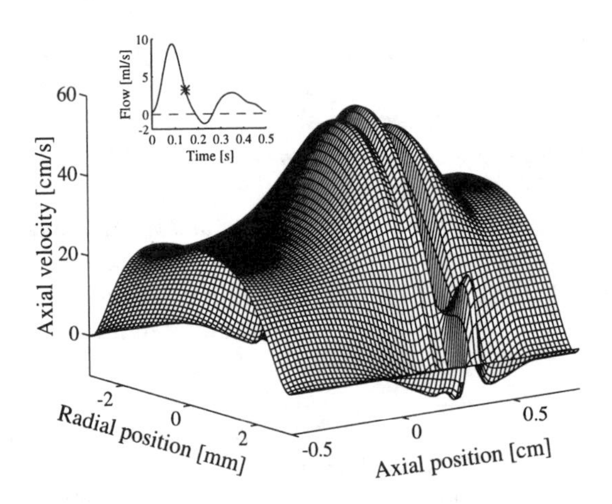

Figure 10: Axial velocity distribution along the vessel at time indicated in the flow curve. In the post-stenotic region, near the wall, flow separation occurs and vortex circulation takes place.

The remaining part of the diastole after the back flow phase is characterised by a period with a forward flow during which a second flow peak occurs. Unlike the systolic phase, the velocity distribution is disturbed near the wall (Fig. 11).

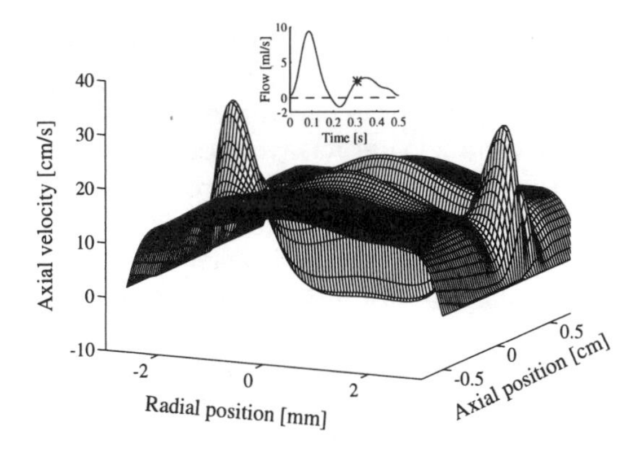

Figure 11: Axial velocity distribution along the vessel at time indicated in the flow curve. In the post-stenotic region, because of the vortex circulation, blood particles near the wall move forwards with high positive velocities, while elements in the axial core move backwards and the velocity profiles become negative.

Owing to the disturbance, the wall stress in the post-stenotic region is higher than in the pre-stenotic one (Fig. 12). During this phase of the diastolic period, the vortex circulation spreads throughout all post-stenotic regions (Fig. 11). The stress on the wall is very high and the peak occurring in the post-stenotic region is four times the shear in the unstenotic region, being even larger than the systolic peak (Fig. 12).

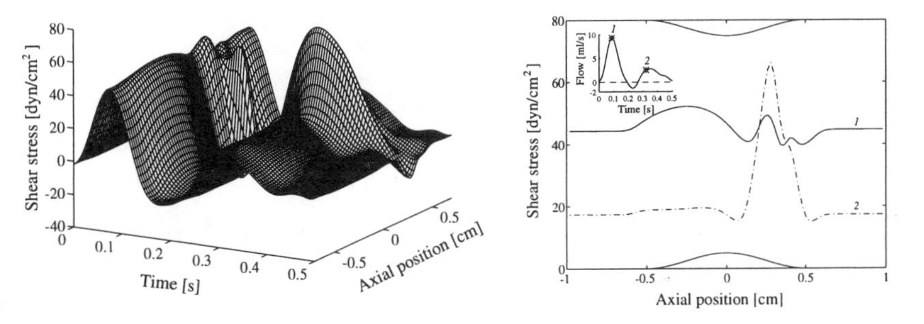

Figure 12: *Left panel.* Wall shear stress along the vessel during a cardiac cycle. The simulated mild stenosis strongly disturbs the stress distribution overall in the diastolic phase. *Right panel.* Wall shear stress distribution along the vessel calculated at the cardiac cycle instants indicated in the flow curve.

Blood flow separation and recirculation, i.e. the vortex circulation, starts during systolic deceleration and occurs in the downstream region of stenosis. This is in accordance with Ref. 38, that visualised a clear zone of flow separation in a mould of iliac artery with a localised atherosclerotic plaque. Sabbah[39] observed in angiographies of coronary arteries just distal to the mild atherosclerotic plaque, a region of relative stasis of contrast material representing a condition of vortex circulation. More recently, Smedby *et al.*[27] using cineangiography with digital analysis of arrival times of contrast material, evidenced clear flow separation zones downstream of early femoral atherosclerotic plaques, so proving that flow disturbances can be usefully employed to detect stenotic sites.

A large range of flow conditions characterises the physiological state of the cardiovascular system, e.g. going from rest to exercise, therefore the local flow pattern in the stenotic region can change dramatically.[19] In fact, flow separation, vortex and the consequent wall stress peak are events characterising conditions of large flow rate. However, other physical and geometrical factors together with the flow rate, condition the appearance of vortex: local tapering and inner radius in the divergent region, amplitude of the flow peak, heart rate, wall elasticity, and blood viscosity.

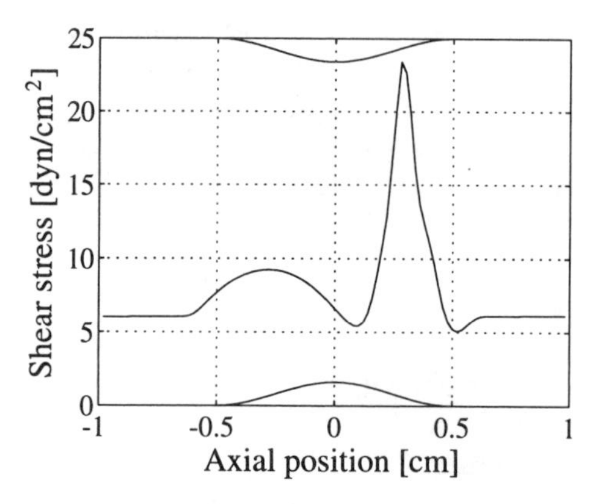

Figure 13: Distribution through the stenotic segment of the time-average over a cardiac cycle of the wall shear stress. A very high peak of stress occurs in the poststenotic region.

Results of numerical simulation make evident that a recirculating flow, though appearing only during a part of the cardiac cycle, induces a high wall stress and above all a disturbed shear gradient whose role in damaging the endothelium is still to be explored. To emphasise the portion of the vessel under the greatest amount of stress, the distribution along the vessel of time-average shear stress over a cardiac cycle was calculated (Fig. 13). The very high peak of shear occurring in the post-stenotic region is surprising. Evidently, the continuous persistence of the vortex circulation close to the wall for almost the entire cardiac cycle considerably influences the time-average value. The highest shear occurring distally of simulated stenosis provides a coherent, theoretical explanation of Languille's experimental observation[40] proving that endothelial cells downstream of a coarctated site have a very different histological aspect from the cells of the rest of the vessel. Moreover, the large shear stress together with the already known shear stress sensitive mechanism causing vascular tone relaxation, may be the cause of the vessel dilatation observed downstream of stenoses.[41]

5 Effects of tapering on pulse propagation

Wave propagation along an arterial vessel has been thoroughly investigated in the last few decades. Propagation phenomenon is intrinsically dependent on the wall elasticity so that it has been reasonable to suppose that information on the state of arterial wall could be indirectly obtained by measuring pressure pulse propagation. To this aim, theoretical analyses based on linearised Navier-Stokes equation were made and relationships linking physical and geometrical vessel properties with wave velocity and attenuation were proposed. Experimental studies have shown significant differences between measurements and theoretical predictions. Milnor & Bertram[23] for instance, have ascertained that pressure wave attenuation measured *in vivo* is considerably greater – about 10 times – than that calculated by linear theory. The results obtained by linear theories are dubious mainly because the equations governing blood motion are highly nonlinear, and also because the invariance of the propagation coefficient along the artery is unreally assumed. In fact, in the linear models the invariance of physical and geometrical properties along the arterial vessel is erroneously assumed. In particular, the natural narrowing of the arterial vessel is neglected, thus assuming the same propagation constant for forward and retrograde travelling waves. A theoretical analysis of the nonlinear effects on wave propagation was produced in Ref. 11. In particular, it was studied how a pressure pulse moves in a Newtonian viscous fluid contained in a deformable, tapered tube. Geometrical and physical parameters of the tube were similar to the physiological ones of a canine femoral artery and a transient pressure perturbation was simulated by applying a single flow pulse in the proximal extremity of the vessel. The flow pulse perturbs an initial condition of steady pressure and during the transient it was possible to distinguish between a direct pulse pressure that moves from the proximal section towards the distal end and a reflected one travelling in the opposite direction. To maximally enhance the reflection, an infinite extremity impedance in the distal section was adopted. In order to evidence the influence of nonlinearities, the case of a vessel with a cylindrical shape and the case of a vessel with a tapering equal to the anatomical one were compared. Results of this study are here briefly summarised.

In the case of a vessel with a cylindrical geometry the pulse shape is almost preserved while travelling towards the distal end, whereas the amplitude slowly decreases (Fig. 14).

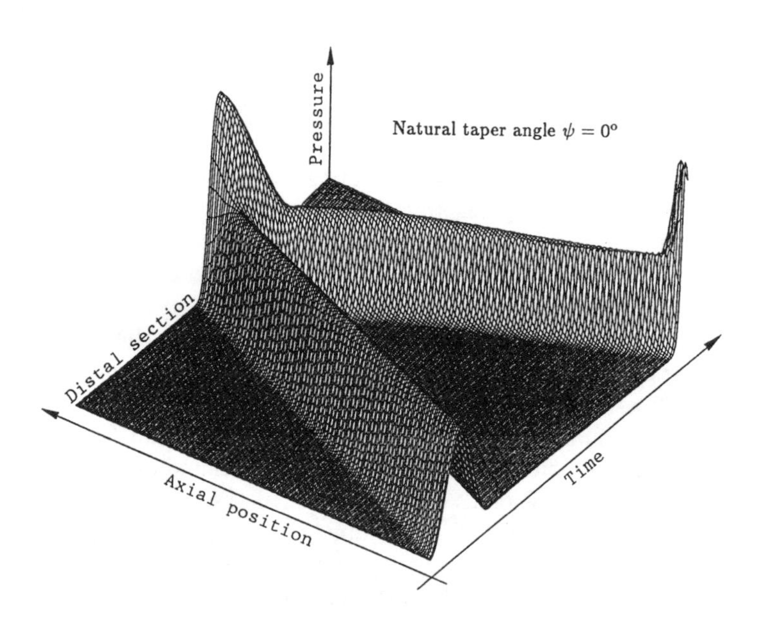

Figure 14: Three-dimensional view of the pressure pulse propagation along an untapered vessel. The pressure pulse, induced by flow perturbation in the proximal extremity, moves along the vessel towards the distal section. Here the pulse is completely reflected, then it returns to the proximal section. (From Reference 11)

Propagation phenomena radically changed when the tapered vessel was simulated (Fig. 15). The pulse amplitude in a tapered vessel increases continuously as it moves along the vessel towards the distal extremity so that when it arrives in this section it is higher than in the untapered case. In particular, at the distal end the pulse is much more than double that at the proximal extremity. On the contrary, as the pulse is reflected and goes backwards, it is greatly damped, so that in the proximal section the pulse peak is considerably lower than in the cylindrical vessel. For a natural taper angle equal to 0.14° the pulse is almost completely cancelled after the first reflection and the perturbation is almost extinct when the pulse returns to the proximal extremity.

Natural tapering of vessels induces then a different behaviour in the pulse propagation according to whether the pulse is moving forward or backward. In fact, during forward travel the amplitude is rising and during backward travel it is reducing. However, reduction is greater than amplification, so that the overall effect is a greater attenuation than the cylindrical geometry.

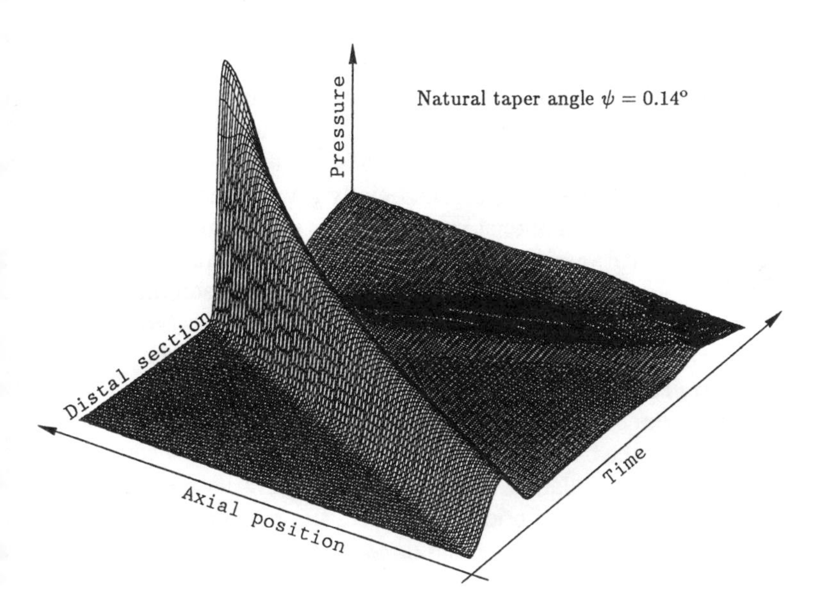

Figure 15: Three-dimensional view of the pressure pulse propagation when the natural taper angle is close to the typical value of a canine femoral artery. (From Reference 11)

These results gave a more exhaustive explanation of some phenomena observed in the arterial system. When pulse pressure induced by the cardiac ejection runs along the arterial tree – moving from the aorta to the peripheral points – its amplitude is amplified. This fact can be explained as an effect of the continuous reduction of cross-sectional area of the single vessels. Moreover, the smallness at the root of the aorta of the reflections can be also interpreted as an effect of the natural tapering of the arterial vessels.

6 Effects of flow rate

One of the fundamental problems in vascular research is the accurate determination of those parameters which characterise the local hemodynamics of main vessels, such as longitudinal impedance, propagation constant, wave propagation velocity and so on.[1] These parameters are generally based on classic Womersley's linear theory. However, several experimental and computational works published in the last two decades[7,9,23,10,11] clearly demonstrate the important role that fluid-motion and wall nonlinearities play in determining the blood velocity field in arteries. In particular, as reported in the previous section, numerous works focus on the effects of blood pressure on vascular hemodynamics, with emphasis on vascular wall motion. On the contrary little attention has been paid to the effects of changes in blood flow, which may be large in relation to variations in tissue metabolic needs.

The purpose of this section is to investigate, by using the mathematical model illustrated above, the effects of blood flow variations on the longitudinal impedance per unit length (Z_l), which is one of the main parameters in determining arterial hemodynamics, especially wave propagation properties.

The proposed computational method allows us to easily change all geometrical and physical vessel properties, as well as the flow rate conditions, and then to obtain accurate numerical solutions in which measurement errors are absent.

Different natural taper angles were simulated to investigate how this arterial property affects the Z_l calculation. Moreover, to evaluate how the flow regime affects the estimation of Z_l, different hemodynamic conditions were simulated. One is the basal condition relating to a rest state, whereas the other relates to a light vasodilation state, which induces in the vessel segment a 50% mean flow increase. Periodic axial pressure gradient (dP/dz) and flow curves necessary for the longitudinal impedance calculation were computed with reference to the middle section of the vessel segment. In particular, axial pressure gradient was evaluated across a 6 mm long segment. Typical time courses, during a cardiac cycle, of blood flow $Q(t)$ and pressure $P(t)$, in the middle section of the vessel, are shown in Fig. 16.

Data obtained by each simulation were subjected to Fourier harmonic analysis. The spectrum was limited to frequencies below 12 Hz, thus considering the first seven harmonics according to Milnor.[1] For each harmonic the complex quantity $Z_l(w)$ was then obtained on the basis of the relation:

$$Z_l(f) = -\frac{dP(f)}{dz}/Q(f) \quad . \tag{34}$$

Normalised values of Z_l were calculated by dividing the Poiseuille resistance $R_p = 8\mu/\pi R^4$ into the real part of Z_l and by dividing the inertance $L_w = 8f\rho/3R^4$ into the imaginary part. L_w represents the theoretical inertance when frequency tends to zero, according to Womersley's theory.[1]

Different hemodynamic conditions have been considered. The flow waveforms in the middle section of the arterial segment, in basal conditions and after peripheral vasodilation are shown in the left panel of Fig. 16. The increase of mean flow rate is 50% with respect to basal value. The pressure waveforms in the same section and conditions are shown in the right panel of Fig. 16.

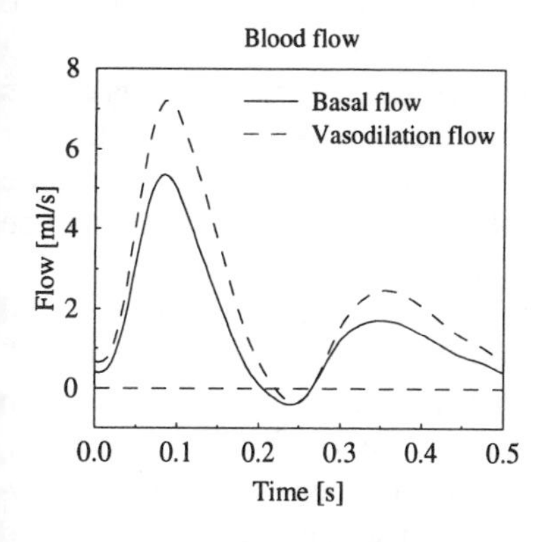

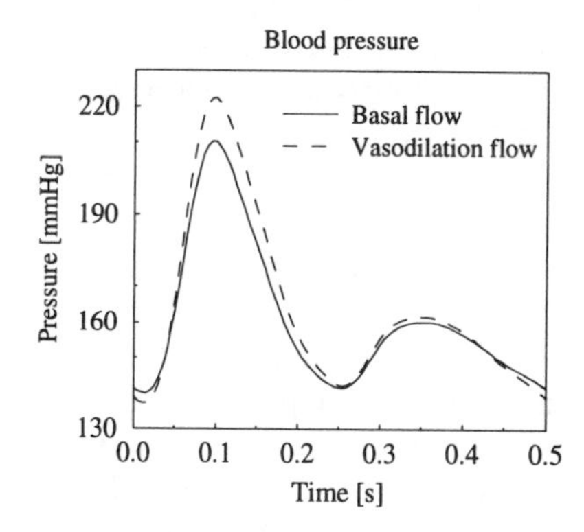

Figure 16: *Left panel.* Flow waveform in the middle section of the arterial segment in basal conditions (solid line, mean value=1.6 ml/s) and after peripheral vasodilation (broken line, mean value=2.4 ml/s). *Right panel.* Pressure waveform in the same section and for the same conditions. (From Reference 42)

The changes in waveform and mean pressure appear not significant and do not appreciably modify the operating region in the wall elastic characteristic. This corresponds to physiological conditions, where pressure tends to be maintained unchanged by control mechanisms, even though flow is largely varied to meet metabolic needs.

Figure 17 shows the longitudinal impedance versus frequency as computed in basal regime for three different values of natural taper angle.

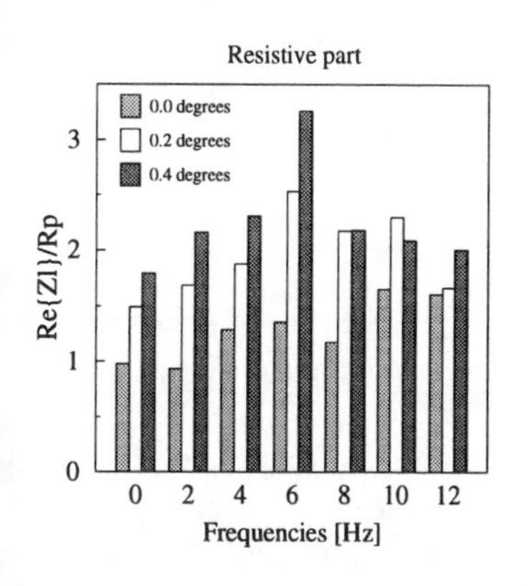

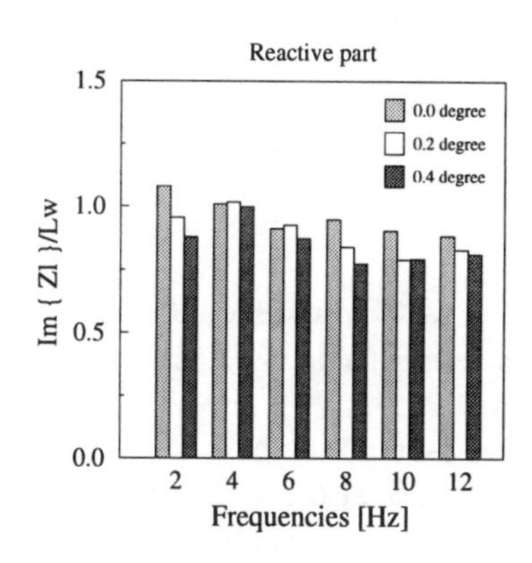

Figure 17: Normalised real (*left panel*) and imaginary (*right panel*) part of Z_l versus frequency, for different values of the anatomical taper angle. (From Reference 42)

The normalised real part (left panel) shows that increased taper angle, in spite of smallness of its absolute value (max 0.4°), significantly affects resistive part of Z_l at most frequencies.

The longitudinal resistance, i.e. impedance at zero frequency, is about 1.8 times compared to no taper case and this difference increases with frequency until 6 Hz. Indeed, the inertial forces associated to the convective terms of fluid motion equations, manifest themselves in an amount that is not negligible when the vessel is narrowing. This finding is in accordance with the literature,[9,23] and confirms that convective acceleration affects hemodynamics in the range of low (not only zero) frequencies. Also the normalised imaginary part appears affected, but to a minor extent.

Figure 18 shows the longitudinal impedance versus frequency as computed in basal and peripheral vasodilation regimes, corresponding to flow curves of Fig. 16. All the geometrical and physical parameters of the vessel segment (e.g., natural taper angle = 0.4°) are unchanged. The real part (left panel) is normalised by dividing by the Poiseuille resistance. It is evident that just a limited increase (50%) in flow corresponds to a not negligible positive step of real part of Z_l for most of the considered harmonics (only at 8 and 12 Hz a small variation is found).

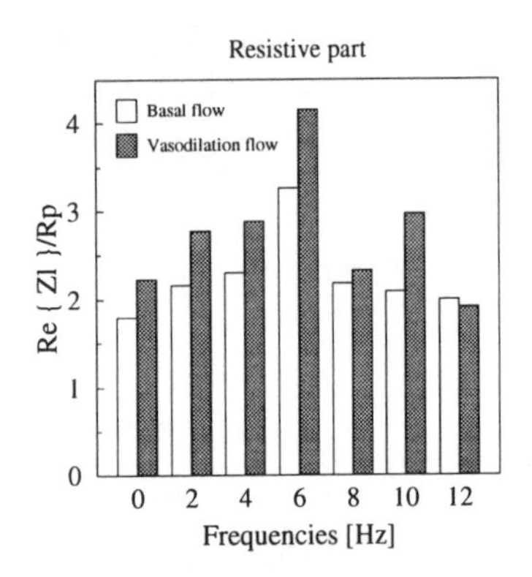

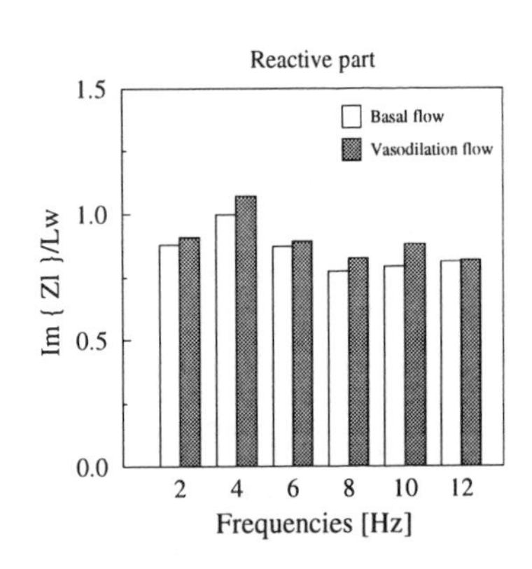

Figure 18: Normalised real (*left panel*) and imaginary (*right panel*) part of Z_l as a function of frequency for the two different flow regimes shown in Fig. 16. (From Reference 42)

The observed increase of the resistive part of Z_l ranges from about 25% at 0 Hz to 27% at 6 Hz. For greater flow rates -however lower than 3 times the basal value- the percent variation of the resistive component of Z_l results approximately proportional to mean flow. Of course variations in the resistive part of the vascular longitudinal impedance has direct effects on the attenuation constant for waves travelling along the vessel. The imaginary part of Z_l

(Fig. 18) does not appear significantly affected by mean flow variations.

It is emphasised that due to the nonlinear behaviour of the system the longitudinal impedance depends on the pulsatile flow components in addition to mean flow. In other words it is a function of flow waveform.

In a steady flow regime the longitudinal impedance becomes purely resistive and its dependence on the flow value as calculated by the simulator is shown in Fig. 19.

As it is evident from Fig. 19, the ratio of mean gradient to mean flow highly depends on the steady flow. The flow resistance becomes, for high flows, 6 times higher than the low flow values, therefore Poiseuille equation largely underestimates the steady resistance when the flow increases.

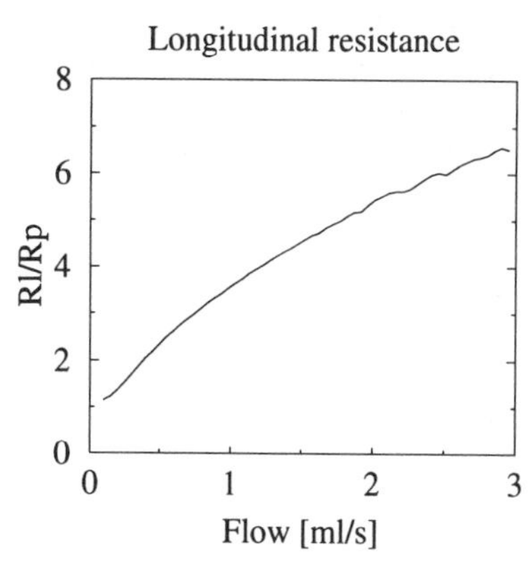

Figure 19: Normalised longitudinal resistance versus flow value in steady conditions. (From Reference 42)

In conclusion, the obtained results show that the vascular parameters estimated by processing arterial pressure and flow data are flow dependent. This is particularly evident for those parameters which are related to the resistive part of Z_l. As a consequence, linear theories that do not take into account this dependence may lead to erroneous conclusions.

7 Concluding remarks

In this chapter a nonlinear model of the arterial blood motion in deformable tapered vessels is presented. The mathematical formulation is developed assuming a two-dimensional blood flow and taking into account the nonlinear terms of Navier-Stokes' equations -convective accelerations- as well as instantaneous taper angles and wall motion. The theory also includes an expression of the axial and radial velocity profiles, which permits their calculation with low computational complexity. Simulations results show that dramatic differences exist with respect to predictions provided by linear theories. This finding is in accordance with the literature and confirms that it is very important to retain the convective accelerations in motion equations for a correct analysis of blood motion.

It is also shown that parietal shear stress is significantly affected by arterial wall elasticity and lumen.

Taper angle, in spite of smallness of its absolute value (max 0.4°), significantly affects local hemodynamics as well as pulse wave propagation. Indeed, the inertial forces associated to the convective terms of fluid motion equations, play a role that is not negligible when the vessel is narrowing. The above-stated results demonstrate that the propagation constant depends on the propagation direction. As a consequence, linear models, which are based on the opposite assumption, may lead to mistaken conclusions when extended to study the hemodynamics of a tapered vessel.

Further, it is shown that a mild stenosis significantly perturbs velocity profiles. The disturbance occurs after the systolic peak and persists for the whole diastolic phase.

It is finally emphasised that due to the nonlinear behaviour of the system the longitudinal impedance depends on the mean flow rate as well as on the pulsatile flow components. Therefore, a similar dependence may be expected for all the derived arterial quantities.

Acknowledgements

This work was supported by the Italian Ministry for the University and the Scientific and Technological Research (MURST).

References

1. Milnor, W.R. *Hemodynamics*, 2nd edition, Williams & Wilkins, Baltimore and London, 1989.

2. Womersley, J.R. An elastic tube theory of pulse transmission and oscillatory flow in mammalian arteries, Wright Air Development Center Technical Report WADC-TR- 56-614, Wright Patterson Air Force Base, Ohio, 1957.

3a. Taylor, M.G. An approach to the analysis of the arterial pulse wave. I. Oscillations in an attenuation line. *Phys. Med. Biol.*, 1957, 1, 258-269.

3b. Taylor, M.G. An approach to the analysis of the arterial pulse wave. II. Fluid oscillations in an elastic tube. *Phys. Med. Biol.*, 1957, 1, 321-329.

4. McDonald, D.A. & Taylor, M.G. The hydrodynamics of the arterial circulation, *Prog. Biophys. Med. Biol.*, 1959, 9, 107-173.

5. Westerhof, N., Bosman, F., De Vries, C.J. & Noordergraaf, A. Analog studies of the human systemic arterial tree, *J. Biomechanics*, 1969, 2, 121-143.

6. Avanzolini, G., Belardinelli, E., Capitani, G. & Passigato, R. Steady state numerical model of the human systemic arterial tree, *Proceedings of the IFAC Symposium on Automatic Control and Computers in the Medical Field*, Brussels, 1971.

7. Ling, S.C. & Atabek H.B. A nonlinear analysis of pulsatile flow in arteries, *J. Fluid Mech.*, 1972, 55, 493-511.

8. Streeter, V.L., Keitzer, W.F. & Bohr D.F. Pulsatile pressure and flow trough distensible vessels, *Circ. Res.*, 1963, 13, 3-20.

9. Ling, S.C., Atabek H.B., Letzing W.G. & Patel D.J. Nonlinear analysis of aortic flow in living dogs, *Circ. Res.*, 1973, 33, 198-212.

10. Belardinelli, E. & Cavalcanti, S. A new nonlinear two-dimensional model of blood motion in tapered and elastic vessels, *Comput. Biol. Med.*, 1991, 21, 1-13.

11. Belardinelli, E. & Cavalcanti, S. Theoretical analysis of pressure pulse propagation in arterial vessels, *J. Biomechanics*, 1992, 25, 1337-1349.

12. Pedley, T. J. *The fluid mechanics of large blood vessels,* 1980, Cambridge University Press, London.

13. Deshpande, M. D., Giddens, D. P. & Mabon, R. F. Steady laminar flow through modelled vascular stenoses, *J. Biomechanics,* 1976, 9, 165-174.

14. Perktold, K., Resch, M. & Florian, H. Pulsatile non-Newtonian flow characteristics in a three-dimensional human carotid bifurcation model, *J. Biomech. Eng.*, 1991, 113, 464-475.

15. Lou, Z. & Yang, A computer simulation of the non-Newtonian blood flow at the aortic bifurcation, *J. Biomechanics*, 1993, 26, 37-49.

16. Carew, T. E., Vaishnav, R. N. & Patel, D. J. Compressibility of the arterial wall, *Circ. Res.*, 1968, 23, 61-68.

17. Patel, D. J., Greenfield, J. C. & Fry, D. L. *In vivo* pressure-length-radius relationship of certain blood vessels in man and dog, In *Pulsatile Blood Flow* (Edited by Attinger E.O.), 1964, 293-302, McGraw-Hill, New York.

18. Manak, J. J., The two-dimensional *in vitro* passive stress-strain elasticity relationship for the steer thoracic aorta blood vessel tissue, *J. Biomechanics*, 1980, 13, 637-646.

19. Cavalcanti, S. Hemodynamics of an artery with mild stenosis, *J. Biomechanics*, 1994, in press.

20. Fung, Y.C. *Biomechanics: Mechanical Properties of Living Tissues.*, Springer-Verlag, New York, 1981.

21. Westerhof, N., Elzinga, J. & Sipkema, P. An artificial arterial system for pumping hearts, *J. Appl. Physiol.*, 1971, 31 776-781.

22. Gear, C.W. Automatic detection and treatment of oscillatory and /or stiff ordinary differential equations, *Numerical integration of differential equations and large linear systems,* ed. J. Hinze, Springer-Verlag, New York, 1982, pp. 190-206.

23. Milnor, W.R. & Bertram, C.D. The relation between arterial viscoelasticity and wave propagation in the canine femoral artery *in vivo*, *Circ. Res.*, 1978, 43, 870-879.

24. Li, J.K.J., Melbin, J., Riffle, R.A. & Noordergraaf, A. Pulse wave propagation, *Circ. Res.*, 1981, 49, 442-451.

25. Hutchison, K.J. & Karpinski, E. *In vivo* demonstration of flow recirculation and turbulence downstream of graded stenoses in canine arteries, *J. Biomechanics*, 1985, 18, 258-259.

26. Boesiger, P., Maier, S. E., Liu Kecheng, Scheidegger, M. B. & Meier, D. Visualization and quantification of the human blood flow by magnetic resonance imaging, *J. Biomechanics*, 1992, 25, 55-68.

27. Smedby, Ö., Högman, N., Nilsson, S. & Urikson, U. Flow disturbances in early femoral atherosclerosis-an *in vivo* study with digitized cineangiography, *J. Biomechanics*, 1993, 26, 1105-1115.

28. Nerem, R. M. Vascular fluid mechanics, the arterial wall, and atherosclerosis, *J. Biomech. Engng*, 1992, 114, 274-282.

29. Siouffi, M., Pelissiser, R., Farahifar, D. & Rieu, R. The effects of unsteadiness on flow through stenoses and bifurcations, *J. Biomechanics*, 1984, 17, 299-280.

30. Withaya, Y. & Young, D. F. Initiation of turbulence in models of arterial stenoses, *J. Biomechanics*, 1979, 12, 185-196.

31. Stein, P. D., Walburn, F. J. & Blink, E. F. Damping effect of distensible tubes on turbulent flow: implications in the cardiovascular system, *Biorheology*, 1980, 17, 275-280.

32. Walburn, F. J. & Stein, P. D. Effect of vessel tapering on the transition to turbulent flow:

implications in the cardiovascular system, *Fed. Proc. FASEB*, 1980, 39, 1068-1070.

33. Daly, B. J. A numerical study of pulsatile flow through stenosed canine femoral arteries, *J. Biomechanics*, 1976, 9, 465-475.

34. Johnston, P. R. & Kilpatrick, D. Mathematical modelling of flow through an irregular arterial stenosis, *J. Biomechanics*, 1981, 24, 1069-1077.

35. Cheng, T., Deville, D., Dheur, L. & Vanderschuren, L. Finite element simulation of pulsatile flow through arterial stenosis, *J. Biomechanics*, 1992, 25, 1141-1152.

36. Back, L. H., Radbill, J. R., Cho Y. I. & Grawford, D. W. Measurement and prediction of flow through a replica segment of a mildly atherosclerotic coronary artery of man, *J. Biomechanics*, 1986, 19, 1-17.

37. Cavalcanti, S. & Carota, L. Velocity profiles distribution along an arterial vessel: a way to improve the detection of stenotic sites, *Med. Biol. Eng. Comp.* 1995, in press.

38. Walburn, F. J., Sabbah, H. N. & Stein, P.H. Flow visualitation in a mold of an atherosclerotic human abdominal aorta, *J. Biomech Engng*, 1981, 39, 1068-1070.

39. Sabbah, H. N., Khaja, F., Hawkins, E. T. *et al.* Flow disturbances in early femoral atherosclerosis-an *in vivo* study with digitized cineangiography, *J. Biomechanics*, 1993, 26, 1105-1115.

40. Langille, B. L., Reidy, M. A. & Kline, R. L. Injury and repair of hendothelium at sites of flow disturbances near abdominal aortic coarctations in rabbits, *Arteriosclerosis*, 1986, 6, 146-152.

41. Roach, M. R. An experimental study of the production and time course of poststenotic dilatation in the femoral and carotid arteries of adult dogs, *Circ. Res.*, 1963, 13, 537-551.

42. Cavalcanti, S. & Gnudi, G. Influence of blood flow on the longitudinal impedance in arteries: a simulation study, *Computational Biomedicine*, ed. K.D. Held, C.A. Brebbia, R.D. Ciskowski & H. Power, Computational Mechanics Publications, Southampton, UK, 1993, pp. 125-132.

Appendix A - On the expression of wall shear stress

The shear stress acting on the arterial wall, τ_w, can be calculated by the projection of the stress vector on the tangent unit vector to the wall:

$$\tau_w(z,t) = \nu\,\rho \left(\frac{1}{R}\frac{\partial w}{\partial \eta} + \frac{\partial u}{\partial z} \right)\Bigg|_{\eta=1} \cos\left(\arctan\left(\frac{\partial R}{\partial z} \right) \right) \quad . \qquad (A1)$$

The expression of the axial velocity derivative with respect to the radial coordinate can be obtained from eqn (16)

$$\left.\frac{\partial w}{\partial \eta}\right|_{\eta=1} = \sum_{k=1}^{N} 2\,k\,\alpha_k \quad . \qquad (A2)$$

From eqns (13) – (14) and the no-slip condition (8.a) it is possible to obtain the axial derivative of the radial velocity

$$\left.\frac{\partial u}{\partial z}\right|_{\eta=1} = \left(\frac{\partial^2 R}{\partial z \partial t} - \frac{3}{R}\frac{\partial R}{\partial z}\frac{\partial R}{\partial t} \right) \quad . \qquad (A3)$$

Finally, the expression of the wall shear is obtained substituting the ($A2$) and ($A1$) into the ($A1$)

$$\tau_w(z,t) = \nu\,\rho \left(\frac{1}{R}\sum_{k=1}^{N} 2k\alpha_k + \frac{\partial^2 R}{\partial z \partial t} - \frac{3}{R}\frac{\partial R}{\partial z}\frac{\partial R}{\partial t} \right) \cos\left(\arctan\left(\frac{\partial R}{\partial z} \right) \right) \quad . \qquad (A4)$$

Chapter 5

A numerical heart and circulation model to simulate hemodynamics for rate-responsive pacing

A. Urbaszek & M. Schaldach

Zentralinstitut für Biomedizinische Technik, Universität Erlangen-Nürnberg, Turnstraße 5, D-91054 Erlangen, Germany

Abstract

A computer model of the human cardiovascular system has been developed for simulating the short-term regulatory processes that adapt the cardiac output of the heart to changing circulatory demands. Both peripheral and cardiac control mechanisms are considered. Therefore, the model has to involve Starling's law as well as chronotropic and inotropic responses, controlled by the autonomic nervous system (ANS). The model serves not only as a tool for investigating the effects of all these control processes under several conditions (physical exercise, posture changes, temperature and other influences) for the healthy subject, the special importance lies in the simulation of cardiomyopathies, chronotropic insufficiency and other functional disturbances typical for pacemaker patients. The goal of the simulations is to develop effective control algorithms for rate-responsive pacemakers based on measured cardiac parameters, especially for those systems providing closed loop control (i. e., the increase of the stimulation rate has a feedback effect on the measured parameter). When natural feedback is still present (which is often the case), simulations serve to adjust the rate response characteristics to the natural feedback. Since these studies depend highly on the particular sensor system, the principle of the pacemaker system considered here, intracardiac unipolar impedance measurements, will also be described briefly.

1 Introduction

Today's multiprogrammable pacemakers provide patients with improved cardiac function by correcting bradyarrhythmias and restoring proper AV synchrony within the heart. However, in the presence of insufficient chronotropic response to exercise, the benefits provided to the patients are limited. Therefore, rate-responsive pacemakers have been designed to permit workload-dependent adaptation of the pacing rate based on the measurement of corporeal and cardiac control parameters.

This is done by sensing one or more parameters being in a certain correlation to the metabolic demand with the help of special sensors. The several principles in use to date for rate adaptation (see Fig. 1 for an overview) can be subdivided into two classes: open loop and closed loop systems, the latter exhibiting a certain heart rate feedback on the sensor signal.[1] Rickards[2] proposed a classification according to the physiological level of sensing into primary sensors detecting the natural heart rate control parameters (i.e., the sympathetic and parasympathetic nerve activities), secondary sensors detecting internal changes caused by exercise and tertiary sensors detecting external changes with exercise (e. g., activity sensors). Obviously, primary and secondary sensors provide closed loop feedback and are closer to the bodies own 'controlling network', the autonomic nervous system (ANS), while tertiary sensor systems providing open loop control often have only a loose correlation to the real hemodynamic demand and are more susceptible to external disturbances.[3] Therefore, aiming at the restoration of natural cardiovascular control and adaptation, closed loop systems are preferable.

On the other hand, the feedback mechanism for the particular sensor, including ANS pathways as well as hemodynamic relationships, must be

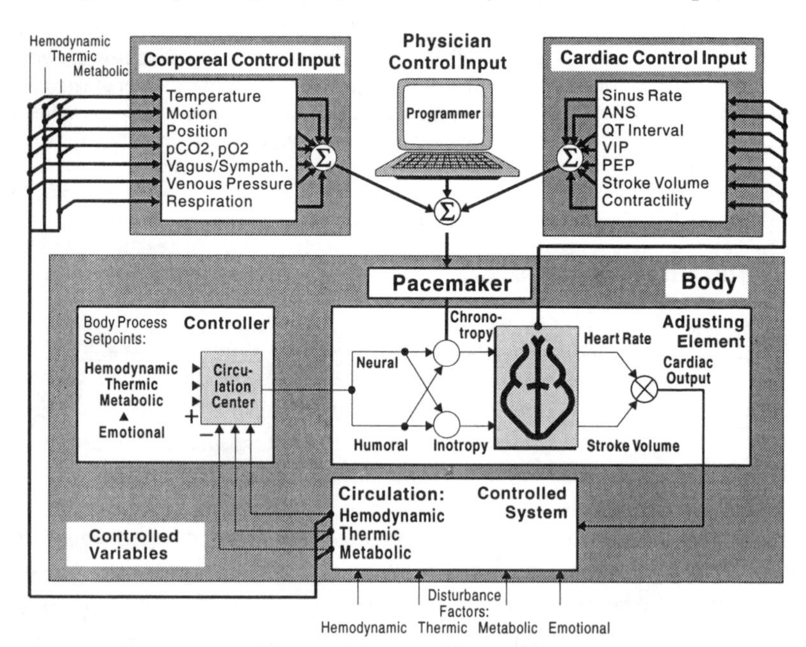

Figure 1: Corporeal and cardiac control parameters employed for rate-responsive pacing

known and, moreover, considered for programming in order to adapt the stimulation frequency properly. While programming of open loop systems requires, in general, the determination of one or more parameters describing the response function curve between the sensor signal and the adaptive rate (e. g., the slope when a linear relationship is assumed), a certain threshold value and the response time,[4] the situation is different in closed loop systems: Less arbitrary assumptions concerning the sensor-rate relationship are necessary; instead, the rate feedback on the sensor signal must be evaluated from open loop (i. e., fixed rate) measurements.

Detailed knowledge of the heart rate feedback on the particular sensor parameter for different load and other challenges to the cardiovascular system (e.g., orthostasis) as a consequence of hemodynamic regulation is a prerequisite for the design of a physiological rate adaptation algorithm. For this reason, a mathematical model of the cardiovascular control mechanisms relevant to this problem is needed, which allows to derive the functional relationships between the measured physiological parameter and hemodynamic quantities in a system-oriented approach.

While the first task can be performed in a general manner, for the second part a special sensor system must be considered. This study is focused on the ANS-controlled pacemaker, which uses intracardiac unipolar impedance measurements to detect changes in the myocardial contraction pattern under the influence of sympathetic nerval activation, i. e., changes in contractility. Since the same nerves control the natural sinus rate in healthy individuals, the obtained information about the ANS tone is a physiological control parameter to adapt the stimulation frequency.

2 Global model structure

In the past, very different hemodynamic circulation models representing the human cardiovascular system or parts of it have been developed on very different levels of abstraction - ranging from very simple, lumped-parameter models neglecting all pulsatile effects up to elaborate nonlinear, time-variant models with distributed parameters (a classification is given in Ref. 5). Obviously, for this analysis a lumped-parameter representation is sufficient, while pulsatile effects should not be neglected, since it is the time course of the mechanical contraction which determines the intracardiac impedance signal. Furthermore, only short-term regulatory processes are of interest - long-term effects such as changing amounts of total blood volume can be neglected.

This "class" of models has been employed in several studies, either for general analysis [5-8] or special purposes such as drug administration [9,10] or total artificial heart applications [11,12] (see Ref. 13 for a review). Although most of these studies deal with similar hemodynamic simulations, the cardiac function must be considered in much greater detail for the objective of this study.

Several physiological subsystems are of importance to the chronotropic and inotropic regulation of the heart: the hemodynamic characteristics of the systemic and pulmonary vessel beds as well as mechanical properties of the myocardium, which determine the dynamics of contraction, have to be considered. Furthermore, fundamental biochemical processes at the cellular level as the underlying mechanisms should also be involved since they can explain the inotropic effects as well as pharmacologic influences (e. g., calcium antagonists, β-blockers), since the analysis must also be applied to typical pacemaker patients which may suffer from other diseases and receive therapeutic agents.

The whole circulation model has, therefore, a hierarchical and modular structure; modules describing the circulatory hemodynamics, heart muscle

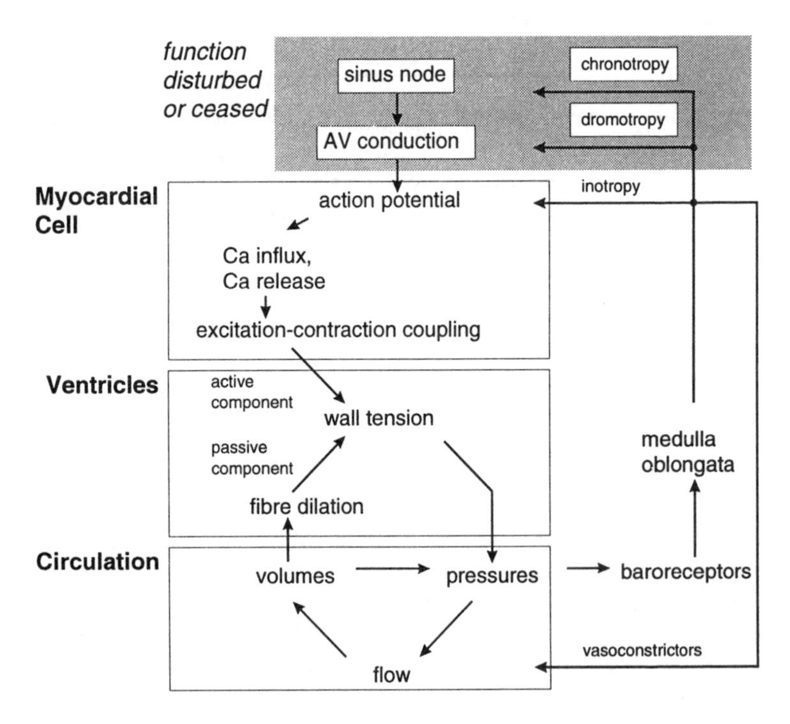

Figure 2: Data dependence graph showing the modular structure of the heart and circulation model.

mechanics and excitation-contraction coupling at the cellular levels are linked via "interfacing" variables such as, e. g., ventricular pressures, which are calculated in the heart mechanics module based upon cardiac muscle relationships (length-tension relation, force-frequency relation) and then promoted to the circulation module (Fig. 2).

3 Circulatory system

The peripheral circulation is represented by a few segments only; this part of the model must reproduce the following effects:

- changes in preload,
- changes in afterload,
- effects of physical exercise,
- effects of postural changes.

This can be resolved by subdividing the whole circulation loop into 10

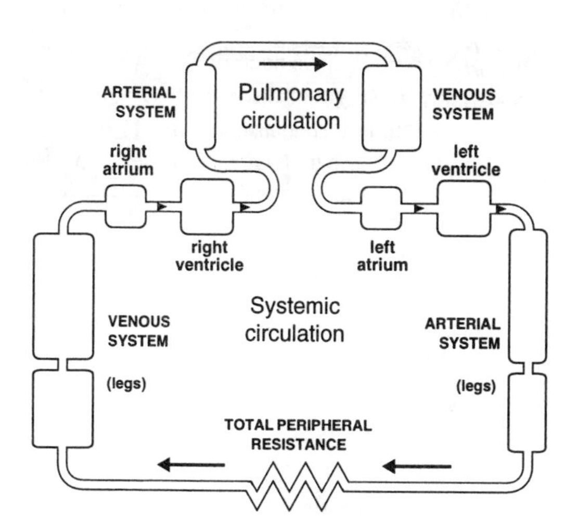

Figure 3: Circulatory loop subdivided into 10 segments for lumped-parameter model representation.

compartments as shown in Fig. 3: The four heart chambers, four compartments representing the arterial and venous systems of pulmonary and systemic parts of the circulation to define pre- and afterload for each side of the heart, and two additional segments for the lower limbs to allow a simplified description of the shifts in blood volume towards the lower body regions during orthostasis.

3.1 Governing equations

The fundamental equations for incompressible, Newtonian fluid flow are given by the continuity and the Navier-Stokes equations (written in vector notation, after Ref. 14):

$$div\,\underline{v} = 0, \qquad \frac{\partial \underline{v}}{\partial t} + (\underline{v} \cdot \nabla)\,\underline{v} = \underline{f} - \frac{1}{\rho}\nabla p + \frac{\eta}{\rho} \cdot \Delta \underline{v} \tag{1}$$

with

$\underline{v}$	velocity vector,
p	pressure,
$\underline{f}$	body force vector (e. g., gravity),
ρ	fluid density,
η	fluid viscosity.

In order to obtain analytical solutions for the computation of blood flow, a number of simplifying assumptions are usually made:

- The blood is considered to be an incompressible, Newtonian fluid,
- the flow is laminar (Poiseuille flow as through stiff tubes with fixed diameter).

In cylindrical coordinates, these general assumptions in addition with further specific premises for particular terms allow eqn (1) to be transformed in a one-dimensional equation for mean (over cross-section) longitudinal flow velocity:

$$\rho \int_1^2 \frac{\partial v}{\partial t}\, dz + \rho \frac{v_2^2 - v_1^2}{2} = \rho\, g\, l \sin\theta - (p_2 - p_1) - \Phi \cdot R \tag{2}$$

with

v	longitudinal flow velocity,
z	longitudinal coordinate,
g	gravitational constant,
l	effective vessel length,
θ	angle between the vessel axis and horizontal position,
Φ	blood flow,
R	viscous flow resistance.

This equation is a kind of energy balance for non-stationary viscous flow (Bernoulli equation in a more general form) considering acceleration work (left), increase in kinetic energy due to decrease in diameter between z_1 and z_2, work against gravitation, pressure difference and friction losses (all terms

represent specific energies, i. e., have the dimension "energy per unit mass").
Solving the integral for a vessel of length l with the substitutions

$$\Phi = \frac{dV}{dt} = A\frac{dz}{dt} = A\,v$$

and

$$L = \rho\frac{l}{A}$$

yields

$$L\frac{d\Phi}{dt} + \frac{\rho}{2}\left(\frac{\Phi_2^2}{A_2^2} - \frac{\Phi_1^2}{A_1^2}\right) = p_1 - p_2 + \rho\, g\, l \sin\theta - \Phi\cdot R \tag{3}$$

with

V	volume,
A	cross-sectional area,
L	inertance, representing the inertial resistance of blood against acceleration.

Although the assumptions for obtaining eqn (3) are contradictory for some terms - e. g., for the inertial term the velocity profile was assumed to be flat, for the viscous term to be parabolic (Poiseuille flow) - this can be compensated by the use of 'corrected' values for L or R.[15,16] For the different regions of the circulation, one or more terms of eqn (3) can be neglected. The flow between atria and ventricles is dominated by the inertial and kinetic terms, while the viscous term and, of course, gravity are negligible:[17]

$$L\frac{d\Phi}{dt} + \frac{\rho}{2}\frac{\Phi_2^2}{A_2^2} = p_1 - p_2\,, \tag{4}$$

with Φ_2 being the flow at the smallest flow cross-section (mitral valve). Between the ventricles and arterial compartments, the same assumptions, except that the viscous resistance must be taken into account here, are made:[6]

$$L\frac{d\Phi}{dt} + \frac{\rho}{2}\frac{\Phi_2^2}{A_2^2} = p_1 - p_2 - \Phi\cdot R. \tag{5}$$

The "flow rectifying" function of the heart valves is simulated by a condition which suppresses negative flows. The kinetic term is negligible for the rest of the circulation[6,9] while the gravitational term plays a significant role in the leg vessels. Furthermore, the inertial effects can be neglected for all other vessels except for the large arteries, so that the rest of the circulation follows the simple equation analog to Ohm's law:

$$\Delta p = \Phi \cdot R , \tag{6}$$

where Δp denotes the sum of the pressure gradient and the hydrostatic difference, the latter being nonzero for the arterial and venous leg vessels:

$$\Delta p = p_1 - p_2 + \rho \, g \, l \sin\theta . \tag{7}$$

Equations (4)-(7) describe the relationships between flow and pressure gradient for the different parts of the cardiovascular system. In addition, volume-flow and pressure-volume relationships are required for a complete flow simulation according to Fig. 2. These relations are obtained from the Windkessel model according to Fig. 3. The volume-flow relation follows simply from continuity for each compartment:

$$\frac{dV}{dt} = \Phi_{in} - \Phi_{out} . \tag{8}$$

The pressure-volume relation is given by the compliance C of the Windkessel compartments, which is defined as follows:

$$C = \frac{\Delta V}{\Delta p} . \tag{9}$$

For most vessels, this value (the reciprocal of elastance) is nearly constant over a "physiological" range of pressures, which leads to the pressure-volume relation

$$p = \frac{V - V_0}{C} , \tag{10}$$

with V_0 being the unstressed volume for every specific vessel compartment. Only for the vena cava, where transmural pressures near 0 can occur, the vessel may partially collapse, which leads to a strongly nonlinear p-V relation in a manner shown in Fig. 4, where C is a function of V.

Furthermore, the veins especially exhibit a time-dependent p-V behavior due to viscoelastic properties of the vessel walls characterized as 'delayed compliance'.[18] This effect is included in eqn (10) by an additional time-dependent term:[6]

$$p = \frac{V - V_0}{C} + R' \frac{dV}{dt} \tag{11}$$

with R' being the wall viscosity coefficient.

The complete equation set for the circulation module (see Fig. 2) is now

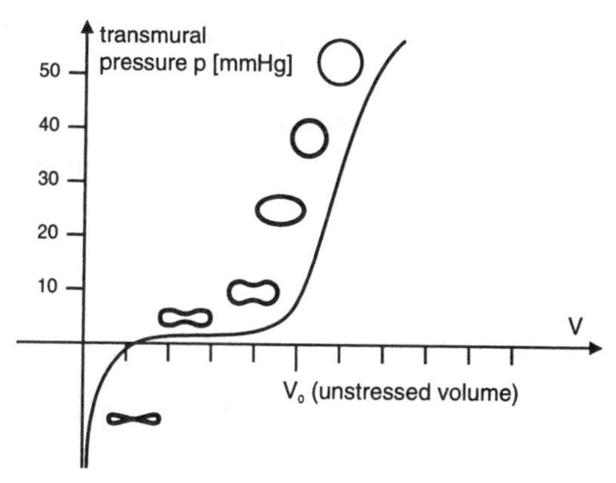

Figure 4: Nonlinear pressure-volume relationship of the vena cava due to collapse at low transmural pressures (redrawn from Ref. 13).

established; the diastolic and systolic pressure-volume relations of the heart chambers which are much more complex are derived in a different way in the next section.

4 The heart

The mechanical properties of the myocardial fibers, which are major determinants of filling and ejection of the heart, are not only complex in itself; furthermore, they are subject to intrinsic regulation (Frank-Starling mechanism) and extrinsic control by the autonomic nervous system (ANS). The coordinated interplay between cardiac and peripheral adaptation allows the cardiac output to be increased about fivefold from 5 l/min for normal adults at rest up to 25 l/min or even more at peak exercise, which underlines the importance of the local and ANS-mediated control mechanisms to maintain a sufficient perfusion pressure when the metabolic demand is increased extensively.

Therefore, the underlying structural characteristics must be represented appropriately in the model. Since ventricular mechanics have a much larger influence on the pumping performance for the heart than the atria, a detailed geometrical model for the ventricles will be developed; in contrast, the atria are more simply modeled in a way similar to the circulatory vessel compartments.

4.1 Atria

During diastolic filling, the pressure-volume relationships of the atria are calculated in the same way as for compliant compartments according to eqn (10). During systole, atrial contraction causes a pressure rise of 4 to 6 mmHg (right) and 7 to 8 mmHg left, respectively.[18] This additional pressure step is incorporated into eqn (10) by an additive term; the systolic pressure rise is controlled by a state variable indicating the active state of atrial contraction.

4.2 Ventricles

A detailed representation of ventricular dynamics is developed in two steps: At first, measurements on isolated papillary muscle strips are utilized to establish relationships between fiber length and tension development. Secondly, the law of Laplace is applied to obtain the cavity pressures from wall tensions, which requires certain geometrical assumptions about the shape of the ventricles.

4.2.1 Single fiber relationships

Fiber stretching and contraction is usually explained using the three-component model developed by Hill in 1939. It is also applicable to cardiac muscle, provided that typical differences to skeletal muscle are taken into account.[19] The major presumption of this model is: A contractile element (CE) develops force upon excitation but is freely extensible at rest, a series elastic element (SE) is in series either with CE (Maxwell version) or with CE and PE in parallel (Voigt version). The parallel elastic element (PE) is in parallel with CE or CE and SE in series and maintains the resting length of CE (Fig. 5).

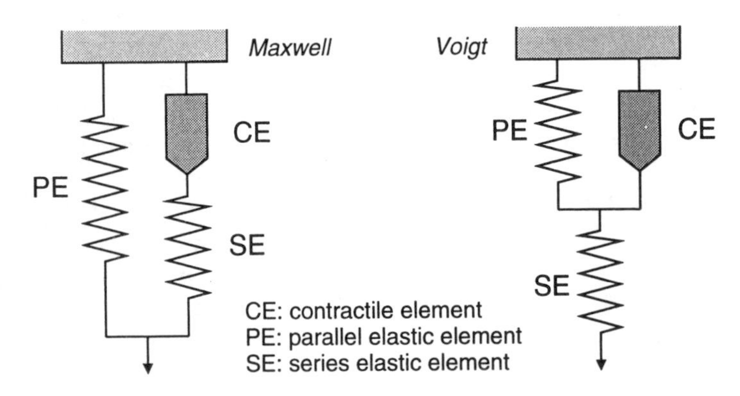

Figure 5: Two versions after Maxwell and Voigt of Hill's three-element muscle model (redrawn from Ref. 21).

Although these models are very simple abstractions (viscous properties are not involved, further shortcomings are discussed in Ref. 20) and the functional elements have no direct morphological correlatives,[21] they have been proven useful to describe the muscle behavior on a macroscopic level. Therefore, this description will be used also for this analysis, at least in principle. The formulas are based on the Maxwell model, since the forces from the passive component (PE) and the active component (CE and SE) can be simply summed up due to their parallel arrangement; furthermore, in the relaxed state (diastole), the passive stretching depends solely on PE; SE has no influence due to the free extensibility of CE. This leads to the following equation (wall tensions instead of forces will be considered from here on to allow the calculation of pressures from wall tension and wall thickness):

$$\sigma = \sigma_{pass} + \sigma_{act}\,, \tag{12}$$

with

σ wall tension,

σ_{pass} passive component,

σ_{act} active component.

The tension developed upon passive stretching during diastolic filling increases nearly exponentially with circumference; the steep slope at higher volumes may be explained not only from tissue properties, but also from the influence of the pericardium. The passive length-tension relationship is therefore expressed by:

$$\sigma_{pass} = \sigma_0 \cdot \left(e^{\alpha \cdot (l - l_0)} - 1\right) \qquad \text{for } l > l_0\text{, otherwise zero} \tag{13}$$

with

σ_0, α constants,

l_0 unstressed fiber length (definition of a 'typical fiber length' follows).

The active component is more complicated, since the actively developed tension depends on:

- the fiber length prior to excitation (preload, given by the end-diastolic volume),
- the myocardial contractility,
- indirectly, the pressure against the ventricle has to eject (afterload).

The first two influences are comprised by the expression:

$$\sigma_{act} = \sigma_{max}(f_s) \cdot f(l) \cdot g\left(\frac{dl}{dt}\right), \tag{14}$$

with

σ_{max}	maximum tension,
f_s	sympathetic nerve firing frequency,
$f(l)$	active length-tension relationship,
$g\left(\frac{dl}{dt}\right)$	function describing the force-frequency relationship (hyperbolic Hill relation),

where $f(l)$ is a function of the fiber length and mimics an important feature of sarcomere dynamics: Force or tension development is maximal (σ_{max}) for a certain optimum length and decreases when the length differs from this value (Fig. 6 shows simulated curves of this relationship). Unlike skeletal muscle,

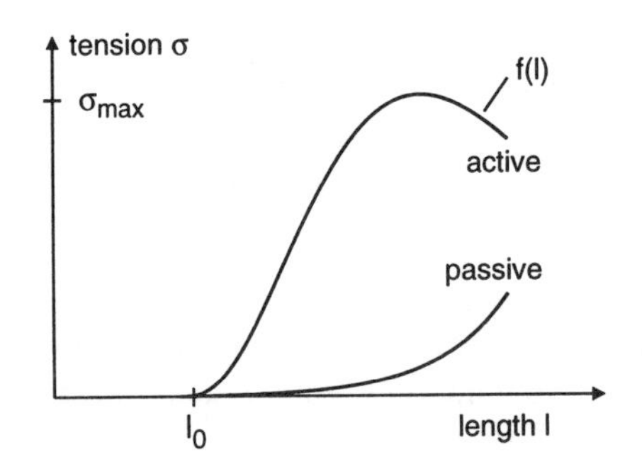

Figure 6: Passive and active length-tension relationships as simulated for the myocardial fibers for the ventricular model; the resulting wall tension σ is the sum of both components.

cardiac muscle operates under physiological conditions exclusively on the ascending limb of this relationship, i. e., a higher fiber stretching, caused by increased diastolic filling, leads in non-failing myocardium to higher tensions during contraction. Including this basic mechanism allows the model to mimic the Frank-Starling mechanism. The hyperbolic relation between velocity and load opposing contraction is introduced by the function g.

The inotropic influence of the ANS, mediated mainly by sympathetic stimulation (the ventricles have only poor parasympathetic innervation), increases the activity of cross-bridge building between the contractile filaments and leads to a rise in σ_{max}, which is therefore a function of the sympathetic nerve firing rate f_s (see below). With the functions f and g, tension development changes also for different afterloads (later valve openings, different pressures against which the ventricles must eject blood).

4.2.2 Ventricular geometry

The geometrical shape hypothesis of the two ventricles must be detailed enough to describe the following phenomena:

- the relationship between wall tensions and cavity pressures,
- mutual influences of both ventricles,
- Frank-Starling mechanism.

A geometrical arrangement fulfilling these requirements is the sphere-ellipsoidal approximation for the left and right ventricles as depicted in Fig. 7. Further assumptions along with this model are:

- All parts of the ventricle are excited at the same time; i. e., the typical conduction pattern is neglected.
- The walls consist of isotropically contracting tissue, wall tension is considered to be constant over the whole cross-section, i.e., different tensions at the inner and outer side of the ventricles are neglected.
- The global spheric shape and ellipsoidal structure are maintained throughout the cardiac cycle, only the dimensions (sphere radius, long ellipsoidal axis, and wall thicknesses) change.
- The myocardium is considered to be incompressible, which allows to calculate wall thickening with decreased overall dimensions during contraction.

The latter assumption of constant tissue volume holds for every of the three wall sections: left free ventricular wall, septum, and right ventricular wall.

At first, sphere radius r and ellipsoidal half axis a, which represent mean values of the inner and outer measure (to be exactly, which divide the wall sections into two subcomponents of equal volumes each), are determined from left and right

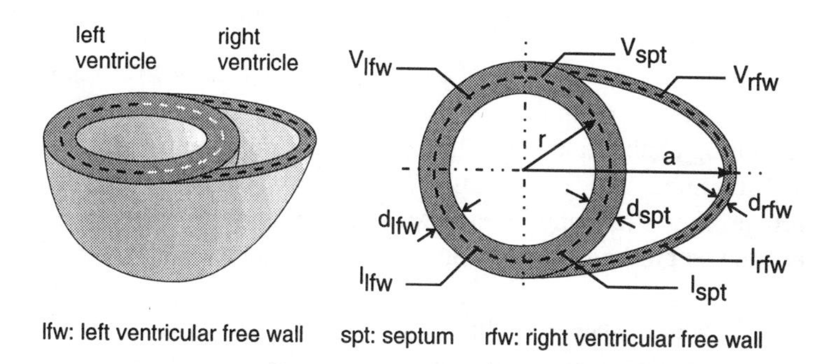

Figure 7: Sphere-ellipsoidal model of left and right ventricles

ventricular volumes according to the following geometric relationships:

$$r \approx \frac{3}{4\pi} \cdot \sqrt[3]{V_{LV} + \frac{1}{2}\left(V_{lfw} + V_{spt}\right)} \tag{15}$$

with

V_{LV}	left ventricular volume,
V_{lfw}	wall volume of the left ventricular free wall,
V_{spt}	wall volume of the interventricular septum,

and

$$a \approx \frac{3}{2\pi r^2}\left(V_{RV} + \frac{1}{2}\left(V_{rfw} + V_{spt}\right)\right) - r \tag{16}$$

with

V_{RV}	right ventricular intracavitary volume,
V_{rfw}	wall volume of the right ventricular free wall.

In these equations, several approximations have been made; e. g. it is assumed that the volume of a sphere with radius r is composed by the left ventricular blood volume, half the left ventricular and septal wall volumes. For different wall thicknesses and wall volumes this is not exactly true, but the error resulting from these approximations is negligible.

From these dimensions, the circumferential measures of each wall section as depicted in Fig. 7 follow from the following relationships:

$$l_{lfw} \approx 2\pi r \cdot \lambda \tag{17}$$

$$l_{spt} \approx 2\pi r \cdot (1 - \lambda) \tag{18}$$

$$l_{rfw} \approx \frac{\pi}{2}\left(\frac{3}{2} \cdot (r + a) - \sqrt{r \cdot a}\right) \tag{19}$$

In eqns (17) and (18) λ is the ratio of the left ventricular free wall circumference to the whole sphere circumference; it is initialized with the value 0.5 and changes during contraction depending on left and right ventricular pressures.

The circumferential length values obtained from eqns (17)-(19) determine the average fiber stretching in each wall section and are used, therefore, to calculate the length-dependent component of the wall tensions (eqns (13) and (14)).

For thin-walled spherical objects, the relationship between wall tension σ, wall thickenss d, radius r and internal pressure p is given by the law of Laplace:[19]

$$p = \sigma \cdot \frac{2d}{r} \tag{20}$$

Table 1: Typical values of ventricular dimensions (from Ref. 22)

	LV wall thickness d_{lfw}	spt. wall thickness d_{spt}	RV wall thickness d_{rfw}	LV minor axis ($\approx 2r - d_{lfw}$)	RV minor axis ($\approx a - d_{spt} - r$)
cm	1.37	1.23	0.39	5.2 (end-diast) 3.4 (end-syst.)	3.4 (end-diast.) 2.6 (end-syst.)

This formula results from a balance between the forces produced by the internal pressure and the wall tensions. Since the assumption of a thin-walled geometry is not acceptable for the ventricles (see Table 1, from Ref. 22), the force equilibrium must be re-evaluated for thick-walled geometries. For a cross-section along the sphere diameter it can be written (see Fig. 8):

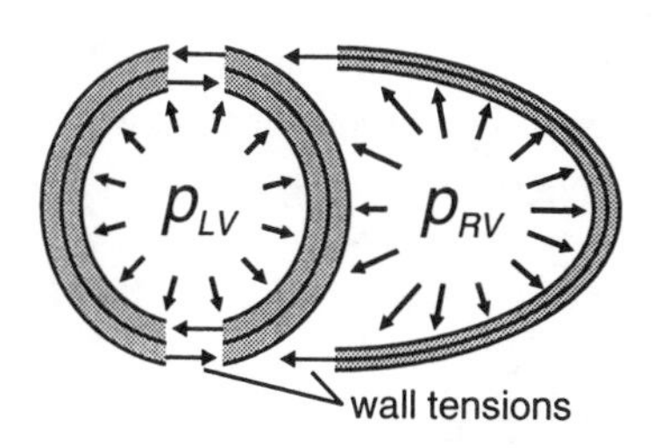

Figure 8: Force equilibrium between ventricular pressures and wall tensions

$$p \cdot A_{cavity} = \sigma \cdot A_{wall}$$

$$p \cdot \pi \cdot \left(r - \frac{d}{2}\right)^2 = \sigma \cdot \pi \cdot \left(\left(r + \frac{d}{2}\right)^2 - \left(r - \frac{d}{2}\right)^2\right)$$

$$p \cdot \left(r^2 - rd + \frac{d^2}{4}\right) = \sigma \cdot 2rd$$

$$\sigma = \frac{p}{2}\left(\frac{r}{d} - 1 + \frac{1}{4}\frac{d}{r}\right)$$

With the values from Table 1, the third term in parenthesis is small compared to the others and will be neglected, which leads to the final relationship between pressure and wall tension:

$$p \approx \frac{2\sigma}{\frac{r}{d} - 1}.$$

This general relationship must now be applied to the different wall sections: For the left and right free walls, this relationship holds directly (the non-spherical shape of the right ventricle is neglected for this purpose):

$$p_{LV} \approx \frac{2\sigma_{lfw}}{\frac{r}{d_{lfw}} - 1}, \tag{21}$$

$$p_{RV} \approx \frac{2\sigma_{rfw}}{\frac{r}{d_{rfw}} - 1}. \tag{22}$$

For the interventricular septum, the opposing effect of both ventricular pressures must be considered. Therefore, the following approximation is used:

$$p_{LV} - p_{RV} \approx \frac{2\sigma_{spt}}{\frac{r}{d_{spt}} - 1}. \tag{23}$$

With eqns (21)-(23), from three different wall tension values σ_{lfw}, σ_{spt} and σ_{rfw} each calculated from the individual circumferential length of the respective wall section, the two ventricular pressures p_{LV} and p_{RV} are obtained. According to Fig. 2, these values are provided for the circulation module.

With eqns (12) through (23), the mechanical part of the ventricular contraction is described; what is still missing is a sufficient description of the 'active state', i. e. the time course and the degree (amplitude) of the muscle fiber activation during systole. This is of special importance, since these cellular and subcellular biochemical processes are responsible for changes in myocardial contractility or inotropic state, which are directly controlled by the ANS efferent nerves and are therefore attractive to be monitored by rate-adaptive pacemakers to obtain a physiological control signal for rate response.

5 The myocardial cell

The basic mechanism of muscle activation is the excitation-contraction coupling in the myocardial cells. In the presence of free Ca^{2+} ions in the intracellular space, the contractile filaments actin and myosin start to build up cross-bridges at special binding sites, which results in a relative shift between these two macro-molecules of about 20 nm for each cross-bridge binding.[23] According to the sliding-filament theory, by repeated building, release and rebuilding of cross-bridge bindings, the sarcomere, which is the smallest unit of a contractile fiber and has a length of $\approx 2.5\ \mu m$, is able to contract about $0.4\ \mu m$ in 100 ms (measured on a frog muscle preparation[23]). The release of Ca^{2+} into the cell as a result of electrical excitation, the excitation-contraction coupling, and the relaxation following the action potential as well as the influence of sympathetic nerve transmitter substances (adrenergic substances) are the processes that the myocardial cell model has to cover.

Figure 9 summarizes the different ion currents contributing to the intracellular Ca^{2+} concentration: With the onset of the action potential Ca^{2+} ions start to move into the cell following the electrochemical gradient through voltage-dependent Ca^{2+}-specific ion channels of several types (L-type and T-type) in the sarcolemma. This Ca^{2+} influx contributes only a small percentage directly to excitation-contraction coupling; its main function is to trigger further Ca^{2+} release from the sarcoplasmic reticulum. This intracellular tubular network serving as an intracellular Ca^{2+} store provides most of the activator calcium from special release sites (terminal cisternae); re-uptake of Ca^{2+} during relaxation takes place on different sites.[24] Therefore, the sarcoplasmic reticulum is represented in the model by two compartments as shown in Fig. 9.

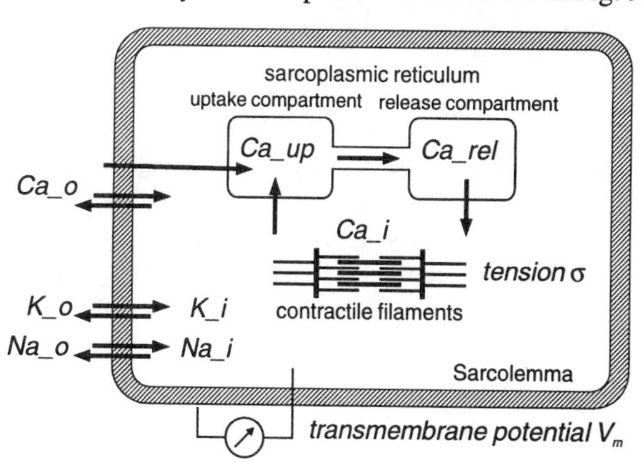

Figure 9: Intracellular Ca^{2+} concentration, which determines force of contraction, results from passive (along the electrochemical gradient) and active (via ion pumps) ion movements between extracellular and intracellular space, depending on the transmembrane potential, and to and from the uptake and release compartments of the sarcoplasmic reticulum.

5.1 Governing equations

In general, intracellular Ca^{2+} concentration results from the sum of all contributing currents:

$$\frac{d\,Ca_i^{2+}}{dt} = \frac{\Sigma\, I_{Ca}}{2V \cdot n \cdot e} \tag{24}$$

with

I_{Ca} any Ca^{2+} current into the intracellular space (negative if directed outwards),

V intracellular volume,

$n = 6.024 \cdot 10^{23}\ mol^{-1}$,

$e = 1.6022 \cdot 10^{-19}\ As$.

Each passive ion current is driven by the electrochemical gradient, which is a superposition of the two driving forces, transmembrane potential (electrical gradient) and diffusion potential (a function of the concentration gradient), and depends on the ion-specific channel conductivity:[24]

$$I_{Ca} = G_{Ca} \cdot (E_{Ca} - V_m)\,, \tag{25}$$

with

G_{Ca} conductivity,

E_{Ca} diffusion potential,

V_m transmembrane potential.

The conductivity values G_{Ca} depending on the ion channel permeabilities of the sarcolemma or sarcoplasmic reticulum are voltage- and time-dependent. The membrane potential itself depends on the membrane capacity and the trespassing ion currents; besides the different types of Ca^{2+} currents, Na^+ and K^+ play a key role for the electrical activity. The following equation is derived from an electrical analog model of the sarcolemma after Hodgkin-Huxley (Fig. 10):

$$\frac{d\,V_m}{dt} = \frac{1}{C} \cdot \sum_i G_i \cdot (V_m - E_i), \tag{26}$$

with

C sarcolemmal membrane capacity ($C/A \approx 1\ \mu F/cm^2$)

$i = Ca^{2+}, Na^+, K^+$.

It is evident that the complex activation and inactivation patterns of the

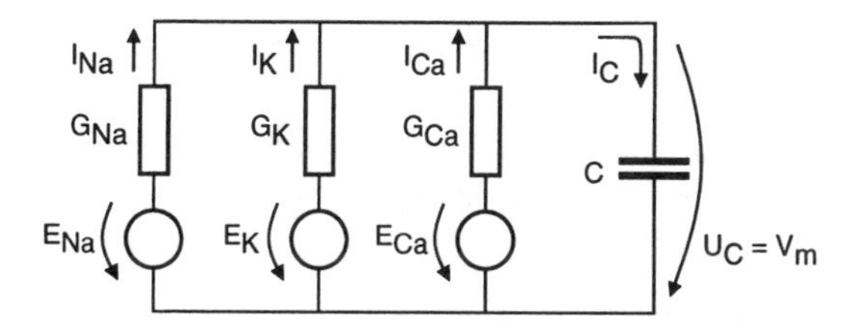

Figure 10: Electrical equivalent of the sarcolemma, consisting of the sarcolemmal capacitance C, ion channel conductivities G and diffusion potentials E (simplified).

different types of Ca^{2+}, Na^{+} and K^{+} channels, which are individually modulated by adrenergic agents,[25,26] have a crucial influence on the transmembrane potential and the excitation process.

The diffusion potential for Ca^{2+} results from the concentration gradient according to the NERNST equation:

$$E_{Ca} = -30.5\ mV \cdot \log \frac{Ca_i^{2+}}{Ca_e^{2+}} \tag{27}$$

with

Ca_i^{2+} intracellular Ca^{2+} concentration (10^{-7} mol/l during resting potential),

Ca_e^{2+} extracellular Ca^{2+} concentration ($2 \cdot 10^{-3}$ mol/l).

While Ca_i^{2+} increases during excitation about two orders of magnitude to approximately 10^{-5} mol/l, the concentration gradients of Na^{+} and K^{+} remain nearly unchanged due to much higher concentrations. Their diffusion potentials are therefore constant:

$$E_K = -95\ mV, \quad E_{Na} = 72 mV. \tag{28}$$

To restore the resting concentration values after each excitation and to 'refill' the sarcoplasmic reticulum with calcium, different types of exchangers (e. g., Na^{+}/Ca^{2+} exchange) and active ion pumps exist to pump ions against the electrochemical gradient out of the cell, which is an energy (ATP-) consuming process. The ion pump current depends on the concentration on the input side of the pump according to first order kinetics (Michaelis-Menten theory):

$$I_{p,\,Ca} = \frac{I_{\max} \cdot Ca_i^{2+}}{K_m + Ca_i^{2+}} \tag{29}$$

with

$I_{\max}$ maximum pump rate,

K_m concentration for half maximum pump rate.

With eqns (24)-(29), the time course of intracellular Ca^{2+} concentration throughout the cardiac cycle is described. The tension developed by the contractile filaments depends directly on Ca_i^{2+}; and since the underlying mechanism is also a chemical reaction (binding of Ca^{2+} to the regulatory protein TnC), the mathematical relationship according to Ref. 27 is similar to eqn (29):

$$\sigma\,(Ca_i) = \sigma_{sat} \cdot \frac{Ca_i^n}{K^n + Ca_i^n} \tag{30}$$

with

σ_{sat} maximum tension for Ca^{2+} saturation,

n number of binding sites,

K Ca^{2+} concentration for half maximum tension.

The number of binding sites n as determined from several investigators ranged from one to "values much larger than two".[27] Therefore, eqn (30) was fitted to experimental data from Fabiato,[24] which is shown in Fig. 11 for the values

$K = 6.23\ \mu mol/l$,

$n = 2$.

The obtained tension value is then transferred to the ventricular mechanics module as σ_{max} in eqn (14), the index 'max' refers here to the fact that this is

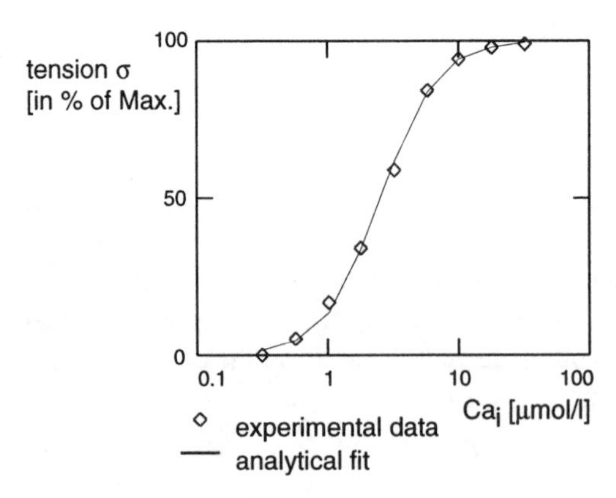

Figure 11: Relationship between intracellular calcium concentration and developed tension, fitted to experimental data from Fabiato.[24]

the value for zero contraction velocity and optimum sarcomere length. In fact, the real value becomes smaller due to the weighting functions *f* and *g* in eqn (14).

5.2 Adrenergic influences

When sympathetic nerve activity is increased, the neurotransmitter norepinephrine is released and binds to α- and β-adrenergic receptors, which are specialized molecules in the sarcolemma. This leads to a change in the conformation of the receptor molecules and results in the activation of the cAMP second messenger system inside the cell, which influences several of the biochemical processes contributing to the excitation-contraction coupling. β-adrenergic agents increase the L-type Ca^{2+} current,[26] enhance the Ca^{2+} uptake of the sarcoplasmic reticulum by stimulating the pump activity and lower the Ca^{2+} sensitivity of the contractile filaments,[25] while α-adrenergic stimulation increases Ca^{2+} sensitivity,[24,25] but knowledge about the involved mechanisms is still incomplete. Taken together, these effects increase significantly the intracellular Ca^{2+} concentration during systole, which is mainly a consequence of the increased presystolic Ca^{2+} content of the sarcoplasmic reticulum and, therefore, larger amount of released Ca^{2+}. With respect to myocardial contraction, these inotropic effects lead to a higher peak tension, higher rate of rise in tension and accelerated relaxation (see Fig. 12).

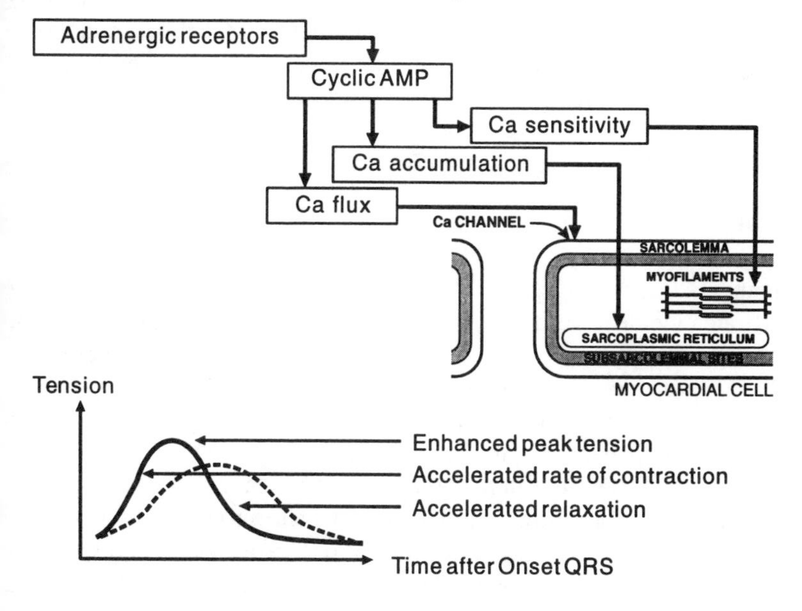

Figure 12: Adrenergic effects on contractility

5.3 Autonomic nervous system control

For the natural blood pressure regulation, the baroreceptors play a key role as pressure sensors. They provide the Medulla oblongata (circulatory control center) with information about the carotid and aortic instantaneous blood pressures via afferent nerves. These signals are integrated with other incoming information, e. g. from the Cortex, and via efferent pathways the function of several organs is controlled. The most important mechanisms regarding circulatory control is the adjustment of the heart function to changing metabolic demands by increasing the heart rate (chronotropic adaptation) and contractility (inotropic adaptation). To maintain proper atrial-ventricular synchrony under significantly higher rates, the AV conduction time is also shortened (dromotropic adaptation). Furthermore, the vascular tone is regulated specifically for different organ systems: While perfusion of muscular tissue is enhanced by vasodilation, vasoconstrictory signals are sent to other organ systems.

The ANS control is incorporated in the model by the following elements: The static and dynamic behavior of the arterial baroreceptors is simulated by an analytical function, which reproduces the sigmoid-shaped static transfer function as well as the dynamic response to pressure changes with different sensitivities to raising and falling pressures.[28] From this information, efferent nerve activities are calculated to increase heart rate (for reference simulations of a healthy subject), to increase the contractile status of the myocardium (see above section) and to adapt the vascular tone, which is simulated by shifting the pressure-volume relationship of systemic vessels in eqn (10), see Fig. 2.

6 Results

The modular structure of the model allows to perform simulations with only one separate module to verify if typical characteristics of a special subsystem, e. g. the mechanical properties of the myocardium, are adequately represented, as well as simulations of the overall behavior. By this procedure, hypotheses of physiological interrelationships which have not been completely uncovered so far can be set up and tested for consistency.

For example, a typical feature of the excitation-contraction coupling is the staircase phenomenon ('Bowditch Treppe'), which means the beat-by-beat increase in amplitude of twitch force after a rest period or when excitation frequency is suddenly increased, until a steady state is reached.[24,25] This effect is explained primarily by a beat-by-beat increase in the sarcoplasmic reticulum

Ca^{2+} content due to higher and more frequent Ca^{2+} influx, which results in a higher Ca^{2+} release. The same effect is to be expected after a rapid adrenergic stimulation; exactly this behavoir is confirmed by the myocardial cell module Fig. 13).

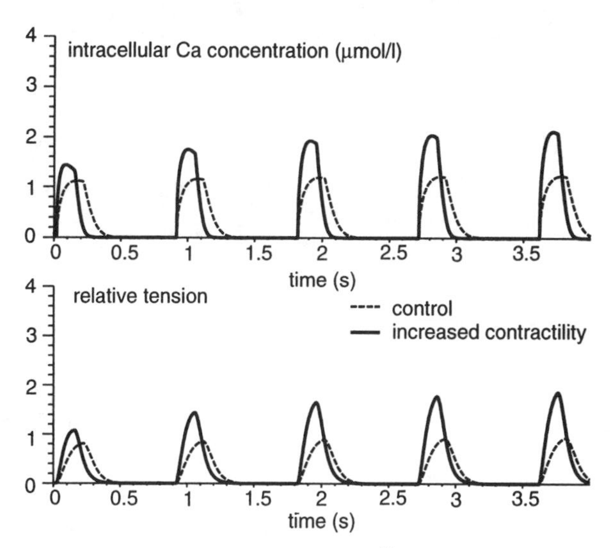

Figure 13: Simulated time courses of intracellular Ca^{2+}-concentration and sarcomere tension under normal and inreased sympathetic activity.

Another result is the realistic approximation of mechanical parameters uch as ventricular wall tensions over the cardiac cycle by the heart mechanics nodule. Figure 14 compares experimentally obtained traces of left ventricular vall tension and thickness from Ref. 29 with simulated tensions of left entricular wall, septum and right ventricular free wall (right).

Pressure-volume loops of the right ventricle are often used for invasive neasurements to evaluate cardiac hemodynamics. The area of the p-V loop orresponds to the stroke work of the ventricle (additional blood acceleration vork is negligible) and allows the quantification of cardiac performance; noreover, the coupling of the left ventricle and arterial system is also analysed n the pressure-volume plane. Thirdly, the slope of the end-systolic pressure-volume relationship correlates to the contractile state of the heart and has been ised as an index of contractility.[30] This sort of analysis can also be performed vith the numerical model, Fig. 15 shows simluated pressure-volume loops for oth ventricles.

Simulated traces of left and right ventricular pressures as well as systemic

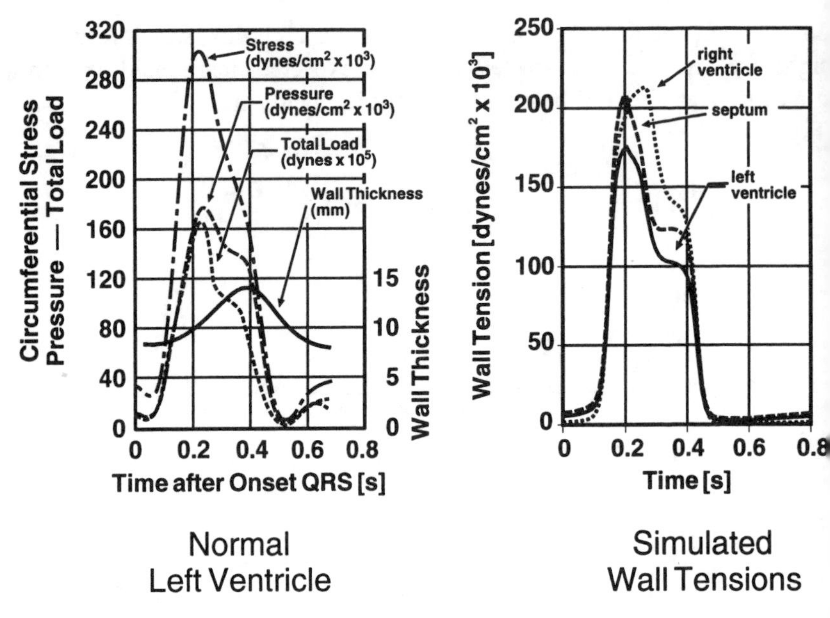

Figure 14: Comparison of measured and simulated parameters. Left: measured left ventricular wall stress (tension), thickness, pressure and total load. Right: simulated wall tension of left ventricular wall, septum and right ventricular free wall.

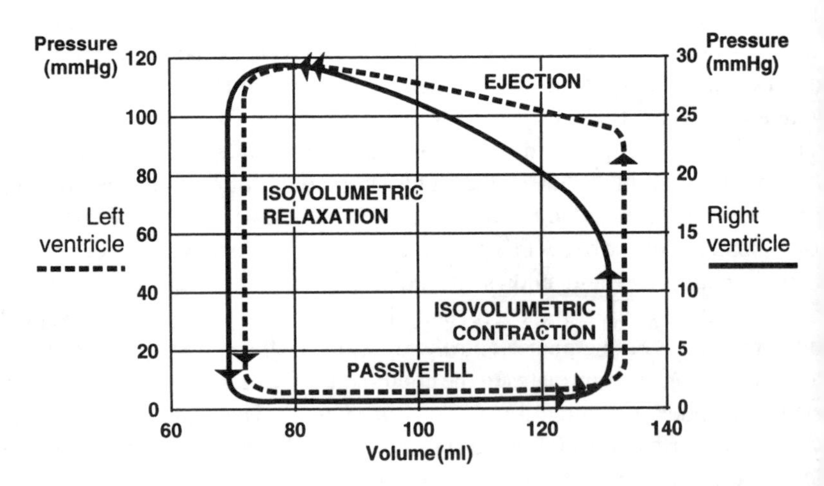

Figure 15: Simulated pressure-volume loops for the left and right ventricles

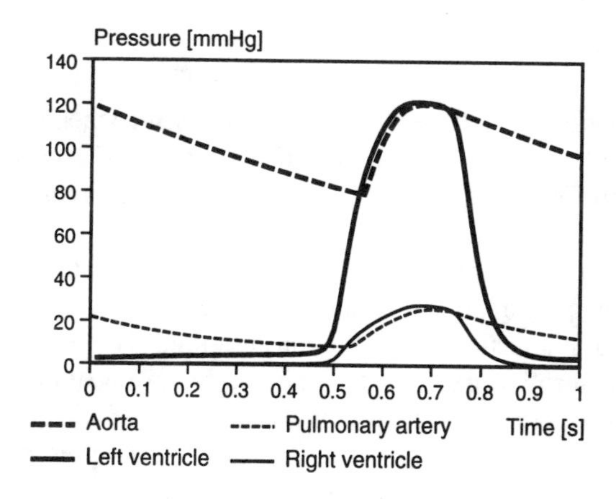

Figure 16: Simulated pressure time courses

and pulmonary arterial pressures are shown in Fig. 16 over one heart period.

Under muscular exercise, the total peripheral resitance is decreased up to 50%,[31] which would lead to a significant drop in mean arterial blood pressure when no control mechanisms were present (see Fig. 17, slim lines). With chrontoropic and inotropic adaptation, mean arterial blood pressure is maintained (bold lines).

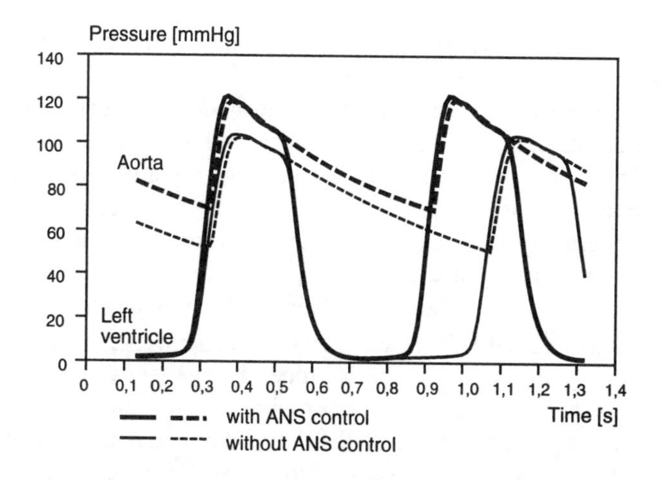

Figure 17: Simulated left ventricular and arterial pressures over two heart periods for exercise conditions (total peripheral resistance reduced to 50%) without any control mechanisms (slim lines) and with ANS control (bold lines).

7 Applications

As the simulation results demonstrate, the presented heart and circulation model is able to mimic the hemodynamic situation of a typical pacemaker patient under different challenges (e. g., physical load, orthostasis) and pacing frequencies (fixed rate or rate-adaptive). The knowledge of the body's response is especially important for closed loop rate-responsive systems, since the heart rate feedback via natural hemodynamic and ANS responses must be taken into consideration for adjusting the adaptation algorithm. This analysis was applied to the ANS-controlled pacemaker, which is based on unipolar intracardiac impedance measurements. The pacemaker is either simulated by software or directly linked to the model via a specialized AD-converter to perform simulation experiments and to evaluate several influences, e. g., the effects of certain myocardial dysfunctions or medication, on the hemodynamic effectiveness of rate adaptation.

7.1 Intracardiac impedance measurement for ANS-controlled pacing

Two intracardiac impedance measurement methods have been established so far: multipolar and unipolar measurements. While mulipolar impedance plethysmography has been used to determine changes in left ventricular volume,[32] this method is not applicable in the right ventricle because of its odd geometrical shape.

Since the normal electrode position is in the right ventricle, it is more favourable to measure the intracardiac impedance by a unipolar electrode against the pacemaker housing as counter electrode. This method has, moreover, the great advantage that the stimulating electrode serves at the same time as a sensor; i.e., the principle is applicable using a conventional pacemaker electrode and no additional sensor is required. On the other hand, the physiological meaning of this signal is different from that obtained with the multipolar conductance catheter.

Unipolar impedance measurement against the pacemaker housing is performed by injecting a 4 kHz square wave constant current of 40 μA through the tip of the implanted stimulation electrode between 100 ms and 300 ms after the pacing stimulus. After signal processing (phase demodulation, Bessel bandpass filtering, AD conversion) the regional effective slope quantity in a preselected ROI (region of interest) is determined using a specialized algorithm (RQ algorithm[1]). After further processing and digital filtering, the rate response control parameter is obtained.[33]

The unipolar impedance signal is mainly determined by conductivity changes in the near vicinity of the electrode. Due to changes in volume percentage of myocardial tissue and blood in the electrode surrounding during isovolumetric contraction and ejection, which have verry different specific conductivities, the impedance signal reflects the geometrical changes of the myocardium during contraction (Fig. 18).

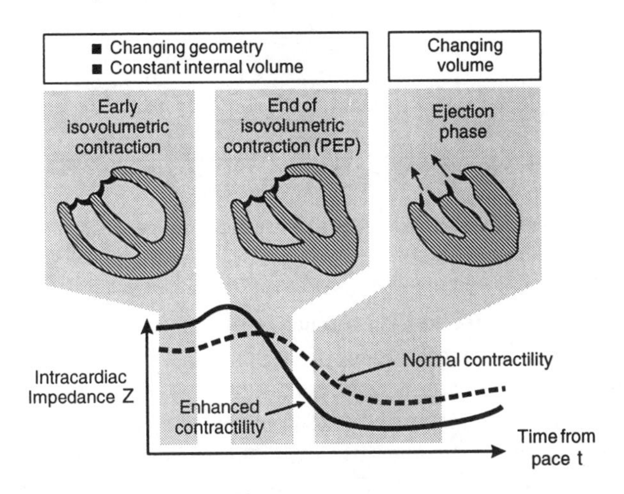

Figure 18: The intracardiac impedance signal during contraction for different contractile states

As discussed in the previous sections, the changes of the single fiber mechanogram under sympathetic stimulation tend to a faster contraction and relaxation as well as to a higher peak tension (see Fig. 12). This influence must also be seen in the contraction pattern of the whole myocardium and, therefore, in the impedance signal. Within the measurement window of 200 ms length covering the late isovolumetric contraction and early ejection phase, a region of interest (ROI) is determined during pacemaker calibration, where the effective impedance signal slope shows the highest correlation with ANS activity (see Fig. 19).

7.2 Clinical experience

In a multicenter study, 210 patients received a pacemaker capable of rate modulation based on the above principle; both single-chamber (BIOTRONIK Neos-PEP, n = 169) and dual-chamber devices (BIOTRONIK Diplos-PEP) were implanted. The evaluation was based on 1. concurrent observation of the spontaneous sinus rate where present, 2. standard exercise protocols, 3. Holter monitoring and 4. echocardiography. The control parameter has been shown

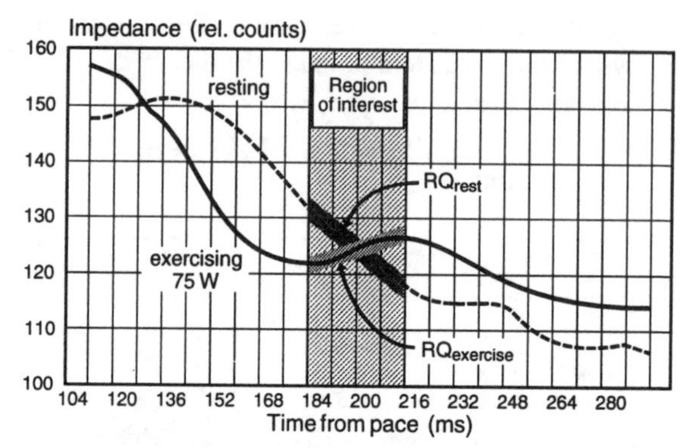

Figure 19: Time course of regional intracardiac impedance, measured under different load conditions with the pacemaker measurement circuit and transmitted via telemetry.

to correlate well with the left ventricular pre-ejection phase, which is considered to be a largely pacing frequency independent measure of cardiac adaptation to circulatory demands. The cardiac output was seen to rise significantly with rate adaptation in response to acute physical exercise, both initially and after several months of this therapy. Exercise protocols (Fig. 20 shows traces of stimulation rate and mean arterial blood pressure during bicycle ergometry)

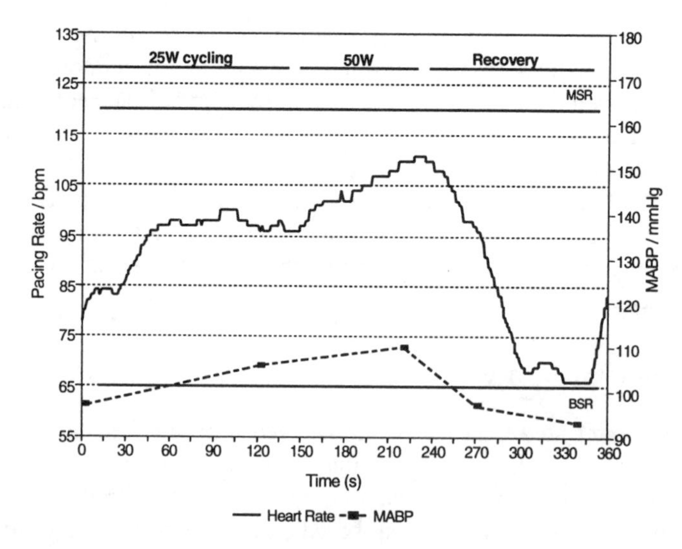

Figure 20: Pacing rate trend over six minutes during bicycle ergometry for a 70 years old patient with sick-sinus syndrome The maintained mean arterial blood pressure (MABP) indicates appropriate rate response.

confirmed the rapid, appropriate rise in stimulation rate to different types of physical exercise as well as to emotional challenges.[34]

8 Conclusions

For those rate-responsive pacing systems which provide closed loop control, the physiological feedback mechanisms must be well understood and taken into account by the adaptation algorithm to ensure a pacing frequency that is most beneficial for the patient's hemodynamic situation. Applied to the impedance-based ANS-controlled pacemaker, simulations with a numerical heart and circulation model could contribute to the identification of sympathetic tone information in parameters related to the myocardial mechanical activity during contraction, such as ventricular wall tension and wall thickness. Furthermore, it provides a tool to optimize this kind of electrotherapy to individual conditions such as impaired ventricular function, diseases or medications.

The model is also applicable to other principles of closed-loop rate adaptation, such as the newly developed technique to utilize the ventricular evoked potential measured with fractally coated, low polarization pacing electrodes for extracting sympathetic tone information.[35] This surface technology drastically reduces the polarization artifact and allows for the first time the undistorted measurement of the evoked potential by an implanted device. This signal is directly influenced by the polarization behavior of the myocardial cells, which is shown to be influenced by adrenergic agents (see above sections). Therefore, the presented model which includes the excitation-con-

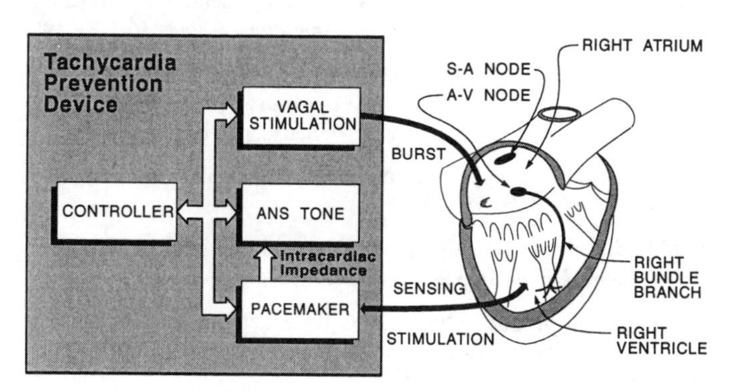

Figure 21: Block diagram of a tachycardia prevention device based on the ANS pacemaker principle: Upon detection of sympathetic hyperactivity at an early stage, electrical bursts are applied to the endocardial vagus nerve endings to prevent tachyarrhythmia.

traction coupling on the cellular level can also serve to explain changes in this signal and contribute to the development of a physiological rate-adaptive pacing system based on the ventricular evoked response signal.
The importance of detecting and utilizing ANS information in electrotherapy goes even beyond rate-adaptive pacing, which requires enhanced understanding and modeling of the neural control mechanisms: Not only bradycardia therapy gains considerable benefit from sympathetic tone information; this signal opens new perspectives also for antiarrhythmic therapy. A new method is the use of the ANS controlled pacemaker to maintain the neural equilibrium (Fig. 21), which serves as a more flexible tachyarrhythmia prevention and, thus, further enhances the benefit of the implanted therapeutic device for the patient.[1,35]

References

1 Schaldach, M. *Electrotherapy of the Heart*, Springer, Berlin, 1992.

2 Rickards, A.F. & Donaldson, R.M. Rate-responsive pacing, *Clin. Prog. Pacing and Electrophysiol.*, **1**:12-19, 1983.

3 Lau, C.P. *Rate Adaptive Cardiac Pacing: Single and Dual Chamber*, Futura Publishing, Mount Kisco, New York, 1993.

4 Kappenberger, L.J. Technical improvements in sensors for rate adaptive pacemakers, *Am. Heart J.*, **127**:1022-1026, 1994.

5 Sandquist, G.M., Olsen, D.B. & Kolff, W.J. A comprehensive elementary circulatory model of the mammalian circulatory system, *Ann. Biomed. Eng.*, **10**:1-33, 1982.

6 Beneken, J.E.W. & DeWit, B. A physical approach to hemodynamic aspects of the human cardiovascular system, *Physical Bases of Circulatory Transport: Regulation and Exchange*, ed. E.B. Reeve & A.C. Guyton, W. B. Saunders, Philadelphia, 1967, pp. 1-45.

7 Coleman, T.G. A mathematical model of the human body in health, disease, and during treatment, *ISA Trans.*, **18**:65-73, 1979.

8 Snyder, M.F. & Rideout, V.C. Computer simulation studies of the venous circulation, *IEEE Trans. Biomed. Eng.*, **16**:325-334, 1969.

9 Leaning, M.S., Pullen, H.E., Carson, E.R. & Finkelstein, L. Modelling a complex biological system: the human cardiovascular system - 1. Methodology and model description, *Trans. Inst. M. C.*, **5**:71-86, 1983.

10 Tsuruta, H., Sato, T., Shirataka, M. & Ikeda, N. Mathematical model of cardiovascular mechanics for diagnostic analysis and treatment of heart failure: Part 1 model description and theoretical analysis, Part 2 analysis of vasodilator therapy and planning of optimal drug therapy, *Med. & Biol. Eng. & Comput.*, **32**:3-18, 1994.

11 Pillon, M., Jufer, M. & Hahn, C. Numerical closed-loop model of a cardiovascular system - application to the development of regulation algorithms for the total artificial heart (TAH), *IFAC Symposium "Modelling and control in biomedical systems" preprints*, 258-264, 1988.

12 Pillon, M., Duffour, H. & Jufer, M. In vitro experiments: Circulatory assist device interaction with a virtual cardiovascular system, *Proc. Ann. Int. Conf. IEEE - EMBS*, **14**:740-741, 1992.

13 Melchior, F.M., Srinivasan, R.S. & Charles, J.B. Mathematical modeling of human cardiovascular system for simulation of orthostatic response, *Am. J. Physiol. (Heart Circ. Physiol.)*, **262(31)**:H1920-H1933, 1992.

14 Schade, H. & Kunz, E. *Strömungslehre*, Walter de Gruyter, Berlin, 1980.

15 Noordergraf, A. Hemodynamics, *Biological Engineering*, ed. H.P. Schwan, McGraw-Hill, New York, 1969, pp. 391-545.

16 Wetterer, E. & Kenner, T. *Grundlagen der Dynamik des Arterienpulses*, Springer, Berlin, 1968.

17 Thomas, J.D. & Weyman, A.E. Numerical modeling of ventricular filling, *Ann. Biomed. Eng.*, **20**:19-39, 1992.

18 Guyton, A.C. *Textbook of Medical Physiology*, W. B. Saunders, Philadelphia, 1981.

19 Detweiler, D.K. Circulation, *Best & Taylor's Physiological Basis of Medical Practice*, ed. J.R. Brobeck, Williams & Wilkins, Baltimore, 1979, pp. 3.1-3.262.

20 Blinks, J.R. & Jewell, B.R. The meaning and measurement of myocardial contractility, *Cardiovascular Fluid Dynamics Vol. 1*, ed. D.H. Bergel, Academic Press, London and New York, 1972, pp. 225-260.

21 Jacob, R. & Gülch, R.W. Kontraktion, *Handbuch der inneren Erkrankungen, Band 1 Teil 1: Herz-, Kreislauf- und Gefäßerkrankungen*, ed. G. Brüschke & B. Heublein, Gustav Fischer, Jena, 1985, pp. 112-151.

22 Lentner, C. *Geigy Scientific Tables Volume 5: Heart and Circulation*, CIBA-GEIGY Ltd., Basel, 1992.

23 Rüegg, J.C. Muskel, *Physiologie des Menschen*, ed. R.F. Schmidt & G. Thews, Springer, Berlin, 1990, pp. 66-86.

24 Isenberg, G. Zellphysiologische Grundlagen der kardialen Sympathikus- und Parasympathikuswirkung, *Autonomes Nervensystem und Herzrhythmusstörungen*, ed. R. Griebenow & H. Gülker, Georg Thieme, Stuttgart, 1990, pp. 17-55.

25 Bers, D.M. *Excitation-Contraction Coupling and Cardiac Contractile Force*, Kluwer, Dordrecht, Boston, London, 1991.

26 Rosen, M.R., Jeck, C.D. & Steinberg, S.F. Autonomic modulation of cellular repolarization and of the electrocardiographic QT interval, *J. Cardiovasc. Electrophysiol.*, **3**:487-499, 1992.

27 Gründeman, R., de Beer, E. & Schiereck, P. Contraction mechanism in striated muscle, *The Physics of Heart and Circulation*, ed. J. Strackee & N. Westerhof, Institute of Physics Publishing, Bristol and Philadelphia, 1993, pp. 121-152.

28 Polley, R., Urbaszek, A. & Schaldach, M. A mathematical approach to short-term baroreceptor behavior, *Proc. Ann. Int. Conf. IEEE - EMBS*, **14**:2697-2698, 1992.

29 Mirsky, I. Elastic properties of the myocardium: a quantitative approach with physiological and clinical applications, *Handbook of Physiology - Section 2: The Cardiovascular System*, ed. R.M. Berne, N. Sperelakis & S.R. Geiger, American Physiological Society, Bethesda, MD, 1979, p. 497.

30 Suga, H. & Sagawa, K. Instantaneous pressure-volume relationships and their ratio in the excised, supported canine left ventricle, *Circ. Res.*, **35**:117-126, 1974.

31 Jacob, R. & Kissling, G. Dynamik des intakten Herzens, *Handbuch der inneren Erkrankungen, Band 1 Teil 1: Herz-, Kreislauf- und Gefäßerkrankungen*, ed. G. Brüschke & B. Heublein, Gustav Fischer, Jena, 1985, pp. 151-188.

32 Baan, J., van der Velde, E.T., Steendijk, P. & Koops, J. Calibration and application of the conductance catheter for ventricular volume measurement, *Automedica*, **11**:357-365, 1989.

33 Schaldach, M. & Hutten, H. Intracardiac impedance to determine sympathetic activity in rate responsive pacing, *PACE*, **15**:1778-1786, 1992.

34 Pichlmaier, A.M., Braile, D., Ebner, E., Greco, O.T., Hutten, H., von Knorre, G.H., Niederlag, W., Rentsch, W., Volkmann, H., Weber, D., Wunderlich, E. & Schaldach, M. Autonomic nervous system controlled closed loop cardiac pacing, *PACE*, **15**:1787-1791, 1992.

35 Schaldach, M. The myocardium-electrode interface at the cellular level, *Cardiac pacing and electrophysiology*, ed. A.E. Aubert, H. Ector & R. Stroobandt, Dordrecht, Kluwer, 1994, pp. 169-188.

Chapter 6

Evaluation of haemodialysis systems using computer simulation

D. De Wachter,[a] P. Verdonck,[a] R. Verhoeven[a]
& R. Hombrouckx[b]

[a]*Hydraulics Laboratory, University of Gent, Gent, Belgium*
[b]*Werken Glorieux, Hogerluchtstraat 6, Ronse, Belgium*

Abstract

To describe haemodialysis systems and therapies a numerical model is made to calculate extra corporeal blood flow characteristics. These results serve to determine the actual kinetic transport parameters in an artificial kidney as they are influenced by the proper system being used and its settings. These calculated transport parameters are used in a kinetic model, to compute dialysis adequacy in a patient independent manner. Finally a comparison is made between different systems and between different dialysis strategies.

Nomenclature

a	wave velocity	[m/s]
A	area	[m^2]
C	concentration	[mg/ml]
D	diameter	[m]
E	elasticity modulus	[N/m^2]
e	wall thickness	[m]
g	gravitational acceleration	[m/s^2]
h	piezometric head	[m]
L	length	[m]
N	number of capillaries	[-]
p	pressure	[Pa]
Q	flow	[m^3/s]
q	flow per unit length	[$m^3/s/m$]
Q_c	clearance	[ml/min]
t	time	[s]
x	coordinate along conduit's axis	[m]
z	vertical coordinate	[m]
α	angle	[rad]
$\Delta\pi$	osmotic pressure difference	[Pa]

λ	pressure loss coefficient	[-]
ρ	fluid density	[kg/m^3]
τ	wall shear stress	[Pa]

subscripts

b	blood
d	dialysate
i	solutes, inflow
k	kidney
m	membrane
o	outflow
u	ultrafiltrate

1 Description of Haemodialysis Systems

1.1 Introduction

The aim of a regular dialysis treatment or chronic haemodialysis is to allow the survival of patients whose chronic renal insufficiency has reached end-stage. Normal kidneys perform three basic functions : (1) excretion of the waste products; (2) regulation of water and electrolyte balance; (3) regulation of endocrine and metabolic functions. Only the first two functions can be replaced by chronic haemodialysis because the latter requires the presence of a functioning renal parenchyma. In a normal subject all the metabolites resulting from nitrogen catabolism are eliminated by the kidneys. In renal insufficiency these metabolites accumulate in the blood and the tissues. This toxicity is responsible for the origin of many clinical disorders related to uraemia. When the number of functioning nephrons falls below 25% of normal the first symptoms of renal disfunction appear. The regulation of the excretion of electrolytes and water fails completely at 5% of normal renal function.

Haemodialysis primarily employs diffusion[1] for solute removal and ultrafiltration for fluid removal. For larger solutes, convective[2] transport by ultrafiltration makes a significant additional contribution to diffusive transport, especially with high permeability membranes. The role of osmosis

---

[1] Diffusion, from the Latin verb "diffundere", meaning "to pour or spread out in different directions"

[2] Convection, from the Latin verb "convehere", meaning "to carry together"

is limited in haemodialysis. In general the artificial kidney must carry out 4 tasks :

1. elimination of uremic toxins (urea, creatinin, uric acid, sulphates, phosphates,...). This is performed mostly by diffusion through the dialysis membrane;

2. elimination of excess water (1 to 4 l per session) since the patient produces little or no urine. This is performed by ultrafiltration through the membrane;

3. control of plasma electrolytes (Na^+, Cl^-, K^+, Ca^{++}). This is performed by adding these ions at the proper concentration in the dialysate;

4. blood pH control. This is performed by adding a buffer to the dialysate (acetate or bicarbonate).

The treatment of renal failure by artificial support has been synonymous with haemodialysis. Nevertheless there are a number of other modalities of treatment which make use of an artificial membrane : Haemofiltration (a process in which uremic blood is cleansed by convective transport (ultrafiltration) after dilution of the blood with a physiological solution) and Plasma Separation (separation of blood in plasma and morphological elements (red and white blood cells, thrombocytes) across a membrane by filtration). This chapter is restricted to extra corporeal haemodialysis systems.

Haemodialysis strategies in Europe have progressively shortened since the early 1970's due to the availability of new techniques, e.g. more performant membranes. Nowadays the majority of patients are on thrice weekly of four-hour treatment sessions. The latter strategy is undoubtedly influenced by reimbursement protocols and staffing convenience. For many patients haemodialysis is not a definitive solution but allows to wait for renal transplantation.

The number of patients alive on different forms of renal replacement therapy in different countries in 1992 are the following :

France	23.000	U.K.	20.000
Germany	30.000	Japan	90.000
Italy	20.000	U.S.A.	110.000

Countries which have a large number of dialysis patients relative to their population such as Italy and Japan, have few or no renal transplantations. Conversely U.K. and Scandinavian countries have a good transplantation program which reduces their dialysed population. Haemodialysis is costly, each session costing from 300 to 750 ECU, three times a week, of which the cost of disposable equipment (haemodialysers, blood lines, dialysate, etc.) represents from 50 to 100 ECU.

Milestones in the history of haemodialysis are :

1943: W. Kolff (Netherlands) first uses with success the "artificial kidney" in a man.

1945-1950: Development of the technique for the routine treatment of acute renal failure.

1959: B. Scribner et al. (United States) perfect the permanent external arteriovenous shunt. This opens the way for periodic haemodialysis.

1966: M.S. Brescia and J.E. Cimino (United States) propose the creation of an arteriovenous fistula in order to perform periodic haemodialysis.

1.2 Haemodialysis treatment

Haemodialysis equipment includes an extra corporeal blood circuit with one or two blood pumps, bloodlines, a haemodialyser, a dialysate delivery system and monitoring devices. Blood is sucked by a peristaltic pump (blood pump) at a flow rate between 200 and 350 ml/min from an arteriovenous fistula (i.e. an arterialised vein by an anastomosis between an artery and a vein) in the arm of the patient through a needle into an extra corporeal circuit. It circulates along a semi-permeable membrane in the haemodialyser and returns to a vein. Simultaneously, a dialysate containing the necessary ions and the buffer circulates on the other side of the membrane mostly at counter current to maintain a concentration gradient across the membrane for the uremic toxins (dialysate delivery system).

1.2.1 Extra corporeal blood circuit

Blood must be safely withdrawn from the patient without damage to its formed elements or to plasma proteins, conveyed at the appropriate blood flow rate to the dialyser, and then back to the patient without the risk of embolism due to air or thrombi.

A typical blood circuit used in haemodialysis (two-needles system) is shown in Fig. 1.1. Blood is withdrawn from the arterial segment by the blood pump (1) and pumped through the dialyser (2) back to the patient via the venous segment. With an arteriovenous fistula or graft, blood access is achieved by the insertion of arterial and venous needles percutaneously into the fistula or graft. Air traps (3) are located in the blood tubing set to trap air or foam and to prevent air entry into the patient. In a single needle system it is also used to temporarily store blood and to measure pressure. An air-foam detector senses the level of air and the build up of foam and will switch off the pump when necessary. An anticoagulant pump (4) delivers an anticoagulant like heparin in the extra corporeal circuit via a T-fitting to prevent clotting of the blood. The rate of infusion can be varied to suit the patient's needs.

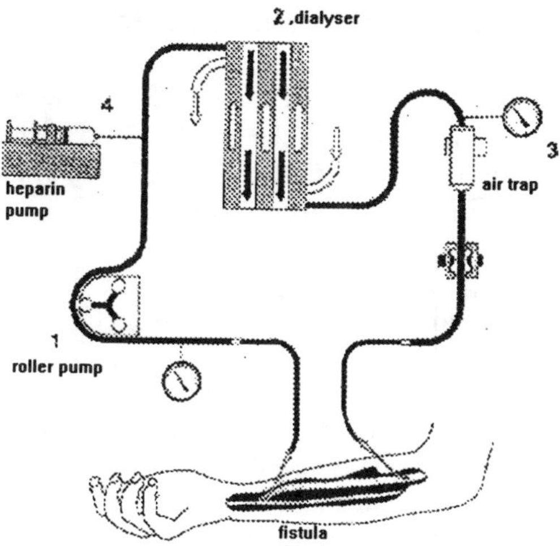

Figure 1.1 Schematic representation of an extra corporeal blood circuit.

The insertion of two needles is world-wide the usual accepted method of vascular access (over 90%). This is often difficult and associated with patient discomfort. Therefore one is using unipuncture or single needle dialysis. This may be achieved by a double lumen needle catheter which combines the conventionally used two needles into one single unit or alternatively pulsatile flow systems may be used. In such a single needle system, only one needle is inserted into the patient's blood access and there is alternation of flow through this needle between flow from the patient into the extra corporeal circuit and flow from the extra corporeal circuit back into the patient. The sequencing and control of these so-called arterial and venous flow phases may be controlled by a pressure-pressure mode, a pressure-time mode, a time-time mode or a volumetric control, De Wachter (1993). A system that is commonly used today relies upon two blood pumps, one upstream and one downstream the dialyser (double-headed pump, Fig. 1.2a). The upstream pump will draw blood into the system until a pre-set high pressure is attained (arterial phase). Consequently the downstream pump returns filtered blood to the patient while the pressure remains higher than the pre-set lower pressure (venous phase). Recently a single-pump device with only one roller pump alternating its pumping direction has been developed that filters the blood twice per pump stroke. The switch between the two phases is determined by the weight of the blood accumulated in the system, which makes it a volumetric dialysis apparatus. Because of its low priming volume, it can be used in paediatric renal treatment. The pump is named a bi-directional pump (Fig. 1.2b), Hombrouckx (1989), after its characteristic reversing of pumping direction.

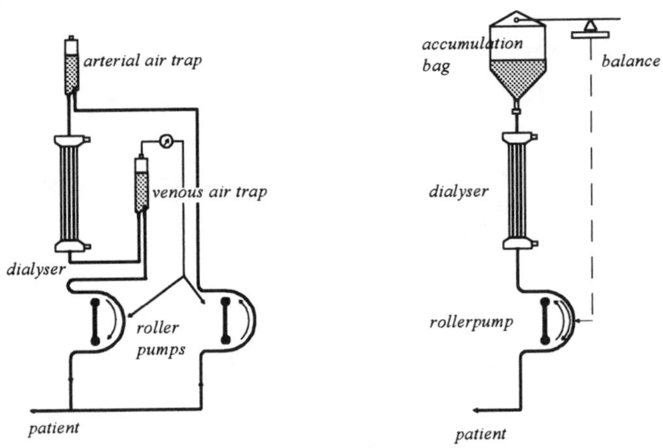

Figure 1.2 Single needle haemodialysis systems: left panel Double Headed Pump (a), right panel Bi-directional Blood Pump (b).

1.2.2 Blood pumps

Blood pumps used for haemodialysis are usually roller pumps based on the peristaltic principle. A pump segment is occluded between rollers on the pump shaft and a curved rigid track. As the pump shaft rotates, the rollers force blood out of the pump segment. After the roller has passed over the tubing, elastic recoil causes blood to be drawn into the pump segment, refilling it. Usually, two or three rollers are mounted on the pump shaft. The pulsatility of pumping decreases with the number of rollers. The flow rate of the pump is a product of the speed of rotation and the stroke volume, that is, the volume pumped per revolution of the pump shaft. The stroke volume depends on the inside diameter of the pump segment, the length of tubing occluded in the curved track, and the degree of occlusion between the roller and the track. The degree of occlusion is usually adjustable. Under occlusion results in backflow, foaming and inadequate dialyser perfusion. Over occlusion can result in haemolysis and damage to the pump segment and eventually to the rollers of the pump as well.

1.2.3 Vascular access

Vascular access is usually created on a medium-size artery such as the radial artery. The first permanent means of easy access to the blood for the treatment of chronic uraemia was the arteriovenous shunt of B.H. Scribner in 1960. Since then the technique has been perfected especially with the arteriovenous fistula introduced by Cimino and Brescia in 1966. Other techniques like arteriovenous grafts are also available.

The arteriovenous grafts (which may be autologous, homologous, heterologous or synthetic) are inserted between an artery (radial artery at its origin, or humoral, axillary or femoral arteries), and a nearby vein. The by-pass can be curvilinear or rectilinear.

1.2.4 Haemodialysers

Haemodialysers in clinical use may be classified as parallel plate, coil or hollow fiber configuration. This text is essentially limited to hollow-fiber devices as other dialyser types are disappearing from use. Figure 1.3 displays a hollow fiber dialyser example.

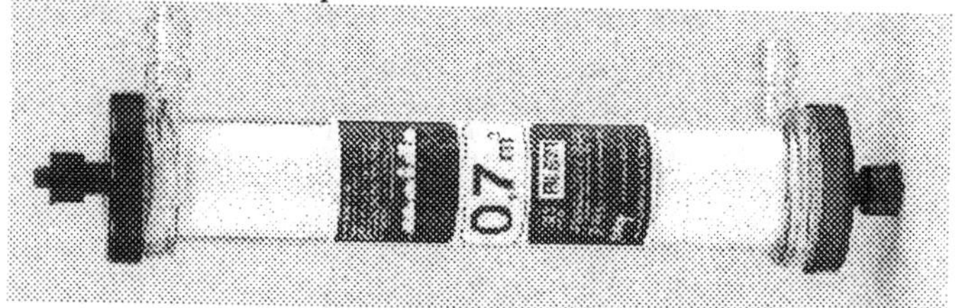

Figure 1.3 Hollow fiber haemodialyser.

Hollow-fiber dialysers are composed of a group of 10.000 to 15.000 hollow fibres (capillaries) with an internal diameter of about 200 microns. A number of manufacturing processes were developed to produce cellulose (a polysaccharide) based membranes like Cuprophan, Cellophone and Cellulate (cellulose acetate). The advent of the middle molecular hypothesis highlighted a failing of these widely used membranes. Development of synthetic membranes to overcome the inability to remove middle molecular weight solutes followed. Although these membranes were originally developed to improve middle molecular clearance, their biocompatibility is also an important benefit : Polyacrylonitrile (PAN 15), Polysulfone (PS), Polymethylmethacrylate (PMMA).

In hollow fiber design the membrane surface area may be changed by altering the number of fibres, the fiber length or the fiber diameter.

1.2.5 Dialysate delivery system

The dialysate delivery system provides dialysate of the desired composition, temperature, pressure, and flow to the dialyser. Dialysate delivery systems may be single-patient systems or central multipatient systems. In a central multipatient system, dialysate is prepared at a central location delivered to individual patient stations where devices for monitoring and controlling temperature, flow rate, pressure are located. Central systems are usually more cost effective, but suffer from the disadvantage that a malfunction of the central proportioning unit will simultaneously affect several patients with discontinuation of dialysis at these patient stations. Also, central systems do not allow for individualisation.

2 Mathematical description of extra corporeal blood flow

2.1 The unsteady flow equations

In haemodialysis systems it is the mass flow that is important for solute transport. As a result the unsteady flow is supposed to be one-dimensional in the direction of the bloodlines. The following equations are derived for a uniform velocity, neglecting the cross sectional velocity differences.
Consequently, the number of unknowns is reduced to two. The flow Q and the pressure p describe the unsteady flow problem completely. By that, only two equations will suffice to solve the problem. The first equation (momentum equation) represents Newton's second law projected on the flow direction, while the second (continuity equation) expresses conservation of mass.

2.1.1 The momentum equation

Equation (1) expresses the force equilibrium of an elementary small fluid particle along the pipe's (x) axis that is oriented under an angle α with respect to the horizontal (Fig. 2.1).

$$pA - \left(p + \frac{\partial p}{\partial x} dx \right) A - \rho g dx \sin\alpha - \tau \pi D dx = \rho dx \frac{dQ}{dt} \quad (1)$$

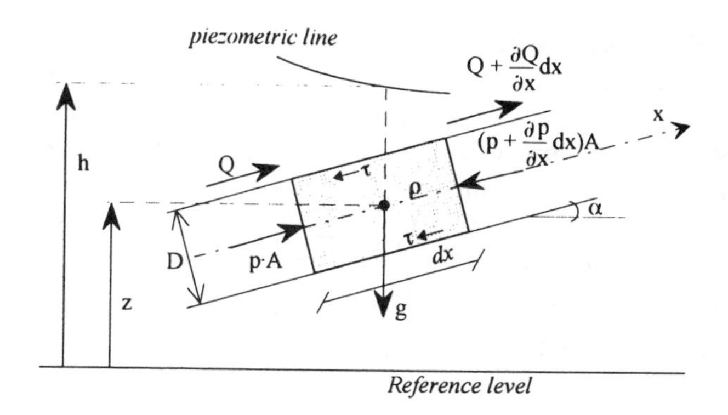

Figure 2.1 The momentum equation for a small fluid particle

The shear stress τ is rewritten using the Darcy-Weisbach formula:

$$\tau = \frac{\partial p}{\partial x}\frac{D}{4} = \frac{\rho\lambda Q^2}{8A^2} \tag{2}$$

The pressure p equals ρg(h-z), a combination of piezometric head h and vertical difference z. If the derivative of the discharge is also expanded, the momentum equation becomes:

$$gA\frac{\partial h}{\partial x} + \frac{Q}{A}\frac{\partial Q}{\partial x} + \frac{\partial Q}{\partial t} + \frac{\lambda Q|Q|}{2DA} = 0 \tag{3}$$

In this equation Q^2 is written as $Q|Q|$ to take into account the flow direction. Remark that this equation is independent of the pipe's slope, because pressurised flow is not affected by gravity.

2.1.2 The continuity equation

Reconsider the elementary small fluid particle in the circular pipe. During a period dt the initial mass ρdV = ρAdx will change due to the varying flow and pressure. This change originates partly from the compressibility of the fluid, partly from the elasticity of the pipe and from fluid in- or outflow along the pipe wall. This can be expressed by eqn (4):

$$-\rho\frac{\partial Q}{\partial x}dxdt - \rho\pi Dqdxdt = \rho A\frac{1}{E_v}\frac{dp'}{dt}dxdt + \rho A\frac{D}{eE}\frac{dp'}{dt}dxdt \tag{4}$$

The left side term is the volume variation minus the mass loss through the wall, while the first right term expresses compressibility according to Hooke's law and the last term gives the mass variation caused by elasticity as formulated with the barrel formula. The parameter E_v is the volumetric elasticity of the fluid. In eqn (4) dp'/dt is only due to dynamic pressure variations, because the unsteady state calculation must be consistent with the steady state. So static pressure (p_0) induced compressibility and elasticity are not included. The dynamic pressure is written as $p' = p - p_0 = \rho g(h - h_0 - z)$. By expanding the term dp'/dt and defining a, the wave velocity, with:

$$a^2 = \left[\rho\left(\frac{1}{E_v} + \frac{D}{eE}\right)\right]^{-1} \tag{5}$$

the continuity equation is expressed as:

$$\frac{a^2}{gA}\left(\frac{\partial Q}{\partial x} + q\right) + \frac{\partial h}{\partial t} + \frac{Q}{A}\frac{\partial h}{\partial x} = 0 \tag{6}$$

2.2 Numerical solution using the method of characteristics

The eqns (3) and (6) form a set of partial hyperbolic quasi linear differential equations, for whom no general solution exists. The method of characteristics, Streeter (1978), will transform them into a set of total differential equations that can be solved analytically. If a linear combination of the two equations is taken: $(2.3) + \xi\,(2.6) = 0$ and ξ is solved as $\xi = \pm g/a$, then two solutions are obtained, one with positive sign (C^+) and one with negative sign (C^-):

$$\left.\begin{aligned} &\pm\frac{gA}{a}\frac{dh}{dt} + \frac{dQ}{dt} + aq + \frac{\lambda Q|Q|}{2DA} = 0 \\ &\frac{dx}{dt} = \frac{Q}{A} \pm a \end{aligned}\right\} C^{\pm} \tag{7}$$

The upper equations in (7) are total differential equations that are only valid along the curves defined by the lower equations, the so called *characteristic lines*. The slope of these lines is $Q/A \pm a$. Since the wave velocity is not accurately known and since the mean fluid velocity Q/A is in practice only a fraction of a (e.g. 1m/s vs. 20m/s), the term Q/A can be neglected in the slope definition.

Now the equations $dx/dt = \pm a$ define straight lines in the x,t field. The corresponding differential equations are solved along these lines. If we split up the pipe length in a number of intervals Δx, a rectangular x,t grid is created. If the boundary conditions are known both Q and h can be solved for every grid point. E.g. in Fig. 2.2 the unknowns in point P are calculated from the parameters in A and B, by integrating the differential equations along the C^+ and C^- lines. This yields two equations with two unknowns from which the results are readily obtained.

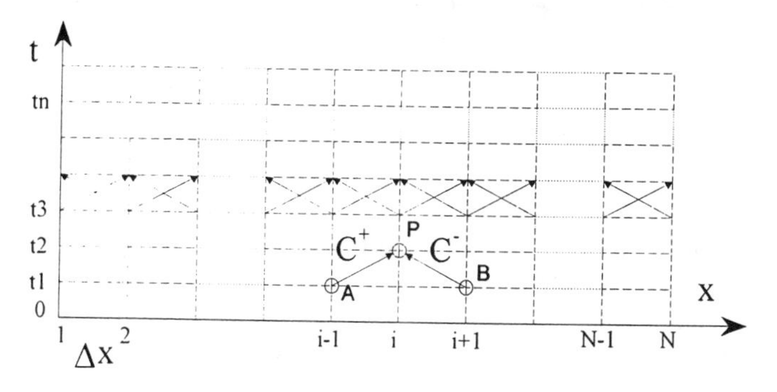

Figure 2.2 Numerical x,t grid for solution along the characteristic lines

Since for dialysis systems the parameter q (see ultrafiltration flow) is proportional to the pressure, eqn (8), the numerical integration of the differential equations in the kidney is performed by a predictor corrector method, Isaacson (1966). In the other parts a second order approximation will do.

2.3 Parameters and boundary conditions

The differential equations (7) must be solved in every grid point of the blood path. To start the calculation sufficient boundary conditions and initial values for every point in the flow path are necessary. The initial values are acquired by a preceding steady-flow simulation with a fixed flow through every part of the system. It will calculate the corresponding pressures. On the other hand the boundary conditions will describe the time dependent behaviour in the first and last grid points of a conduit. These boundary conditions most generally are different for every component of the system. Dialysis systems discussed in this work can be easily subdivided in distinct manageable parts.

The different parts are partitioned into two groups: one group that consists of the physical blood paths requiring boundary conditions, the other consists of devices supplying boundary conditions (Table 2.1). Boundary conditions must supply a value to either the pressure (P) or flow (Q) or state a relation (R) between these two.

Table 2.1 System subdivision and boundary conditions (BC)

requiring BC	supplying BC	type
plastic tubing	roller pump	Q
artificial kidney	expansion chamber	R
	accumulation bag	P
needle	fistula or vein	P

Because some devices require special handling, a detailed discussion of parts and their proper boundary conditions and parameters follows.

2.3.1 Bloodlines

Since the finite difference equations describe the flow characteristics in every blood path, the constants in these equations must be defined. They originate from the geometrical properties of the bloodline itself (length, diameter, height difference and wall roughness) and from the physical fluid properties (kinematic viscosity, specific mass). The celerity is a parameter influenced by both bloodline and fluid. The wall roughness is used to calculate the friction losses (λ parameter).

A bloodline cannot specify any boundary conditions. They are supplied by its neighbouring components. In the special case of two directly connected bloodlines the pressure and discharge in the last grid point of the first will equal those in the first grid point of the second bloodline.

2.3.2 Hollow fiber artificial kidney

The interior of a hollow fiber kidney can be considered as a bundle of N parallel bloodlines, the total number of capillaries. The same parameters as in 2.3.1 are required for each capillary. Besides those a parameter is needed that describes total ultrafiltrate flow Q_u across the dialysate membrane, the ultrafiltration coefficient K_u [in ml/min/mmHg].

$$Q_u = K_u(P_b - P_d - \Delta\pi) \quad (8)$$

To calculate the local ultrafiltration in each grid point q_u, K_u is divided by the capillary length and N.

Special handling is needed for the in- and outflow sections. Here the flow separates into the capillaries or converges from them. This causes pressure losses which are introduced between the last (first) point of the plastic tubing and the first (last) grid point of a capillary. These losses are determined by experiment. The flow in the first (last) point of the capillary equals one Nth of the in(out)flow. In this way the boundary conditions for the capillaries are established.

An artificial kidney has two flow paths: the one just described for the blood and at the other side of the membrane a dialysate circuit. In this description, it is assumed that the dialysate flow is constant or at least quasi-static, i.e. the flow variation is slow compared to the variations in the blood circuit. Consequently the dialysate flow is characterised by a mean flow rate, the pressure head at the dialysate outlet and the friction losses in the kidney.

2.3.3 Angioaccess devices

The patient's blood can be accessed by either a needle or a catheter. Both devices are easily described as common bloodlines. The only differences are the extra pressure head losses at the needle or catheter tip. Generally these are flow dependent and non-equal for in- and outflow.

As a boundary condition at the patient's side the time dependent pressure either in the fistula or arteriovenous shunt (needle) or in the large vein or atrium (catheter) can be given. This pressure must be measured in vivo.

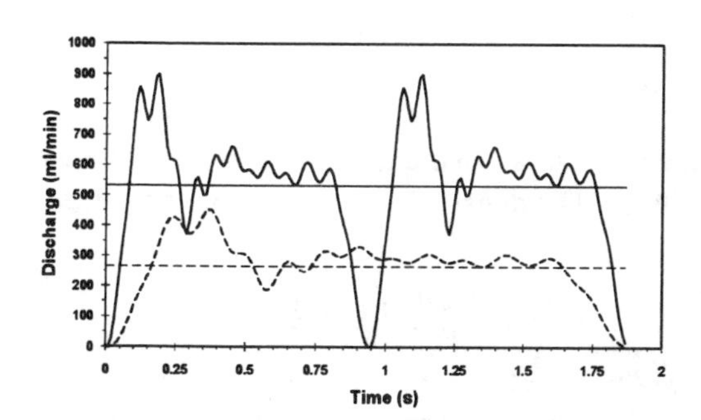

Figure 2.3 Time dependent pump discharges with mean value indicated

2.3.4 Roller pump

This device can supply as boundary condition for the connected bloodlines the discharge as a function of time. This is most easily handled with a Fourier series using a variable frequency and scalable constants to simulate different mean flow rates (Fig. 2.3). The Fourier coefficients are determined from laser Doppler anemometry measurements.

2.3.5 Air trap or expansion chamber

The device is very similar to an air vessel, applied to damp flow oscillations in order to protect systems from waterhammer. To provide the boundary conditions to the connected bloodlines, the storage function is implemented according to the following (Fig. 2.4)

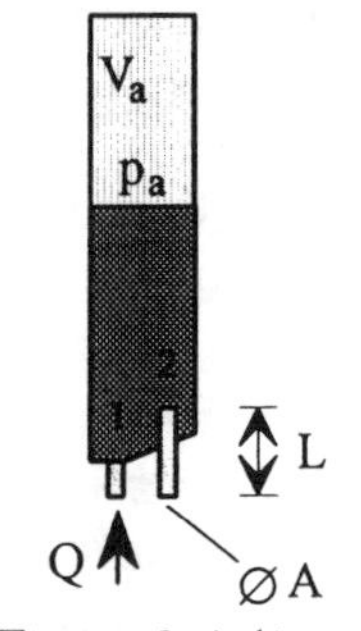

Figure 2.4 Air trap

The same two equations govern the expansion chamber dynamics: the continuity equation:

$$\frac{dV_a}{dt} = Q_1 - Q_2 \tag{9}$$

where V_a is the volume of air in the chamber and indexes 1 and 2 refer to the connected bloodlines. The dynamic equation can be expressed as:

$$p_{1,2} = p_a + \Delta p_{loss} + \frac{G_{1,2}}{A_{1,2}} + \rho \frac{dQ_{1,2}}{dt} \frac{L_{1,2}}{A_{1,2}} \tag{10}$$

which states a relation between pressure and flow in the last point of any connecting bloodline as a function of the air pressure (p_a) increased by the friction losses (Δp_{loss}) and the extra pressure caused by the fluid weight $G_{1,2}$ between air trap chamber and the connected bloodline. The pressure p_a is calculated from the polytropic gas compression law:

$$p_a \cdot V_a^{1.2} = \text{constant} \tag{11}$$

For this description the following parameters are needed: the diameter, height and wall roughness of the chamber and the vertical placement in the system and also length and cross-sectional area of the inflow tubes. As an initial value the fluid height in the chamber is to be known.

2.3.6 Accumulation bag

The bag is also used to temporarily store blood. It can be described as a reservoir with variable level. The differences with the expansion chamber are that there is only one bloodline and that no air is above the fluid surface, hence there is always atmospheric pressure acting on it. This simplifies the previous equations although they remain equally valid.

2.3.7 General system settings

Finally the system pre-set parameters are needed: for the single needle systems this includes the control parameters to switch filling and emptying phases: a minimum and maximum pressure or weight. As a last input one needs the total calculation time and the time step to be used. The latter will determine calculation accuracy and must be small enough to precisely describe fast parameter variations.

2.4 Derived quantities: Ultrafiltration and clearance

Ultrafiltration is the process of water displacement across the membrane, eqn (8). Since the local ultrafiltration q_u is calculated in every grid point, the global ultrafiltration Q_u in the kidney is obtained by merely summing them all along one capillary and multiplying by N.
In order to calculate solute removal (Section 4) in the artificial kidney one must know the clearance value, which is defined as

$$Q_c = \frac{Q_{bi} C_{bi} - Q_{bo} C_{bo}}{C_{bi}} \qquad (12)$$

or the fraction of solutes removed multiplied by the blood flow.
In case of no ultrafiltration (only diffusive transport) Q_{c0} can be solved as a function of dialysate and blood flows and of membrane characteristics. For counter current flow, one finds, Maher (1989):

$$Q_{c0} = Q_b Q_d \frac{1-\exp\left[-kA_m\left(\frac{Q_d - Q_b}{Q_d Q_b}\right)\right]}{Q_b - Q_d \exp\left[-kA_m\left(\frac{Q_d - Q_b}{Q_d Q_b}\right)\right]} \tag{13}$$

with k the diffusive membrane permeability [cm/min]. Co-current flow will yield:

$$Q_{c0} = Q_b Q_d \frac{1-\exp\left[-kA_m\left(\frac{Q_d + Q_b}{Q_d Q_b}\right)\right]}{Q_d + Q_b} \tag{14}$$

These formulas, if plotted would show at first a clearance linearly increasing with blood flow. Above a certain blood flow, dependent on both dialysate flow and the factor kA_m, the clearance will asymptotically saturate towards a constant value.

In haemodialysis ultrafiltration is generally non-zero but small compared to blood flow. In this case the total clearance (both diffusive and convective transport) can be approximated by, Maher (1989)

$$Q_c \approx Q_{c0} + S\frac{Q_u}{2}\frac{C_{bo}}{C_{bi}} \approx Q_{c0}\left(\frac{2Q_{bo} - SQ_u}{2Q_{bo}}\right) + S\frac{Q_u}{2}\frac{Q_{bi}}{Q_{bo}} \tag{15}$$

for moderate values of blood flow. Herein S is the sieving coefficient of the membrane for the solute. All the other values in this last equation are known from the dynamical study. The result will be a time dependent clearance value for a given solute.

Remark in eqn (15) that ultrafiltration augments solute transport, but because of the interaction between the diffusive and convective processes, the total clearance is less than the addition of the two.

3 Evaluation of haemodynamics

3.1 Calculation parameters

Three presented systems are simulated: a regular two needle system, a pressure (cfr. double headed pump, Fig. 1.2) and a weight controlled (bi-directional pump) single needle system. Dialysate pressures are adapted to have comparable ultrafiltration flows. Because of the bi-phasic flow, the pumps in the single needle systems turn at twice the speed of that in the two needle system. As artificial kidney, the parameters of a Fresenius F6 dialyser are used, which has a fitted kA_m of 683 ml/min for urea and 470 ml/min for creatinin. The simulated air trap has the characteristics of the Sorin Bellco BL048, except for the one on the arterial side in the pressure controlled system which is a BL049.

For the pressure controlled system minimum and maximum pressures are set at ±6.67 kPa (±50 mmHg), while weight settings are 0 and 50g for the weight controlled system. This latter is not a realistic figure, but it allows for shorter simulation time before the second phase is started and does not significantly alter results. The dialysate flow was fixed at 500 ml/min.

3.2 Pressures

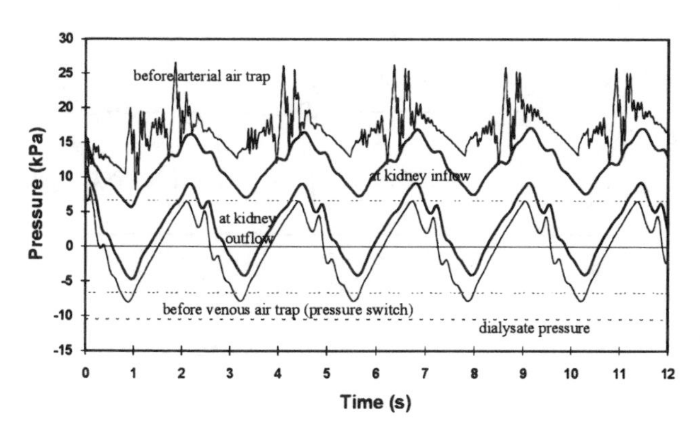

Figure 3.1 Pressures in a pressure controlled system

The pressures in the pressure controlled system show some of the system's characteristics. First it is seen that the pressure signal that appears in front of the arterial air trap (i.e. the air trap at the filling side) is effectively damped by the air trap when it reaches the kidney. Also it is clear from the system's operation that the pressures in the kidney will oscillate between two levels. These levels are determined by the pre-set pressures (dot lines). When the pressure in the venous air trap will cross any of these two levels, the pump that was turning will stop and the other will start (compare this also with the flow picture, Fig. 3.4). As is clearly seen the pressure will undershoot the lowest level because of the distance between the filling pump and the pressure measurement. For view of ultrafiltration flow the dialysate pressure at the kidney blood inflow side is shown. This system allows easy determination of ultrafiltration since it is directly computed from the different pressure settings.

In a weight controlled system (Fig. 3.2), the overpressure falls to zero from the pump towards accumulation bag. When the system is emptied, the pressure in the system becomes negative. At this time the pressure in the kidney can become more negative than the dialysate pressure forcing the ultrafiltrate to flow retrograde from dialysate to blood side of the membrane. It is important to observe this fact for it lowers net water removal from the patient. Ultrafiltration flow during the emptying phase is in any case lower as during filling. This fact demands that the dialysate pressure should be set cautiously for a sufficient mean ultrafiltration flow.

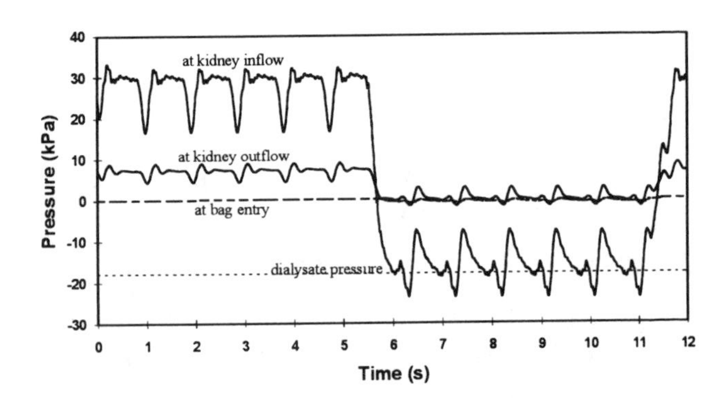

Figure 3.2 Pressures in a weight controlled system

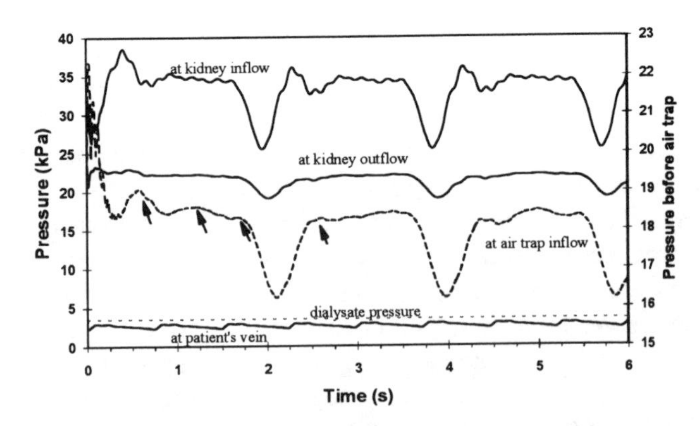

Figure 3.3 Pressures in two needle system

The pressures in a two needle system (Fig. 3.3) are determined by the pressure at the patient's vein. Unlike the presented single needle systems which are always isolated from the patient by a roller pump, a two needle system is directly connected to the patient by a needle or catheter. The immediate consequence of this is that the ultrafiltration is highly dependent on the venous pressure. If this pressure increases, total pressure in the system will shift upwards and in case of a constant dialysate pressure, ultrafiltration is augmented. The pressure at the kidney side of the air trap is shown at a larger scale (right side) to demonstrate the delayed influence of the venous pressure signal (arrows) on top of the pump induced pressure wave.

3.3 Flows

The mean pump speed in the two needle system is half that of the single needle systems, because single needle systems are bi-phasic. The result is that the total blood volume filtered during a fixed time is almost constant for the three systems.

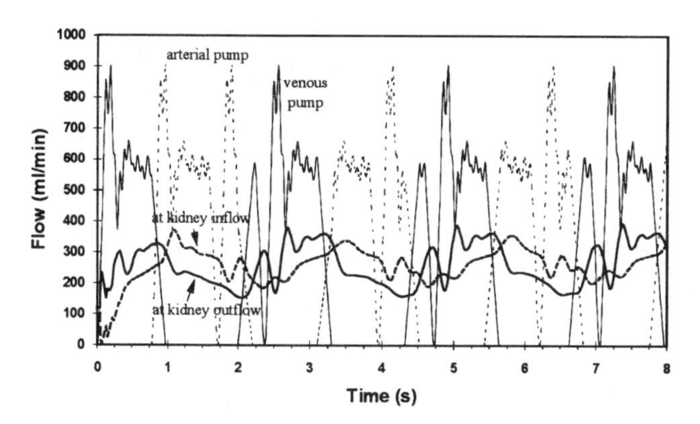

Figure 3.4 Discharges in a pressure controlled system

Figure 3.4 shows this bi-phasic operation of the pressure controlled system. At the start the venous pump (at the system's return side) is turning, because the pressure was higher than the high pressure pre-set (Fig. 3.1). When the pressure falls below the low pressure pre-set, this venous pump stops and the arterial pump starts turning. Since blood is intermittently stored and removed from the air traps between pumps and kidney (Fig. 1.2), the mean blood flow in the kidney is effectively half of the pump's discharge (Table 3.1). The operation of the nearest pump enhances blood flow at each side of the kidney after a certain time lag. The flow in the middle of the kidney (not shown for clarity) is rather constant.

In the weight controlled system (Fig. 3.5), the absolute mean flow through the kidney is at every moment equal to the mean flow delivered by the pump. This will enlarge the clearance values in the kidney since they are flow dependent, eqns (13)-(15), (Fig. 3.7-8). The flow signal from the pump is propagated and damped through the system, although it is still significant at bag side of the kidney. This flow variation is thought to affect fluid exchange between the core blood flow in the capillaries and purified blood near the membrane, thus enhancing diffusion. In the pressure controlled system this variation is much lower (Fig. 3.4).

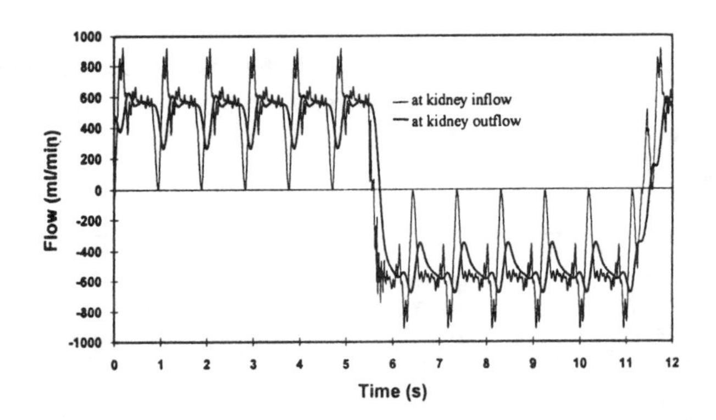

Figure 3.5 Discharges in a weight controlled system

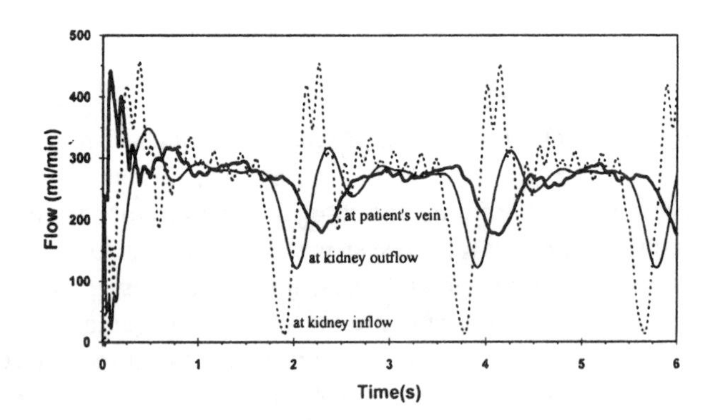

Figure 3.6 Discharges in a two needle system

The flow in a two needle system (Fig. 3.6) is very similar to the flow in the weight controlled system. The difference is that the flow will always be in the same direction and that the mean is halved. The blood flow into the patient's vein (thick line) is influenced by the pump characteristic and by the local pressure .

3.4 Clearance

As already explained with the discharges, the effective clearance of the artificial kidney in the weight controlled system is much higher than in the other systems, especially during the filling phase. Almost no flow signal is preserved in the time dependent clearance as the value saturates at such high blood currents. During the emptying phase the clearance lowers because of co-current dialysate flow and also somewhat by the smaller ultrafiltration flow. Here the flow signal is much more pronounced.

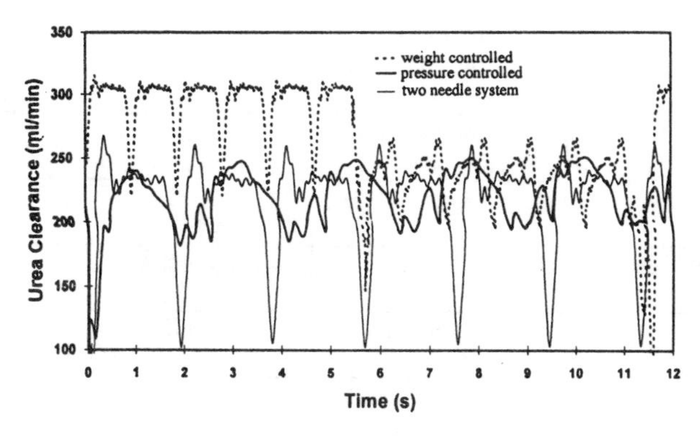

Figure 3.7 Urea clearances

The clearances in the pressure controlled system and the two needle system are about the same, because the mean blood flows through both systems are equal. Only the signal wave form is different as there is more damping in the pressure controlled system (compare with flow variation Figs. 3.4,6).

The difference between the creatinin clearance in the weight controlled system and that in the other systems is smaller as it is for urea. This comes directly from the fact that the creatinin clearance saturates more rapidly at higher blood flows.

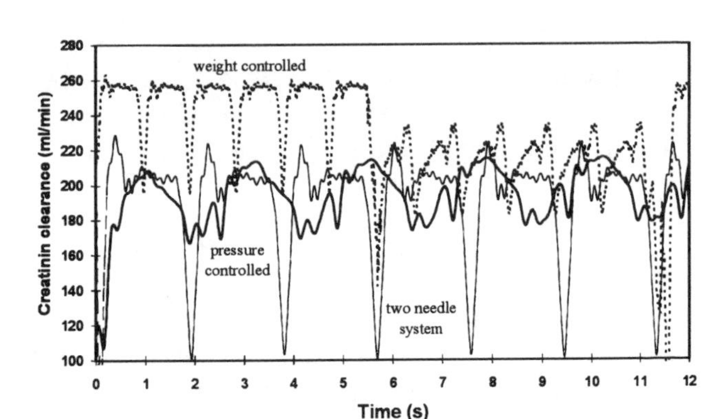

Figure 3.8 Creatinin clearances

To summarise in Table 3.1 the mean flows and clearances are presented.

Table 3.1 Mean values in the different systems

	pressure controlled system	weight controlled system	two needle system
Kidney blood flow	264 ml/min	528 ml/min -528 ml/min†	265 ml/min
Ultrafiltration	440 ml/h	1129 ml/h 94.7 ml/h†	702 ml/h
Urea clearance	221 ml/min	294 ml/min 237 ml/min†	217 ml/min
Creatinin clearance	195 ml/min	248 ml/min 214 ml/min†	192 ml/min

† blood return phase with co-current dialysate flow

4 Solute and water removal in haemodialysis systems

4.1 Introduction

In order to assess the efficiency of haemodialysis therapies, a compartimental model is needed that describes the removal of toxins and excess water from the patient's body. Each compartment contains a certain percentage of the total body water volume with different solutes and each compartment is divided from the other compartments by a transport limiting factor. A common example is the double pool model that is characterised by an intra- and extracellular compartment. The transport limiting factor in this case is the biological membrane between the human cell's interior and exterior space. By identifying the time dependent changes in both water volume and toxin concentration for each compartment, the global dialysis efficiency can be monitored for a given therapy.

4.2 Numerical description of kinetics.

4.2.1 Double pool model

The kinetics in the double pool model can be described by a mass balance equation, giving both the influence of a change in water contents and of solute particle concentration changes in the different compartments, Abbrecht (1971). From Fig. 4.1 the following equations can be derived:

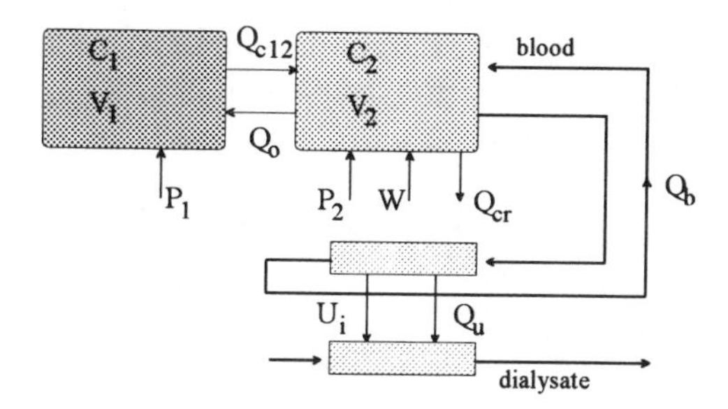

Figure 4.1 Double pool model kinetics

Compartment 1:

$$\frac{dV_1}{dt} = Q_o = L_p A_m \sum_i \Delta\pi_i \sigma_i \tag{16}$$

with V_1 volume of compartment [ml]
Q_o net water flux across the membrane [ml/min]
$\Delta\pi_i$ osmotic pressure gradient for solute particle i [Pa]
L_p membrane permeability for water [cm/min/Pa]
A_m membrane area [cm^2]
σ_i Staverman's reflection coefficient for solute i [-]

$$\frac{dC_{1i}}{dt} = \frac{1}{V_i}\left[P_{1i} - C_{1i}Q_o - Q_{c12}\left(C_{1i} - C_{2i}\right)\right] \tag{17}$$

with C_{ji} solute i concentration in compartment j [mg/ml]

P_{1i} generation rate for solute particle i [mg/min]
Q_{c12} intercompartimental clearance [ml/min]

<u>Compartment 2</u>:

$$\frac{dV_2}{dt} = -Q_u - Q_o + W(t) \tag{18}$$

with W(t) net water intake of patient [ml/min]

$$\frac{dC_{2i}}{dt} = \frac{1}{V_2}\left[P_{2i} - C_{2i}\left(\frac{dV_2}{dt} + Q_{cri}\right) - U_i + Q_{c12}(C_{1i} - C_{2i})\right] \tag{19}$$

with Q_{cri} endogenous clearance of patient for solute i [ml/min]
U_i solute particles removed per unit of time [mg/min]

4.2.2 Discussion

Equations (17) and (19) must be solved for every solute particle in any compartment. To make things not too complicated, only up to five substances are taken into account: urea, creatinin, phosphor, vit. B12 and also an impermeable solute, which is the total of all solutes to which the internal membrane is truly impermeable. All these solute concentrations will change on their turn the total osmotic pressure gradient and consequently have their influence on eqns (16) and (18), as is clear from the following equation:

$$\Delta\pi_i = (C_{1i} - C_{2i})/C_w \tag{20}$$

with C_w the concentration of water [mmol/ml]

To solve this set of equations a Euler algorithm (trapezoidal numerical integration rule) is applied, Isaacson (1966).

4.2.3 Dialysis system

While most parameters in these equations are patient dependent and must be obtained by measurements, three parameters are calculated: Q_o, Q_u and U_i. For a discussion of Q_u, the ultrafiltration flow, see Section 2.4. Q_o, the water flux between compartments is mathematically expressed in the same manner although it is governed by different physical membrane properties, eqn (16). U_i equals the volume of solute particles that is removed from the blood in the artificial kidney system per unit of time and it is dependent on the system' operational characteristics. For both the pressure-pressure controlled system and the two needle system U_i simply equals:

$$U_i = Q_c(C_{bi} - C_{di}) \tag{21}$$

Herein is Q_c the total mean clearance as previously calculated in the fluid dynamical study. C_{bi}, the concentration of solute i in the blood, equals C_{2i} as the blood is considered part of the second compartment.

For the bi-directional weight controlled system the situation is more complex. In this system's kidney the same eqn (21) applies, but the blood concentration in the kidney varies in time because blood is filtered twice and there also is some local recirculation. In Fig. 1.2b four parts can be distinguished, each having their proper concentrations: the bloodline between patient and kidney (index l), the artificial kidney (index k), the bloodline between kidney and accumulation bag (index n) and finally the accumulation bag (index a). For ease of calculation the total pump cycle (i.e. complete blood intake and return phases) is split in different characteristic periods.

If we define:

$t_{k,l,n}$: the time to flow through kidney or bloodline (l or n)
$t_{f,e}$: the time to fill/empty the accumulation bag

it is possible to describe solute transport. At the onset of the filling phase, partly purified blood (inlet concentration C_{ai}) is expelled from the kidney and subsequently replaced by previously filtered blood from the bloodline (concentration C_{li}) and patient's fresh blood (concentration C_{2i}). Thus, only after a period t_k+t_l the kidney is completely filled with fresh blood and thereafter eqn (21) can be applied with $C_{bi} = C_{2i}$.

When the blood will be returned to the patient, the same transition occurs, during a period t_k+t_n and with equivalent concentrations C_{2i}, C_{ni} and C_{ai}. If for ease of calculation, solute removal is expressed per pump cycle (t_f+t_e), the total solute volume removed can be approximated as:

$$\begin{aligned}
U_i(t_f+t_e) = {} & Q_{cf}\,C_{2i}\left(t_f-(t_k+t_l)\frac{V_k+V_l}{2\,V_k+V_l}\right)+Q_{ce}\,C_{2i}\left(\frac{t_k+t_n}{2}\frac{V_k}{2\,V_k+V_n}\right)\\
& +Q_{cf}\,C_{li}\left(\frac{t_k+t_l}{2}\frac{V_k+2\,V_l}{2\,V_k+V_l}\right)\\
& +Q_{ce}\,C_{ni}\left(\frac{t_k+t_n}{2}\frac{V_k+2\,V_n}{2\,V_k+V_n}\right)\\
& +Q_{ce}\,C_{ai}\left(t_e-(t_k+t_n)\frac{V_k+V_n}{2\,V_k+V_n}\right)+Q_{cf}\,C_{ai}\left(\frac{t_k+t_l}{2}\frac{V_k}{2\,V_k+V_l}\right)\\
& -Q_{cf}\,C_{di}\,t_f-Q_{ce}\,C_{di}\,t_e
\end{aligned} \tag{22}$$

In this equation a mean concentration is used during the transition phases. Therefore the bloodvolumes V_k, V_l and V_n are introduced. They are the contents of respectively kidney (k) and bloodlines (n and l). This formula is only valid if V_l and V_n are smaller than V_k. Remark also the two different clearance values Q_{cf} and Q_{ce} for respectively counter current and co-current flow of blood and dialysate in the hollow fiber dialyser.

The set of eqns (16)-(19) can now be solved by applying either eqn (21) or (22).

4.3 Determination of parameters

Before being able to calculate the solution of eqns (16)-(19) the different simulation parameters must be known. They can be divided in two groups: parameters related to the system and therapy and parameters related with the patient. While the former can easily be held constant, the latter are different for each dialysis session and for each patient, although some parameters should be kept within viable limits. Also these patient parameters can only be measured.

4.3.1 System and therapy parameters

4.3.1.1 The two main system parameters ultrafiltration flow and solute clearance are well known from the haemodynamic study (Section 2) as a time dependent quantity. In order to speed up the kinetic simulation, these quantities are averaged in time to allow for larger time steps during calculation. In the case a bi-directional weight controlled system is simulated, the volumes of tubings, artificial kidney and full accumulation bag are needed. The calculation time step should equal the total pump cycle to permit the application of eqn (22).

4.3.1.2 The only therapy related parameter is the duration of the dialysis session. If one also wants information about the rebound phenomenon (see Section 5.2), the total simulation time should be longer than the duration.

4.3.2 Patient parameters

4.3.2.1 Clearances describe solute transport. The intercompartimental clearance is difficult to measure. One of the possible methods consists of doing a curve fit on concentration data, measured during a regular dialysis session. The same operation can also yield data for osmotic flow Q_o.

The remaining endogenous clearance for solute i Q_{cri} (i.e. by the natural kidney) is obtained from a 24h urine sample, Maher (1989).

$$Q_{cri} = \frac{C_{i\,urine} V_{24h\,urine}}{C_{bi}(1440\,min / day)} \tag{23}$$

4.3.2.2 Solute production and water intake are two parameters closely related to the patient's diet and consequently will vary with it. However we can take a mean value. As for water intake this is approximated by

$$W = \frac{G_{pre} - G_{post}}{t_{inter}} \tag{24}$$

where G_{pre} is the patient's weight before the dialysis session and G_{post} the weight after the previous session (dry weight) and t_{inter} the interdialysis time. Remark that this formula will also take into account fluid removal by remaining urine excretion.

If the patient still produces some amount of urine, the solute production can be derived from P_i=$Q_{cri}C_{bi}$. If not, figures are derived from information about the patient's diet. E.g. generation of urea can be calculated from P_{urea} = (PCR - 0.294V)/9.35 with V the patient's dry weight and PCR stands for net protein catabolic rate. Since the ratio of urea and creatinin endogenous clearances is fairly constant, creatinin concentration can be estimated. For other solutes, information is more difficult to acquire, Maher (1989).

4.3.2.3 Compartimental volumes are derived from the patient's dry weight, which can be taken as the weight at the end of a regular dialysis session. Hence, intracellular volume is assumed to be about 38.7% of dry weight and extracellular volume 19.3%, Maher (1989).

4.3.2.4 Initial values for kinetic simulation are the measured blood concentrations of the different solutes before the start of the session.

5 Simulation of dialysis therapy

Dialysis treatment is now simulated with realistic parameters to compare different therapy parameters. First the two presented single needle systems, i.e. the pressure and weight controlled system, are compared for their dialysis efficiency. The next example shows the difference between a short two-hour treatment and a slow treatment using the two needle system. Finally the study is concluded with a comparison between a UDAD therapy, Hombrouckx (1989), and a standard therapy strategy using the weight controlled system.

5.1 Comparison of single needle systems

In this example the single needle systems are compared for their filtering ability for the two main toxic substances: urea and creatinin. A standard 4h dialysis treatment is performed on a 60 kg patient. The respective clearances are shown in Table 5.1. for the patient and for the two systems with a Sorin Bellco BL612-M dialyser. The ultrafiltration flow is 825 ml/h.

Table 5.1 Clearances

	Urea	Creatinin
endogenous	0.3 ml/min	0.5 ml/min
pressure controlled system	206 ml/min	171 ml/min
weight controlled system	278 ml/min 257 ml/min†	222 ml/min 204 ml/min†

† co-current flow

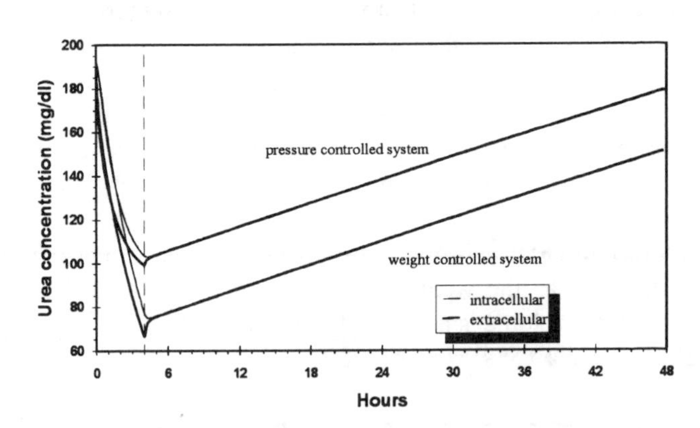

Figure 5.1 Urea concentrations in two single needle systems

In Fig. 5.1 the concentration change of urea is shown during dialysis and the following interdialysis time. It is clearly seen that the weight controlled system has a better removal of urea. The blood concentration level is far below that which is purified in a pressure controlled system. The main reason is the higher mean clearance of the system, caused by the higher blood flow in the weight controlled system (see Section 3.4).

Although this reason seems very plausible, the same is not true for creatinin filtering (Fig. 5.2). Here we find that the creatinin concentration is higher in the weight controlled system, though only slightly. The system's proper characteristics give an unexpected result. Because of the local recirculation (in the system) and the less efficient second filtering of blood acquired in the blood bag (with dialysate flowing co-currently), total purification appears to be less efficient. The importance of this result will become clear from the following. In clinical practice dialysis efficiency is mainly monitored with urea blood levels. Since our result proves the fact that urea and creatinin are not always removed to the same relative extent, this practice should be questioned.

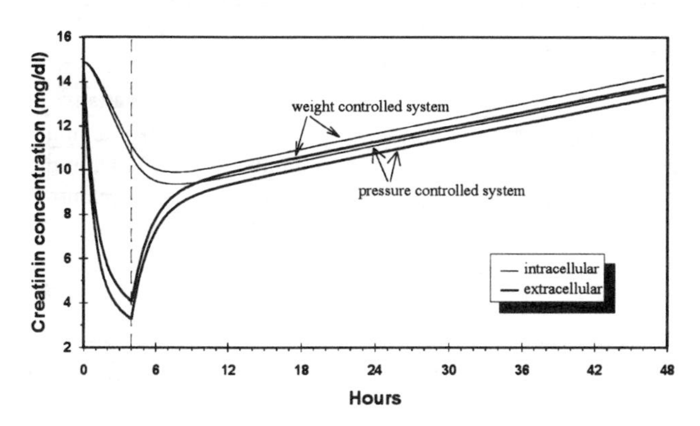

Figure 5.2 Creatinin concentrations in two single needle systems

5.2 Comparison of short and slow therapies

Because a normal dialysis session lasts four to six hours, many researchers have tried to use shorter treatment schedules, mainly because some patients desire it. If the same result, i.e. same level of toxin and fluid removal, is obtained in less time, it is less annoying for the patient and also has economic impact. Yet, some patients suffer too much with such a short or even with a standard four hour dialysis session. For these patients slow therapies are used that last more than six hours and use lower blood flows.

In this example the two needle system, as previously discussed is used with two different pumping speeds: one as fast as in single needle systems (i.e. a mean blood flow of 532 ml/min) and the other at one fourth of this (i.e. mean of 132 ml/min). The clearances, as calculated with the equations from Section 2, are for urea: 362 ml/min and 129 ml/min and for creatinin: 324 ml/min and 123 ml/min respectively. The dialysate pressure was kept at 8 kPa (50 mmHg), which greatly enhanced the ultrafiltration flow in the case of fast pumping. This is necessary to remove a sufficient amount of water during the shorter treatment. Ultrafiltration flow is respectively 1183 ml/h and 465 ml/h, to remove the water uptake of 2.35 kg/day.

The simulated patient weighs 72 kg and initial urea concentration was 171 mg/dl. His renal function has completely ended. The short treatment lasts two hours, while the slow treatment has a six and a half hour duration to obtain about the same level of urea removal.

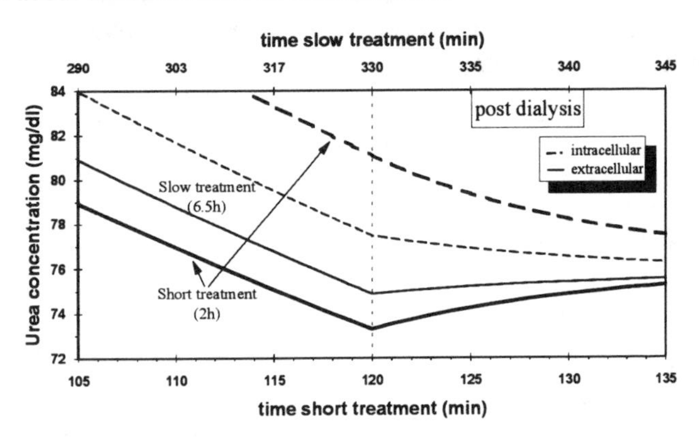

Figure 5.3 Slow and short therapies with a two needle dialysis system

In Fig. 5.3 the end of the dialysis session is shown on the time scale of the short treatment to more easily compare results. To accomplish this the slow therapy's time scale is scaled during dialysis therapy, and is shifted during the postdialysis period. When comparing the urea concentration shifts, it is readily seen that the intercompartimental difference is much enhanced in case of short treatment. This comes from the intercompartimental membrane which hinders urea removal from the intracellular compartment. Since the slow treatment lasts much longer, urea has more time to diffuse from the intracellular to the extracellular compartment, thus keeping the difference effectively smaller. One of the consequences of this difference is the rebound phenomenon, the rapid increase in extracellular concentration directly after treatment. It is explained from the re-equilibration between the two compartments. Both compartimental concentrations will converge to the same concentration in case of zero solute generation. Consequently it is the final concentration, i.e. after the rebound, that will determine real dialysis efficiency. When comparing the rebound phenomena of both therapies, it is clear from Fig. 5.3 that the rebound is higher in case of the short treatment because the intracellular compartment was cleansed less.

5.3 Comparison of UDAD and standard treatment

UDAD stands for Ultrashort Daily Auto Dialysis. It is a therapy which is prescribed for young healthy patients that are capable to dialyse themselves (auto) at home, Hombrouckx (1989). It is a therapy that is performed six times a week (daily) and only lasts one and an half to two hours (ultrashort) with standard blood flow in the system. This therapy is best performed with the weight controlled system because of its ease of operation.

The following simulation shows the creatinin concentrations in a patient during a complete week on the same therapy. The same initial simulated patient is used for the two different treatments: 6 times 1½ h (UDAD) vs. 3 times 4h (standard). In reality no identical patient is possible because a patient's condition is always varying. This patient weights 54 kg, his endogenous clearances are 1.5 and 2.5 ml/min for urea and creatinin respectively. The water intake is 633g/day and urea and creatinin generation are 3.76 and 0.26 mg/min. As artificial kidney the Sorin Bellco BL621 is used with urea clearances of 255/190 ml/min and for creatinin 210/144 ml/min. (The second figures are for co-current flow). Ultrafiltration flow is 738 and 448 ml/min for UDAD and standard therapy for the week's first dialysis session, while it is 480 and 290 ml/min elsewhere. The first session needs an augmented water removal because of the two (UDAD) or three (standard) day interdialysis period.

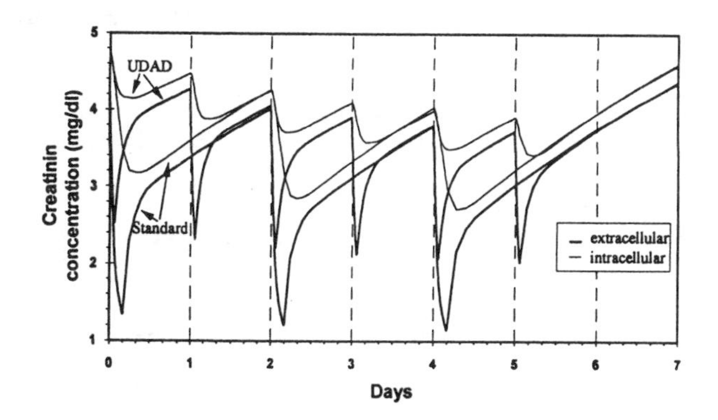

Figure 5.4 UDAD and standard therapies

As is clear from Fig. 5.4, the patient's pre-dialysis creatinin levels remain the same for both therapies. After two days both extracellular and intracellular concentrations are almost equal. The main difference is found in the post-dialysis concentrations. These are much higher in the UDAD patient. It is not the minor toxin removal of one dialysis cycle that is important, but the fact that the difference between pre- and post-dialysis concentrations is kept smaller in the UDAD patient, rendering a more physiological concentration profile, Hombrouckx (1989).

Conclusion

In this chapter the haemodynamics in different haemodialysis systems are calculated. From the pressure and flow characteristics, the operational parameters clearance and ultrafiltration are determined. These parameters serve to compute dialysis efficiency in different systems and using different kinds of dialysis therapies.

References

Abbrecht P.H., Prodany N.W., *A model of the Patient-Artificial Kidney System*, IEEE Trans. Biom. Eng., pp257-264, V18-4,1971

Cambi V., *Short Dialysis*, Martinus Nijhoff Publishing, 1987

De Wachter D., Verdonck P., Verhoeven R., Hombrouckx R., *Comparison of a new and standard single-needle-dialysis system using a mathematical model*, Artificial Organs, 17 (5) : 328-338; 1993

Henderson L., Quellhorst E., Baldamus C., Lysaght M., *Haemofiltration*, Springer Verlag, 1984

Hombrouckx R., Bogaert A., Leroy F., Verhoeven R., Verdonck P. et al., *Limitations of Short Dialysis are the indications for Ultrashort Daily Auto Dialysis*, Trans. ASAIO, 35 (3) : 503-505; 1989

Isaacson E., Keller H., *Analysis of numerical methods*, John Wiley & Sons, New York, 1966.

Jaffrin M., *Organes Artificiels*, Université de technologie de Compiègne, 1991

Jungers P., Zingraff J., Man M., Drueke T., Tardieu B., *The essentials in haemodialysis*, Martinus Nijhoff Medical Division, 1978

Legrain M., Suc J.M., *Nephrology*, Year Book Medical Publishers

Maher J. (Ed.), *Replacement of Renal Function by Dialysis*, Kluwer Academic Publishers, 1989

Nissenson A., Fine R., Gentile D., *Clinical Dialysis*, Prentice-Hall International, 1984

Streeter V., Wylie B., *Fluid Transients*, McGraw-Hill Company, 1978.

Chapter 7

Folding motifs, kinetics, and function in the proximal convoluted tubule

C.J. Lumsden

Institute of Medical Science, Department of Medicine, University of Toronto, Toronto, Canada M5S 1A8

Abstract

Simulation is used to study the dynamics of indicators interacting with the renal proximal convoluted tubule (PCT). The structural model of the PCT emulates the tubule's complex pattern of three-dimensional folding and vascular perfusion. Diffusional shunting among nearby folds strongly influences the washout and regional retention of secreted and reabsorbed indicators. Comparison to unfolded systems shows striking differences that are most pronounced in the early part of the washout process. Visualization of the tracer clouds in the blood, interstitial, and urinary compartments allows the regional pattern of indicator progression through the system to be followed in detail.

Nomenclature

MID multiple indicator dilution

k_1 $k_1'\phi_1/B$

k_2 $k_2'\phi_1/B$

k_3 $k_3'\phi_2/B$

k_4 $k_4'\phi_2/B$

The k_j are the unidirectional flux coefficients for indicator entry into and exit from the tubular epithelial cell (see Fig. 8).

ϕ_1 surface area per unit length of the basolateral [blood-facing] plasma membrane of the cell.

ϕ_2 surface area per unit length of the brush-border [urine-facing] plasma membrane of the cell.
B volume of distribution per unit length along the nephron for indicator exchange into and out of the cell.
k_j' permeability for indicator flux in the direction of k_j $j = 1, \ldots, 4$.
k_5 rate of irreversible sequestration per unit accessible intracellular space.

k_I, k_O effective unidirectional flux coefficients for indicator exchange between the interstitial space and the urine space of the proximal convoluted tubule in the case where the action of the tubular epithelial cell is modeled by a single biological interface, rather than by two interfaces [basolateral and tubular surfaces] acting in series and separated by an intracellular space

k_+, k_- unidirectional flux coefficients for indicator exit from and entry into the peritubular capillary bed. For flow-limited indicators $k_+, k_- \rightarrow \infty$

$W_{B\ell}$ [constant] velocity of blood flow in the one-dimensional model.
W_{Tu} [constant] velocity of urine flow in the one dimensional model.
$W_{B\ell'}$ effective velocity of flow for flow-limited indicators within the compartment formed by the capillary and interstitial spaces.
$\mathbf{W}_{\mathrm{B}\ell}(\mathbf{x})$ blood velocity at position $\mathbf{x}$ in the three dimensional model. The velocity has constant magnitude.
$\mathbf{W}_{\mathrm{Tu}}(\mathbf{x})$ urine velocity at position $\mathbf{x}$ in the three dimensional model. The velocity has constant magnitude.

γ ratio of intracellular to effective vascular volume per unit length.
θ ratio of tubular to effective vascular volume per unit length.
D diffusion coefficient for indicator Brownian mobility in the interstitial space.

$\ell(\mathbf{x})$ space curve defining the central axis of a generalized cylinder in $\mathbf{R}^3$
$\mathbf{t}(\mathbf{x})$ tangent vector to $\ell(\mathbf{x})$ at $\mathbf{x}$.

$u(\mathbf{x},t), w(\mathbf{x},t), v(\mathbf{x},t), z(\mathbf{x},t)$ concentrations per unit time of indicator in the capillary, interstitial, epithelial cell, and urine spaces of the nephron, respectively, at a position $\mathbf{x}$ along the exchange unit at time t. In the case of flow-limited indicators $u(\mathbf{x},t)$ denotes the concentration in the effective vascular space formed by the capillary and interstitial volumes.

1 Introduction

Each nephron of the kidney consists of a spheroidal ultrafilter, the glomerulus, joined to a long, unbranched tubule lined with epithelial cells. A human kidney contains approximately one million such nephrons, each richly perfused with blood, operating in parallel.

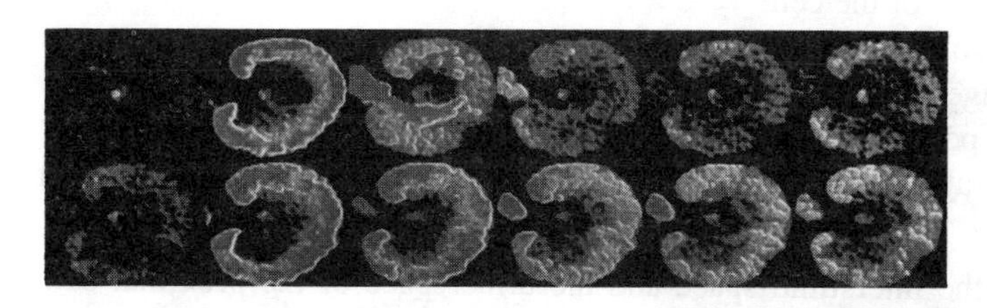

Figure 1: Indicator dilution in a dog kidney. The passage of an X-ray opaque vascular (Top) and interstitial (Bottom) indicator is followed in a series of CT images taken every 0.6 sec following pulse injection into the renal artery (see ref.[3]). Physical models of the nephron and renal microcirculation are required to explain the retention and washout patterns specific to different indicators. The kidney is imaged transversely by the CT scanner, slice thickness 3 mm.

The tubule of the nephron processes the glomerular ultrafiltrate to form concentrated urine. Under normal conditions more than 80 per cent of the water, blood sugars, and amino acids filtered at the glomerulus is reclaimed by the initial segment of the tubule. Secretion of organic acids and organic bases from the postglomerular microcirculation also occurs in this portion, which loops tortuously out from the glomerulus to form a coiled mass called the proximal convoluted tubule or PCT.[1]

The PCT's key role in managing water, ions, and small organic molecules has led to many studies of its solute transport kinetics. Recently, it has become possible to examine these transport mechanisms quantitatively in the intact functioning kidney *in vivo* using the method of multiple indicator dilution (MID). In the MID approach to the kidney, a small bolus (typically 0.5 ml) of saline containing one or more types of indicator molecule is rapidly injected into the renal artery. Because the bolus volume is small and the indicator molecules are present in tracer concentration, the organ is not disturbed. The technique can be applied at the level of the whole organ,[2] via timed serial samples collected from the catheterized renal vein and ureter, or regionally at the level of small numbers of nephrons and their blood supply,[3] via ultrafast dynamic tomography (Fig.1).

The dynamics of indicator transit through the kidney as observed in these procedures cannot be understood without the aid of a quantitative physical model. The model expresses the underlying physical mechanisms of indicator transport by the renal microcirculation and the nephrons. Previous mathematical models of indicator transport by the PCT have not been able to include the tubule's complex pattern of three- dimensional folding. Simplifying assump-

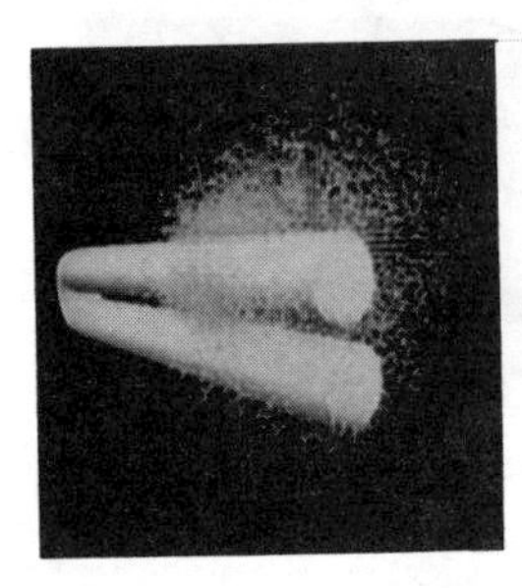

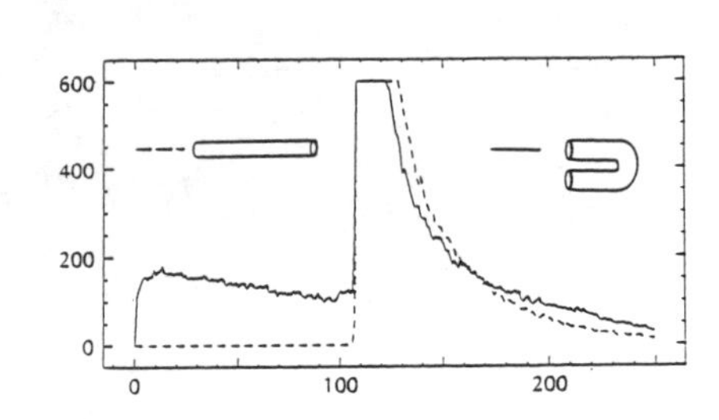

Figure 2: Diffusional shunting in a hairpin fold. Left. Particles from an indicator bolus confined initially to the top arm of the hairpin escape and are captured by the lower arm. Right The characteristic signature of a shunt in a bolus-injection experiment is the appearance at the outflow of shunted indicator particles before the convective transit time of the tubule. The hairpin is a basic folding motif of the proximal convoluted tubule. The PCT is to first orde a set of hairpin folds that are themselves packed closely together in space. Reproduced fron Li and Lumsden[4] with permission.

tions about the folding and its relations to the regional blood supply have bee applied. Most commonly, the PCT is treated as a straight tube. Indicator pass across its surface and into an adjacent straight tube modeling a capillar situated close to the PCT.[2]

Such straight-geometry models may be adequate for interstitial indicators which are confined either to the urine space of the nephron, or to the renal cap illaries and their sheaths of surrounding extracellular matrix.[3] Indicators tha are secreted or reabsorbed, however, cross the tubule convolutions and ma in principle move between otherwise distant tubule segments that are folde close together. This process is called diffusional shunting. Previous studie have shown that diffusional shunting can be an important factor in the indicatc dynamics if the exchange regions are close enough (Fig. 2).[4–6] On the basi of general anatomical observations it is evident that the folds of the proxima convoluted tubule are numerous and closely packed.[1,4] Diffusional shuntin may therefore figure prominently in the normal function of the PCT. Quanti tative models of indicator transport by epithelial surfaces folded in comple three-dimensional patterns resembling those found in the proximal nephro have not, however, been available.

We have now modeled the complex folding pattern of the PCT in three d mensions and treated the organization of its perfusing blood supply explicitl Comparison of indicator transit through the folded model with the behavior

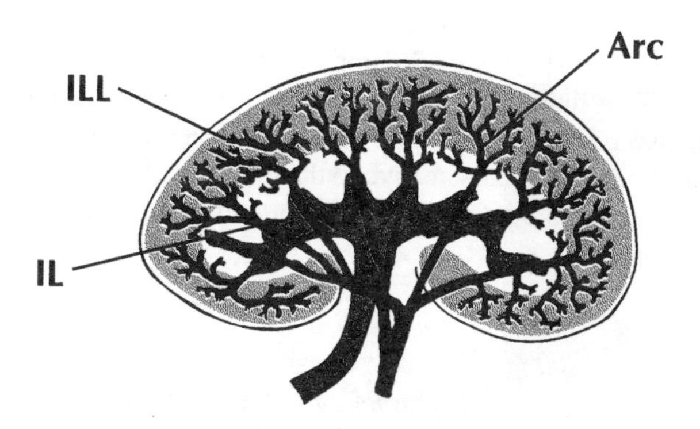

Figure 3: General anatomical organization of the renal blood supply in the dog. IL, interlobar artery; ILL, interlobular artery; Arc, arcuate artery. After Arnautovic.[7]

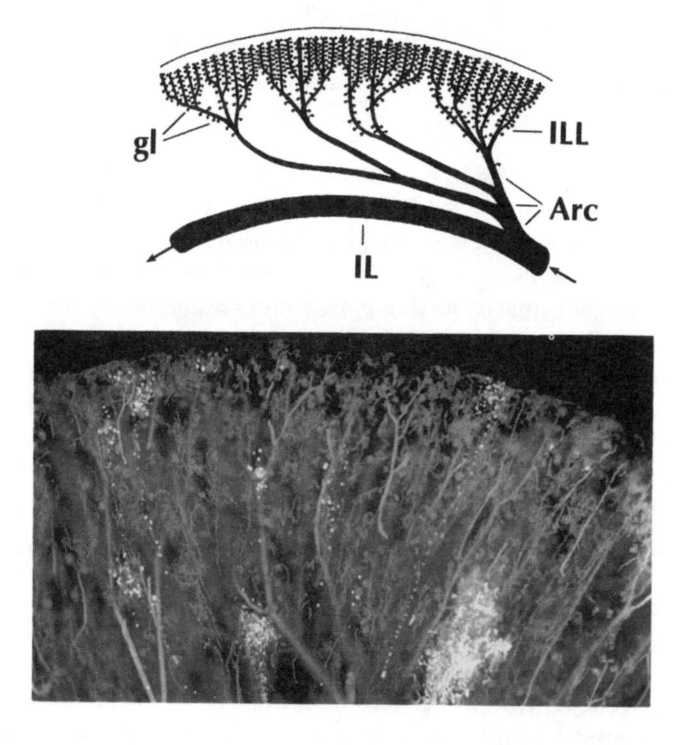

Figure 4: Top. Blood supply to the dog renal cortex. gl, glomeruli. After Selkurt.[8] Bottom. Latex injection of the vascular field illustrated above, showing the complex arrangement of arterial vessels. The light blotches mark zones occupied by selectively injected PCTs. Courtesy of R. Beeuwkes.

of matched straight-tube configurations shows striking differences, suggesting that more complex structural assumptions than previously applied may be needed to accurately interpret the indicator response patterns of the kidney. Our findings also raise the intriguing possibility that certain types of solute molecules may be particularly sensitive to the motifs of PCT folding, and may therefore be applied in the kidney as *in vivo* probes of nephron geometry and its changes in disease.

2 Background

It is the perfusion of specific nephrons by specific blood vessels that determines how much injected indicator reaches the tubular-vascular interface in each small region of kidney tissue, and when. A precise specification of these tubular-vascular relations is therefore needed to predict the fate of indicator entering the system. In this brief overview I will concentrate on the current understanding of the dog kidney. Convenient size and surgical siting, combined with structural similarities to human material, have made the dog kidney the favored system for renal MID.

2.1 Anatomy

The renal artery bifurcates into dorsal and ventral divisions almost immediately upon entering the hilum (Fig. 3). The interlobar branches arising from the two divisions partition the dog kidney into symmetrically arranged vascular fields. Each arterial vessel is part of a true end artery; unlike the renal venous system inter-arterial anastomoses (and arterial-venous anastomoses) are unknown. From the interlobar branches, arcuate vessels coursing at the corticomedullary border carry the renal blood flow to the interlobular arteries (also termed cortical radial arteries) and from there to the afferent arteriole of each glomerulus (Fig. 4). Blood flow to both the cortex and the medulla passes through the glomerular vessels, all of which are located in the renal cortex. The major venous branches parallel the course of the arterial vessels.

2.2 Nephron organization

In addition to blood vessels, the renal cortex of a dog contains some 200,000 nephrons packed side by side, surrounded by the cortical microvasculature and disposed orthogonally to the gently curving surface defined by the renal capsule. The proximal convoluted tubule [PCT] loops tortuously out from the parent glomerulus to form a coiled, vertically extended mass (Fig. 5).The PCT averages 14 mm in length with a luminal diameter of 30 microns. The complex folding pattern packs the PCT and its perfusing capillaries into a cylindrical region about 1 mm long and 250 microns across.

2.3 Tubular-vascular relations

Glomerular efferent vessels, forming the peritubular capillary system, closely invest the tubular convolutions. Beeuwkes and Bonventre[9] found that the proximal convoluted tubules arising from the midcortical and juxtamedullary glomeruli are either partially or completely dissociated from the efferent blood vessel leaving the parent glomerulus. These tubules interact with efferent vessels arising from neighbouring glomeruli (Fig. 6). In contrast, the proximal convolutions of superficial glomeruli are associated predominantly or completely with the efferent vascular network of their parent glomerulus. The general pattern of perfusion is axial, with multiple branches of the peritubular capillary system paralleling the long axis of the compact cylindrical mass formed by the tubule folds (Fig. 7).

Although more limited, the data on rat, monkey, and human material indicate tubular-vascular relations similar to those observed in the dog.[10,11]

Figure 5: Schematic illustration of a typical PCT folding pattern (black line) in relation to the parent glomerulus (grey ellipsoid).

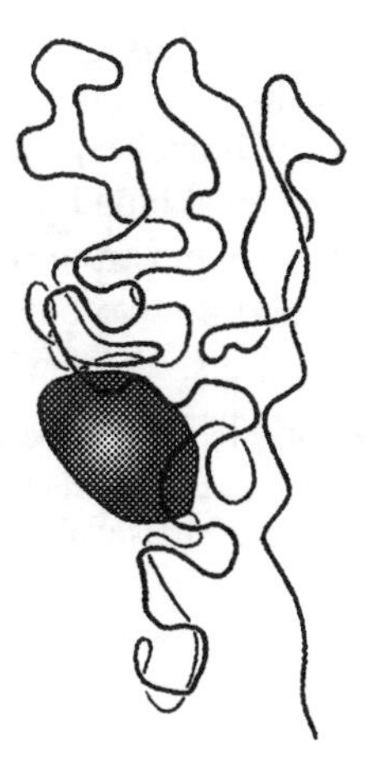

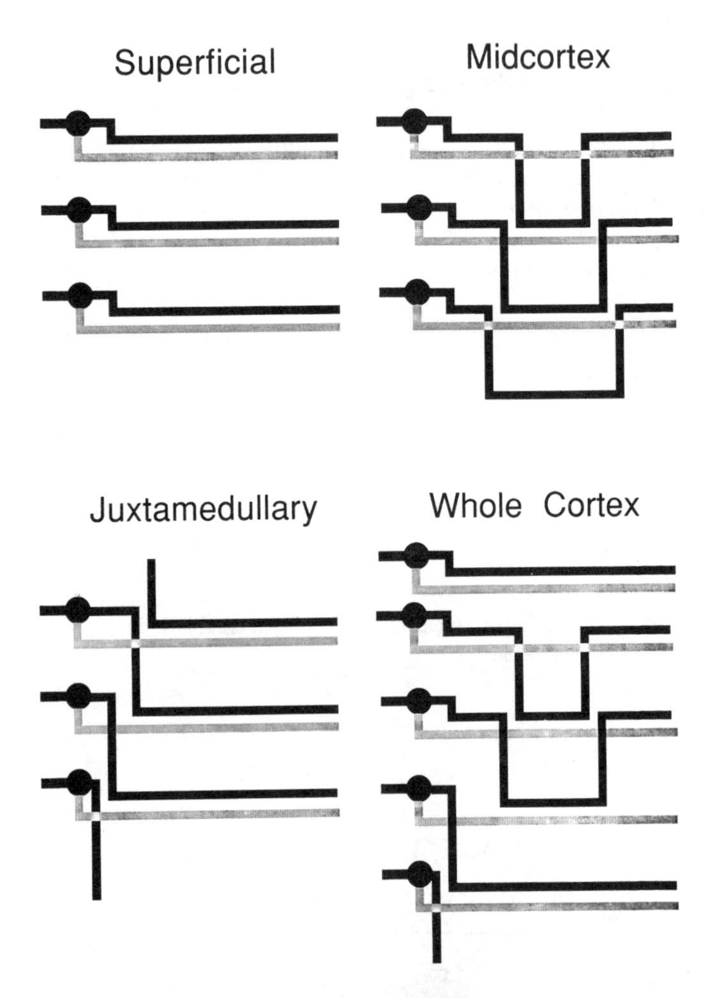

Figure 6: Schematic of the tubular-vascular relations in the dog renal cortex. The dark circles are the the glomeruli. The dark tubes denote the afferent and efferent (peritubular capillary) glomerular vessels. The light tubes denote the proximal convoluted tubule.

2.4 Tissue compartments

Indicator molecules handled by the proximal convoluted tubule may access four compartments (the volumetric and kinetic contributions of the renal cortical lymphatics and nerves is minor and will be ignored here): the intravascular space $B\ell$ of the peritubular capillary network, the interstitial space Int, the tubular epithelial cells Ep, and the tubular lumen Tu (Fig. 8). Indicators capable of permeating the fenestrated endothelium of the peritubular capillary wall may cross from $B\ell$ into Int. Once in Int the ligands may re-enter $B\ell$ or, if recognized by carriers on the basolateral surface of Ep, enter the epithelial cell. Indicators taken up by the epithelial cell have access to the tubular lumen Tu via the luminal membrane of Ep. Alternatively, they may return to the interstitium by moving back across the basolateral membrane. By the same routes, indicators in the tubular lumen have access to the remaining three compartments. Irreversible sequestration may occur in Ep (e.g. uptake into vesicles) and, if present, lead to reduced recovery of indicator relative to the net mass injected. The case of irreversible sequestration will not be considered further here.

Figure 7: Latex cast of a mid-cortical glomerulus in the dog and its efferent vessels forming the complexly branched peritubular capillary tree. The axial pattern of the association with the PCT is evident. Courtesy of R. Beeuwkes.

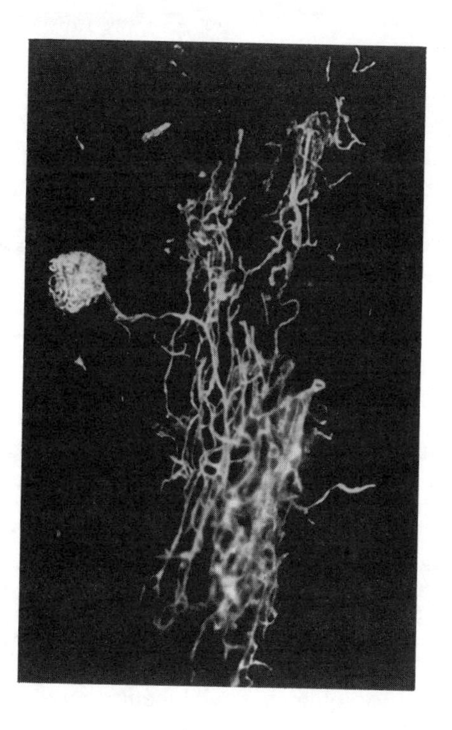

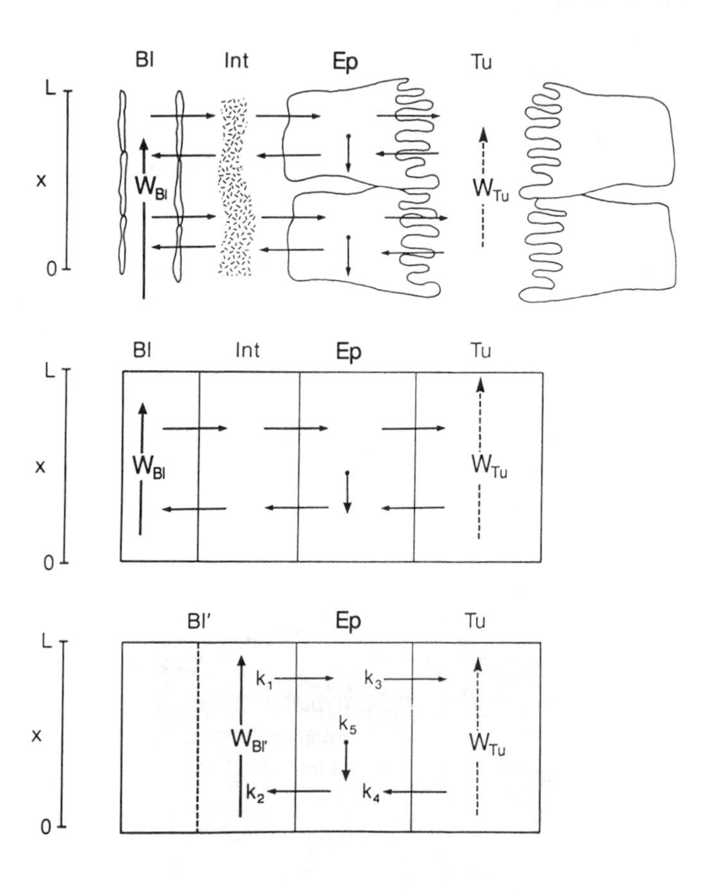

Figure 8: Steps in modeling tracer exchange between the tubule epithelial cells and the peritubular capillary. Top. The local relationships between $B\ell$ and Tu and the tracer fluxes joining them. The flux arrows mark pathways of transport across the basolateral (blood-facing) and luminal (urine-facing) plasma membrane of the epithelial cells Ep lining the proximal convoluted tubule. Middle. Compartmental representation. Bottom. The exchange of low molecular weight tracer molecules between the blood and the interstitium Int of the renal cortex may be flow-limited. The system is then equivalent to three adjacent compartments that run the length of the tubule: Tu, Ep, and $B\ell' = B\ell + Int$.

2.4.1 Flow-limited exchange

The permeability of the peritubular endothelium to the low molecular weight indicators used as interstitial markers in renal MID (e.g. creatinine, L-glucose) may reach large values. Under these conditions the combined space $B\ell' \equiv B\ell + Int$ acts as one effective space, through which indicators move as if convecting at a velocity related to the true velocity of blood flow in the efferent vascular network. The relationship between the true and the effective velocities is governed by the relative volume of distribution per unit length of peritubular vessel. For flow-limited indicators exchanging beween the blood and urine across the tubular epithelium, the number of relevant compartments is therefore three [$B\ell', Ep, Tu$; Fig. 8] rather than four [$B\ell, Int, Ep, Tu$], with the interstitial marker acting as the reference indicator to which the exchange dynamics of a test substance can be compared. The interstitial marker, although filtered at the glomerulus and freely permeable across the peritubular capillary wall, is ignored by the epithelial cells and remains on the extracelluar side of both the luminal and basolateral plasma membranes over the time course of the experiment.

3 The structural model

3.1 Physical assumptions

Following whole-organ theory,[12] the dynamics of indicator within a region of interest are assumed to arise from indicator interaction with the exchange units in the region. Here, the exchange units are the PCTs and their perfusing blood supply. At this stage the model treats only the case seen in the superficial layer of the cortex (Fig. 6), where each proximal nephron is perfused independently of the others. Within each nephron and each capillary, convective transport of the indicator bolus is assumed to be adequately represented by plug flow. Transverse mixing within each flow channel is assumed to keep the concentration of indicator well stirred and uniform lateral to the direction of flow, so that diffusion in the plane lateral to the flow can be ignored. Flow rates are assumed sufficiently brisk relative to the time course of the experiment that diffusion parallel to the flow direction can also be ignored. Diffusion of indicator within the interstitium and the intracellular spaces may occur. Permeabilities and unidirectional flux coefficients are for simplicity assumed to be constant (the simulation method itself, described below, is applicable to variable transport rates). Indicator fluxes across each interface in the system vary linearly with the indicator concentration difference across the interface. Aside from the indicators, the system is in steady state. The physical rationale behind these assumptions and their applicability to kidney studies is presented elsewhere.[2,12]

3.2 Indicator exchange equations

The importance of a computational approach to indicator dynamics in the proximal nephron is now readily seen. For simplicity we make this point with the case of flow-limited indicators. Since the exchange of indicator between $B\ell'$, Ep, and Tu is a process that neither creates nor destroys mass, standard conservation requirements suffice to give the equations of motion. If the tubule and perfusing capillary are, moreover, straight rather than coiled (e.g. Fig. 6), and diffusion within Int and Ep is fast enough to be ignored, the physical assumptions we have made imply that $u(x,t)$, $v(x,t)$, and $z(x,t)$ change in time according to

$$\begin{aligned} \frac{\partial u}{\partial t} &= -W_{B\ell'}\frac{\partial u}{\partial x} + k_2\gamma v - k_1\gamma u \\ \frac{\partial v}{\partial t} &= k_1 u - k_2 v - k_3 v + k_4 z - k_5 v \\ \frac{\partial z}{\partial t} &= k_3\frac{\gamma}{\theta}v - k_4\frac{\gamma}{\theta}z - W_{Tu}\frac{\partial z}{\partial x} \end{aligned} \qquad (1)$$

For an impulse injection experiment the initial conditions are $u(x,0) = c\delta(x)$, $v(x,0) = 0$, and $z(x,0) = c'\delta(x)$ where $\delta(x)$ is the Dirac delta function and the lowercase c, c' are normalization constants defined in terms of the tracer masses and flows entering the peritubular capillary and the proximal tubule respectively.

Application of the Laplace transform

$$\mathcal{L} : f \longrightarrow \mathcal{L}[F(x,t)] \equiv \int_0^\infty f(x,t)e^{-st}dt \equiv \overline{F}(x,s) \qquad (2)$$

and rearrangement gives the exact solution

$$\begin{pmatrix}\overline{U}\\ \overline{Z}\end{pmatrix} = \frac{1}{\delta - (\lambda_1 - A)(\lambda_2 - D)} \cdot$$

$$\left\{ \begin{array}{l} Be^{\lambda_1 x}\left(\frac{Cc}{W_{B\ell'}} + \frac{(D-\lambda_2)c'}{W_{Tu}}\right) + (\lambda_2 - D)e^{\lambda_2 x}\left(\frac{(A-\lambda_1)c}{W_{B\ell'}} + \frac{Bc'}{W_{Tu}}\right) \\ (\lambda_1 - A)e^{\lambda_1 x}\left(\frac{Cc}{W_{B\ell'}} + \frac{(D-\lambda_2)c'}{W_{Tu}}\right) + Ce^{\lambda_2 x}\left(\frac{(A-\lambda_1)c}{W_{B\ell'}} + \frac{Bc'}{W_{Tu}}\right) \end{array} \right\} \qquad (3)$$

in Laplace space, where $\delta = BC$,

$$\begin{aligned} \lambda_1 &= (A + D - \sqrt{(A-D)^2 + \delta}\,)/2 \\ \lambda_2 &= (A + D + \sqrt{(A-D)^2 + \delta}\,)/2 \end{aligned} , \qquad (4)$$

and $\overline{V} = (k_1\overline{U} + k_4\overline{Z})/\Gamma$ with $\Gamma = (s + k_2 + k_3 + k_5)$ and A, B, C, D polynomials in the k_j. These simple, one-dimensional solutions are of considerable use in developing preliminary models of indicator exchange across the tubular epithelial cell.[2]

When the tubule and blood vessels are folded and interwoven, each small volume of renal interstitium may in principle exchange indicator molecules with more than one segment of tubule or capillary. These segments, otherwise distant from each other in the unfolded, straight geometry, lie in close proximity in the folded nephron. The local conservation equations must then be generalized to account for the complex spatial organization of tubule and capillaries, as well as the possible exchange of indicator among multiple nephron segments:

$$\begin{aligned} \partial u(\mathbf{x},t)/\partial t &= -\nabla_{\text{plug}} \cdot [\,u(\mathbf{x},t)\,\mathbf{W}_{\mathbf{B}\ell}] \\ &\quad + [-k_+ u(\mathbf{x},t) + k_- w(\mathbf{x},t)]\ \delta(\mathbf{x} - \Sigma_{B\ell}) \\ \partial w(\mathbf{x},t)/\partial t &= -\nabla^2 w(\mathbf{x},t) \\ &\quad + \sum_{\text{sources}(\mathbf{x})} [k_+ u(\mathbf{x},t) - k_- w(\mathbf{x},t)]\ \delta(\mathbf{x} - \Sigma_{B\ell}) \\ &\quad + \sum_{\text{sources}(\mathbf{x})} [-k_1 w(\mathbf{x},t) + k_2 v(\mathbf{x},t)]\ \delta(\mathbf{x} - \Sigma_{\text{Baso}}) \\ \partial v(\mathbf{x},t)/\partial t &= -\nabla^2 v(\mathbf{x},t) \\ &\quad + [k_1 w(\mathbf{x},t) - k_2 v(\mathbf{x},t)]\ \delta(\mathbf{x} - \Sigma_{\text{Baso}}) \\ &\quad + [-k_3 v(\mathbf{x},t) + k_4 z(\mathbf{x},t)]\ \delta(\mathbf{x} - \Sigma_{\text{Antilum}}) \\ \partial z(\mathbf{x},t)/\partial t &= -\nabla_{\text{plug}} \cdot [\,z(\mathbf{x},t)\,\mathbf{W}_{\mathbf{Tu}}] \\ &\quad + [k_3 v(\mathbf{x},t) - k_4 z(\mathbf{x},t)]\ \delta(\mathbf{x} - \Sigma_{\text{Antilum}}) \end{aligned} \qquad (5)$$

for the vascular space $B\ell$, the interstitium Int, the intracellular space Ep, and the tubular space Tu respectively. The symbols $\Sigma_{B\ell}$, Σ_{Baso} and Σ_{Antilum} denote the capillaries and the exterior and interior surfaces of the PCT. ∇_{plug} represents the transport operator for plug convection. The summation operator indicates the need to include the contribution from all capillary and tubule segments exchanging tracer with the interstitium at position $\mathbf{x}$.

Equations (5) may be simplified somewhat by noting that for the nonmetabolized indicators of interest here, diffusion through the cell adds only a time delay in the transit of the ligand through the nephron. We may therefore approximate eqns(5) with a reduced model in which the surfaces Σ_{Baso} and Σ_{Antilum} are replaced by one surface, Σ_{Tu}, and two kinetic constants, k_I and k_O, representing the passage of indicator molecules between the interstitium and the lumen of the tubule. In order to preserve the structure of the urine space Tu we take $\Sigma_{Tu} = \Sigma_{\text{Antilum}}$. The volume of distribution of the chemically inert cell interior is then added to Int, so that

$$
\begin{aligned}
\partial u(\mathbf{x},t)/\partial t &= -\nabla_{\text{plug}} \cdot [u(\mathbf{x},t)\,\mathbf{W}_{\mathbf{B}\ell'}] \\
&\quad + [-k_+ u(\mathbf{x},t) + k_- w(\mathbf{x},t)]\,\delta(\mathbf{x} - \Sigma_{B\ell}) \\
\partial v(\mathbf{x},t)/\partial t &= -\nabla^2 v(\mathbf{x},t) \\
&\quad + \sum_{\text{sources}(\mathbf{x})} [k_+ u(\mathbf{x},t) - k_- v(\mathbf{x},t)]\,\delta(\mathbf{x} - \Sigma_{B\ell}) \\
&\quad + \sum_{\text{sources}(\mathbf{x})} [-k_I v(\mathbf{x},t) + k_O z(\mathbf{x},t)]\,\delta(\mathbf{x} - \Sigma_{Tu}) \\
\partial z(\mathbf{x},t)/\partial t &= -\nabla_{\text{plug}} \cdot [z(\mathbf{x},t)\,\mathbf{W}_{\mathbf{Tu}}] \\
&\quad + [k_I v(\mathbf{x},t) - k_O z(\mathbf{x},t)]\,\delta(\mathbf{x} - \Sigma_{Tu})
\end{aligned} \qquad (6)
$$

However, even in this partially reduced form, the exchange equations are not amenable to exact solution for the complex geometries that describe capillary and tubule folding. A computational method was therefore developed to estimate $u(\mathbf{x},t)$, $v(\mathbf{x},t)$, and $z(\mathbf{x},t)$ by simulating the motion of indicator molecules through the vascular, interstitial, and tubular spaces described by eqns (6).

3.3 Simulation method

3.3.1 Geometry modeling

The first step was to develop digital representations of the biological interfaces $\Sigma_{B\ell}$ and Σ_{Tu}. Three-dimensional structural databases for the surfaces were built as described previously.[4] Briefly, each vessel was treated as a generalized cylinder with central axis defined by an $\mathbf{R}^3$ space curve $\ell(\mathbf{x})$ with tangent vectors $\mathbf{t}(\mathbf{x})$. The cross section of the vessel in the plane normal to $\mathbf{t}(\mathbf{x})$ was a circle of radius r for all $\mathbf{x}$ on the domain of definition. The cross section represented the vessel interior where convective flow of blood (capillaries) or urine (tubule) occurred. For the simulations discussed here the radius was set at 10 microns for the capillaries and 30 microns for the tubule.

An interactive visualization program was written to allow each vessel to be built under user control. An arbitrary number of vessel start points could be selected. Beginning at a vessel's start point, the construction moved in fi-

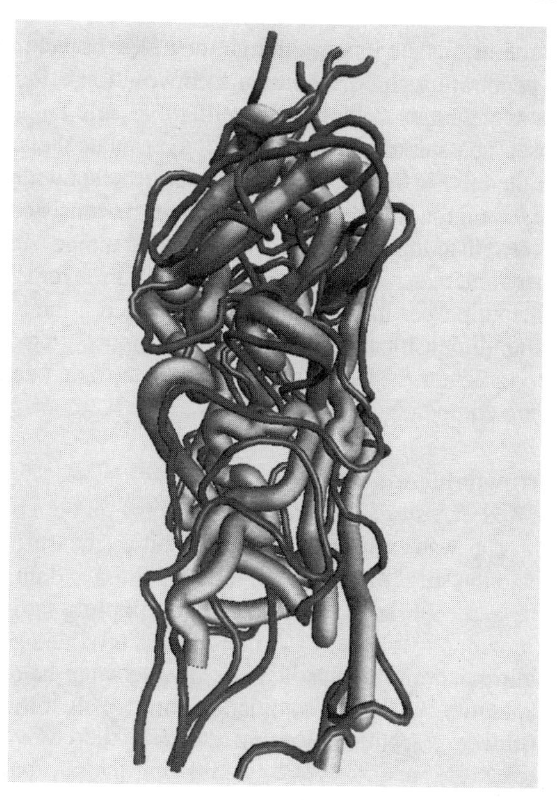

Figure 9: Solid surface rendering of the proximal convoluted tubule model and the perfusing capillaries.

nite steps along the center axis $\ell(\mathbf{x})$. The step size was 5 microns for the capillaries and 10 microns for the tubule, This produced fine resolution of the folding patterns with a construction task that required several thousand steps per vessel. For each step, user input determined the direction of $\mathbf{t}(\mathbf{x})$ and thus the local axial curvature of the vessel. The vessels were displayed as solid surfaces (Silicon Graphics Crimson VGX, Silicon Graphics Inc., Mountain View, CA) in order to check for self-intersections during construction and vessel-vessel intersections when multiple vessels were simulated.

Because of its compact folding pattern the generalized cylinder representing the PCT was built first and the microcirculation added. Σ_{Tu} was elaborated to achieve a qualitative resemblance to a dog mid- cortical nephron imaged by Beeuwkes and Bonventre.[9] Since studies documenting exact packing relations between PCT and peritubular capillary vessels have not yet been reported, no attempt was made at this stage to model the complex branching of the capillary network penetrating among the tubule convolutions. Perfusive coverage of Σ_{Tu} was achieved by threading 5 capillary vessels $\Sigma_{B\ell_k}$, $k = 1,\ldots,5$ among the folds. The capillaries were oriented to emulate the axial pattern of perfusion illustrated in Fig 7. The resulting structure is shown in Fig 9.

A second exchange unit was also constructed. It consisted of a straight tubule, equal in length to the axial length of the folded tubule, surrounded by 5 straight capillaries and matched to the folded configuration for PCT/interstitial volumes of distribution per unit length. This provided a three- dimensional version of the one-dimensional straight-geometry models (e.g. Fig.) studied previously. The indicator exchange properties of the folded and straight exchange units were compared.

3.3.2 Compartment identification

For both the folded and straight units, the geometry model comprising $\Sigma_{Tu} \cup \Sigma_{B\ell_k}$, $k = 1,\ldots,5$ was embedded in a cubic lattice measuring 132 x 87 x 76 lattice points to the side (872,784 total lattice points) and the lattice points inside the tubule and capillaries identified. The sampling lattice sufficed to model the tubule with a resolution of 7 microns and provided good representation of its folding pattern. The lattice coordinates were held in the workstation's main memory during the simulation runs. This allowed real-time visualization of tracer molecule movement.

3.3.3 Simulation method

The indicator dynamics represented by eqns (6) were modeled by placing a cloud of indicator particles on the lattice. Following tracer conditions, the indicator particles did not interact with each other except to exclude simultanteous occupancy of one lattice site by two or more particles. At each time step, the position of each particle was updated by a protocol representing the transport equations (6) for the compartments and interfaces. Inside Σ_{Tu} and the $\Sigma_{B\ell_k}$, plug flow was modeled by deterministic displacement of each tracer particle down the vessel axis. Instant radial diffusive equilibration inside the vessels was implemented by randomly and homogeneously resorting the tracer particles among available lattice sites on each occupied axial transect through Σ_{Tu} and the $\Sigma_{B\ell_k}$. Particle motion in the interstitium was three dimensional translational diffusion following the procedure of Dufort and Lumsden.[13] Interstitial particles immediately adjacent to Σ_{Tu} or one of the $\Sigma_{B\ell_k}$ had finite probabilities per unit time of crossing to an unoccupied site in the axial slice through the vessel interior. Interior particles had related rates for crossing to any available (unoccupied) site in the site ring immediately circumscribing the vessel wall in the normal plane occupied by the particle. Indicator parti cles had equal probabilities per unit time of crossing a capillary wall in either direction(out or in) per unit area, i.e. $k_{+} = k_{-}$.

3.3.4 Transport regimes

Three classes of tubular response were considered by adjusting the corresponding surface penetration rates to strongly favor tubule entry (secretion), tubule exit (reabsorption), or impermeable (interstitial marker). Each simulation was run 5 times using a bolus of 100,000 particles confined to the start of the capillaries and PCT and the runs averaged. The nephron filtration fraction was 0.3, i.e. 30 per cent of the particles were assigned to the origin of the PCT to represent the effects of glomerular filration and the rest divided equally among the five capillary vessels.

Outflows for each vessel were examined separately, along with the net response of all five capillaries. Indicator retention dynamics were examined in two volume elements comprising respectively the entire lattice or a small volume embedded in the center of the tubule folds. All simulations were run under three sets of boundary conditions – periodic, to simulate an infinite lattice of diffusively coupled PCT/capillary units; periodic on the lattice sides and reflecting at the ends (= the lattice faces perpendicular to the general capillary direction), simulating a layer of diffusively coupled units (e.g. the supeficial layer of the cortex, Fig. 6), and reflecting on all faces, simulating a single PCT/capillary unit embedded in a large interstitial region.

Of the 2^5 possible patterns of direction of capillary flow relative to the direction of urine flow, the present study was restricted to all-concurrent flows. The sensitivity of the indicator dynamics to alterations in the perfusion pattern as represented by direction of capillary flow will be reported subsequently.

4 Results

4.1 Effect of boundary conditions

The fully periodic boundary condition runs gave straight tubule responses significantly affected by diffusional propagation of indicator from the initial bolus in the nephron just across boundary interface from the urine outflow and will not be considered further here. For our parameter selections the responses of individual nephron in the effective "infinite layer" and "isolated nephron" cases were similar and so only the case for layer will be shown in the Figures.

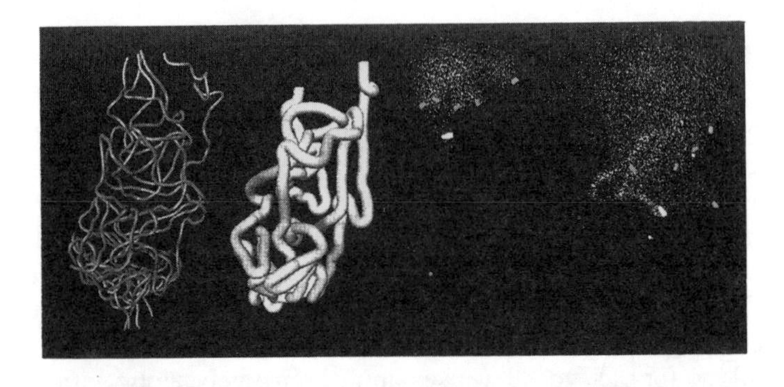

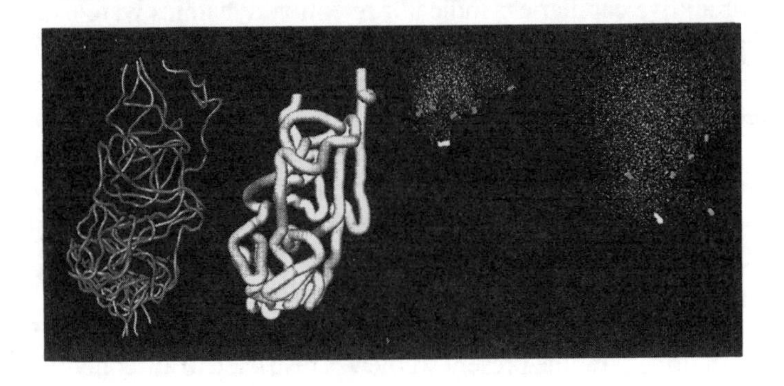

Figure 10: Indicator transit through the PCT and peritubular capillaries following bolus injection. The spatial organization of the tracer particle cloud is shown for a time early and midway through the transit of the injection bolus down the tubule. Top. Secreted indicator. Bottom. Reabsorbed indicator.

4.2 Visualization of indicator exchange

Visualization of the indicator particle clouds demonstrated complex, heterogeneous patterns of particle density sweeping through the system (Fig. 10). Secreted indicators presented a dense band of concentration that swept out the PCT folding pattern as time progressed. The shape of the PCT was less clearly demarcated for reabsorbed indicators but the pattern of interstitial cloud was more homogeneous, compatible with the increased relative probability for particles to be in the interstitium than the tubule.

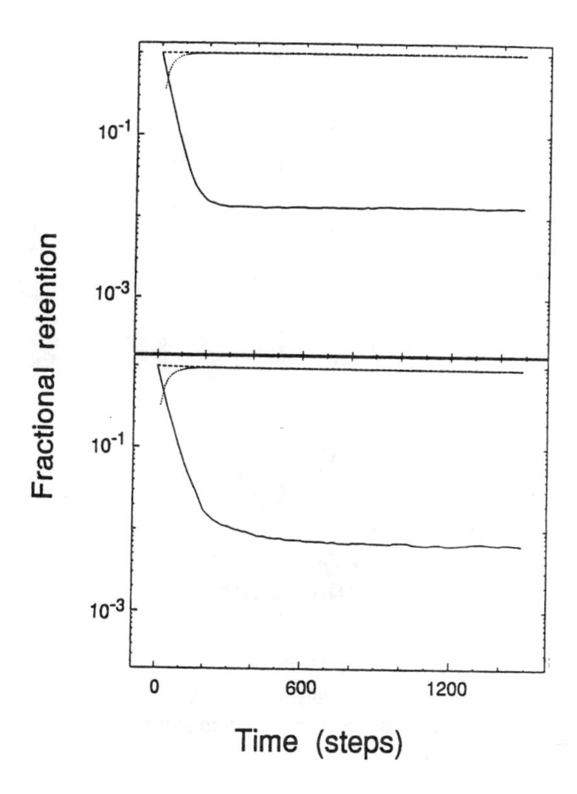

Figure 11: Whole-compartment retention curves for interstitial (tubule-inert) indicator in the straight tubule (Top) and folded tubule (Bottom) models. Solid line, blood compartment; dashed line, entire simulation volume; dotted line, interstitium.

4.3 Comparison of straight and folded geometries

The behavior of the straight and folded PCT models differed strikingly, and these differences were evident in both the fractional outflow and regional retention curves across the types of indicators considered. Regional retention values for the interstitial marker were higher in the interstitial compartment of the straight tubule than in the corresponding compartment of the folded tubule (Fig. 11). Particle washout from the vascular region was slower for the straight tubule case, so that fewer particles remained in the folded tubule simulation volume at the end of the simulation run.

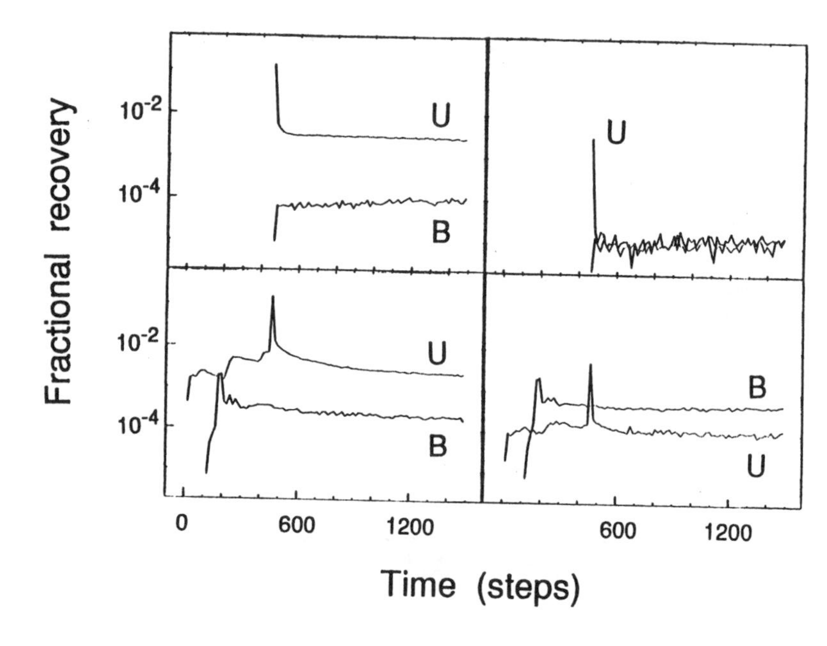

Figure 12: Indicator response behavior of the straight (Top) and folded (Bottom) tubules f secreted (Left) and reabsorbed (Right) indicators compared in terms of the fractional reco eries per time step at the tubule and vascular (= sum of all 5 capillaries) outflows. B, bloo compartment; U, tubule (urine).

The outflow curves for secreted indicator in the straight tubule showed n evidence of diffusional shunting; after passage of the indicator bolus, bloo and urine-borne particles left the system at a relatively constant rate (Fig. 12 The folded tubule shows a marked outflow of particles preceding the arrival o the indicator bolus at the outflow and suggesting strong diffusional shunting The rate of change in fractional recovery was elevated, suggesting a faster rat of removal of indicator from the simulation volume than could be achieved b the straight geometry. For reabsorbed indicator the behavior was similar, bu with urine fractional recoveries decreased an order of magnitude, consisten with the preferential movement of indicator out of the urine space.

The regional retention curves underscore these differences between th straight and folded geometries (Fig. 13). There is again higher retention i the straight system compared with faster washout from the folded geometry At the end of the simulation run, total regional retention (sum of all three

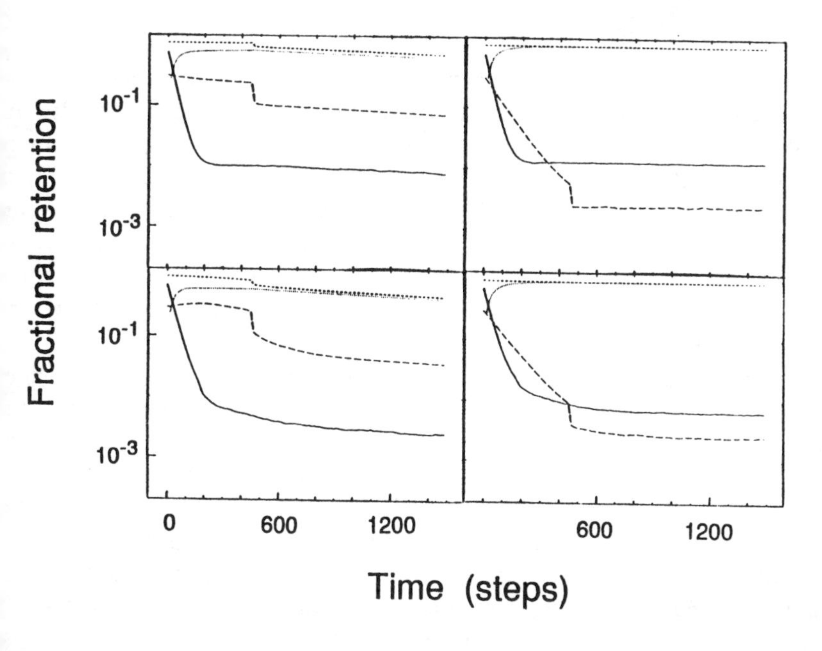

igure 13: As for Figure 12 but comparing the fractional retention of the indicators per time tep in the whole region and each exchange compartment. Solid line, blood compartment; ong dashes, PCT; short dashes, entire simulation volume; dotted, interstitium.

ompartments) was 0.7 for secreted indicator the straight geometry and 0.45 or the folded configuration. The importance of the particle scavenging action f shunting in the folded tubule was demonstrated by the retention patterns for he reabsorbed indictor. With simulated reabsorption decreasing the expected bundance of indicator particle at all points in the tubule per unit time, regional etention was 0.99 for the straight tubule and 0.9 for the folded configuration t the time the simulation terminated.

Examination of the individual capillary responses gave information on the ndicator dynamics in each vascular micro-region. For the straight geometry Fig. 14) no particles were detected in advance of the intracapillary indicator olus. The capillary lengths were equal, with equivalent bolus transit times. n the folded system the capillary lengths differed somewhat and so did the ascular transit times; shunting between capillary segments allowed indicator egress in advance of the original particle bolus. At longer times particle utflows from the various capillaries were approximately equal, but differed narkedly earlier in the exchange process.

5 Discussion

The kidney is one of the most complex organs in terms of its anatomy an its functional relations between microcirulation and regional cell activity. A in the brain, function in the kidney is built around the massive distribution (activity across many interacting elements (nephrons). But unlike the brai which relies on electrophysiology for its intercellular signal transduction, th kidney uses biofluid mechanics to regulate ultrafiltration and the transport (solutes by the renal cells. The proximal nephron, comprising the glomerul ultrafilter and the reabsorptive/secretive capacity of the proximal convolute tubule, is a key component of the renal fluidic circuitry.

Our simulations suggest that the processing of indicators and other solute by the proximal convoluted tubule is a spatially complex process in which th folding pattern of the tubule cannot be ignored. The two striking features (the particle dynamics in the simulated PCT to emerge from the present stud are (i.), the significant early alteration of the indicator dilution signal from th nephron as a result of diffusional shunting, and (ii.), an increased washout (particles from the simulation volume. The multiple tubule folds increased th probability that a solute particle would return to the urine space after it lef and therefore also increased the likelihood of particle removal from the sys tem over the time periods simulated (equivalent to several convective trans times of the PCT). Both of these effects of folding were most pronounced fc secreted indicators, which have the greatest likelihood of entering the urin space and remaining there.

These observations may have interesting implications for understandin the role of complex folding in normal PCT function. Although conclusior about possible roles of *individual* folds *within* any one PCT would be prema ture, it is interesting to note the increased efficiency with which the folde tubule as a whole, as compared to the straight exchange unit, scavenges ind cator particles from the interstitium and removes them in the urinary strean The scavenging effect is greater for secreted than reabsorbed solutes, whic is compatible with the PCT's normal function of directing secreted produc toward the final urine stream while returning reabsorbed solutes to the bloo Our findings raise the possibility that scavenging efficiency is a more sens tive function of tubule folding, compaction, and shunting for secreted particle than for reabsorbed particles. This suggests that PCT folding may enhance se cretion efficiency with minimal degradative impact on reabsorption.

The structural complexity of the folded PCT has lead to the consideratio here of a three dimensional model of the tubule folding pattern and perfu sion. Work with additional three- dimensional models is now needed in orde to properly assess the generality of our findings and to study effects of sys tematic variations in the folding and perfusion patterns and in the transport kinetics. Our present findings strongly suggest that an accurate understanding of three-dimensional folding and blood supply may be essential to the accurate interpretation of function in the proximal nephron.

Acknowledgments

It is a pleasure to thank Tony Zielinski and Christiopher Li for their assistance with the software development and modeling. The work was supported by funds from the Medical Research Council of Canada (MRC) and the Kidney Foundation of Canada. CJL is the recipient of MRC Scholar and MRC Scientist career awards.

References

1. Koushanpour, E. & Kriz, W. *Renal Physiology*, Springer-Verlag, New York, 1986.

2. Lumsden, C. J. & Silverman, M. Multiple indicator dilution and the kidney: Kinetics, permeation, and transport in vivo, *Methods in Enzymology*, 1991, **191**, 34-72.

3. Lumsden, C. J., Silverman, M., Zielinski, A., Potts, D. G., Shafik, I., Brody, A. S., Whiteside, C. I. & Violante, M. Vascular exchange in the kidney: Regional characterization by multiple indicator tomography, *Circulation Research*, 1993, **72**, 1172-1180.

4. Li, C. W. & Lumsden, C. J. The proximal nephron in three dimensions: A transport simulation approach, *Computers in Biomedicine: Proceedings of the First International Conference, Southampton, UK, September 24-26, 1991*, ed. K. D. Held, C. A. Brebbia & R. D. Ciskowski. Computational Mechanics Publications, Boston, 1991, pp. 121-131.

5. Bassingthwaighte, J. B. Physiology and theory of tracer washout techniques for the estimation of myocradial blood flow: Flow estimation from tracer washout, *Progress in Cardiovascular Diseases*, 1977, **20**, 165-189.

6. Rasio, E. A., Bendayan, M. & Goresky, C. A. Diffusion permeability of an isolated rete mirabile, *Circulation Research*, 1977, **41**, 791-798.

7. Arnautovic, I. The distribution of the renal artery in the kidney of the dog, *British Veterinary Journal*, 1959, **115**, 446-453.

8. Selkurt, E. E. The renal circulation, *Handbook of Physiology. Circulation*, ed. W. F. Hamilton & P. Dow, American Physiological Society, Washington, D. C., 1963, sect. 2, vol. II, pp. 1457-1516.

9. Beeuwkes III, R. & Bonventre, J. V. Tubular organization and vascular-tubular relations in the dog kidney, *American Journal of Physiology*, 1975, **229**, 695-713.

10. Beeuwkes III, R. The vascular organization of the kidney, *Annual Reviews of Physiology*, 1980, **42**, 531-542.

11. Beeuwkes III, R. Vascular-tubular relationships in the human kidney, *Renal Pathophysiology*, ed. A. Leaf, Raven Press, New York, 1980, pp. 155-163.

12. Bassingthwaighte, J. B. & Goresky, C. Modeling in the analysis of solute and water exchange in the microvasculature, *Handbook of Physiology. The Cardiovascular System. Microcirculation, Part 1*, ed. by E. M. Renkin & C. C. Michel, American Physiological Society, Bethesda, Maryland, 1984, sect. 2, vol. 4, pp. 549-626.

13. Dufort, P. A. & Lumsden, C. J. Cellular automaton model of the actin cytoskeleton, *Cell Motility and the Cytoskeleton*, 1993, **25**, 87-104.

Chapter 8

Coupled behaviour of lung respiration: computational respiratory mechanics approach

S. Wada[a] & M. Tanaka[b]
[a]*Research Institute for Electronic Science, Hokkaido University, North 12, West 6, North District, Sapporo 060, Japan*
[b]*Department of Mechanical Engineering, Faculty of Engineering Science, Osaka University, 1-3 Machikaneyama, Toyonaka, Osaka 560, Japan*

Abstract

A simulation approach is developed in terms of computational biomechanics to understand a complex mechanical system of the lung respiration from the elemental phenomena. The respiratory model is made up from the mechanical elements such as the gas flow, the tissue defamation, the blood flow, the gas diffusion and so on. The mathematical model is directly provided for each of mechanical phenomena and interactions among them. The concept of analysis by synthesis exerts in order to analyze the system characteristics by the simulation. The breathing dynamics is simulated to exhibit the surface tension characteristics which could not be observed directly in the lung. The pulmonary blood flow affected by the breathing is investigated by means of the simulation supported by the experimental observations. Taking account of the coupled behavior among the pulmonary circulation, the breathing dynamics and the gas exchange, we attempt to seek an appropriate pressure condition in artificial ventilation.

1 Introduction

1.1 Biomechanical problems in lung respiration

The lung respiration is a large-scaled mechanical system having complexity which stems from the structural hierarchy, the geometrical and material nonlinearlities, and the multi-field interactions.[1,2] As a basic function of the lung, the breathing ventilates the alveolar gas diffusing into the capillary blood, and the arterialized blood is carried by the pulmonary circulation. Several mechanical phenomena such as the air and blood flows, the tissue deformation, and the gas diffusion, govern this process in a coupled manner. In the clinical field, phenomenalism is conventional in evaluating the respiratory function physiologically.[3] However, the mechanical approach is also an important alternative toward the lung respiration, because the respiratory function is a result of mechanical phenomena.

Efforts have been devoted to mechanical aspects of the lung respiration. Continuum mechanics are applied to express the lung deformation,[4,5] and several mathematical models are proposed including the parenchyma structure.[6-9] The constitutive relations of the lung tissue elasticity are formulated based on the strain energy function.[10-13] The mechanical characteristics of the surface tension under the lung surfactant are examined by experimental approaches,[14-18] and are discussed with the parenchyma stability.[19,20] Fluid mechanics is employed to evaluate the gas flow in the airway[22-24] considering the elasticity of airway wall[25,26] and to investigate the pulsatile blood flow in the pulmonary circulation.[27] The blood flow in the capillary includes rheological phenomena.[28] The geometrical data are prepared for the airway[29-31] and for the pulmonary vessel tree.[32-35] The gas transport in the airway is investigated incorporating the theoretical studies[36-38] and experimental analyses[39-42] to resolve the mechanism of high frequency ventilation.[43-45] The gas exchange between the alveolar gas and capillary blood across the alveolar membrane is studied on basis of the diffusion phenomena.[46-50]

1.2 Elements to system

From the viewpoint of mechanics, the analysis starts from the element and moves on the system, because the element-wise characteristics and mechanism lead the overall behavior of the system. Therefore, most of the analyses concerning on the lung mechanics have focused on the elements and subsystems isolated from the respiratory system. However, the concept of system synthesis is also necessary because the respiratory function should be evaluated as the system behavior in which the elements and subsystems are assembled and coupled as they are. In fact, the lung respiratory system has a complex interaction among the element phenomena, and the behavior of an isolated element does not always give that in the system. For example, the sheet-flow model proposed by Fung & Sobin[51] shows the relation between the blood flow and the capillary collapsing, but it deeply depends on the blood pressure distribution that must come from the overall behavior of the blood flow in the pulmonary circulation system including the capillary subsystem.

The concept of synthesis is also needed to verify the element or subsystem in the total respiratory system, since the measuring *in vivo* are almost limited to the system behavior which is observed at the outside of lung. For example, there is no technique to confirm the capillary blood pressure directly in the breathing lung. In spite of many experiments for the alveolar surface tension,[52] its real properties in the lung have not been known yet because of the difference between *in vivo* and *in vitro* environments. In these cases, the consideration at the total system enables us to identify indirectly unknown factors at the elements or subsystems. This is an analysis by synthesis.

1.3 Computational viewpoint

Computational approaches have been applied to understand the mechanical behavior of the lung respiration. Apart from the traditional approaches by indirect analog models as electric circuits[53-56] and bond graphs[57], the relatively recent simulation-like approaches by direct physical models are found in the literature by Hoshimiya *et al.*,[58] Schimid-Schoenbein *et al.*,[59] Zhuang, *et al.*,[60] and they have a potential to investigate the mechanical process based on the physical structure in the system. The authors and colleagues have also adopted this approach in the context of computational respiratory biomechanics.[61-67] The concept of analysis by synthesis is a key to analyze the complex respiratory system by the simulation. The model-based approach assisted by the computational mechanics gives an insight into mechanical phenomena in the lung which could not be observed directly by experiments. At this point, the simulation described here should be distinguished from the numerical calculation on the basis of a complete theory.

A certain idealization is necessary in the modeling, although the recent progress of computational technology accepts the real modeling.[68,69] It is critical that a mathematical model simplified heuristically, like a lumped parameter model for the distributed parameter system, has a risk that may hide the elemental detail and lead to only evident results. In order to accomplish the purpose of the model-based approach by simulation, the distributed parameter model will be required at least for the gas flow and transport in the airway and for the blood flow and gas exchange in the pulmonary circulation, even though it is reduced to one-dimensional behavior.

In this chapter, we introduce a respiratory model described from the mechanical elements including their interaction. The simulation approach based on the concept of analysis by synthesis is applied to predict the surface tension behavior from the breathing dynamics. The coupled behavior of the blood flow with breathing is simulated by using the total respiratory model. As the application of simulation approach to clinical field, we attempt to prove an optimal condition of the artificial ventilation.

2 Biomechanical model

2.1 Breathing model

The inflation and deflation of lung are caused by the tissue deformation and the gas flow in airway. In this section, the breathing dynamics are described from these mechanical elements.

2.1.1 Lung deformation

There are several mathematical models for the lung deformation.[4-9] Here, we pick up the model proposed by Fung[4,5] on the basis of continuum mechanics. As the lung parenchyma is decomposed from the micro-scale alveoli, the macroscopic stress is defined as the sum of the forces acting on the alveolar tissue per unit area of the cross-section. Two stresses, the elastic stress σ_e due to tissue elasticity and the stress σ_s due to surface tension on both side of the tissue membrane at alveoli, make the macroscopic stress σ as

$$\sigma = \sigma_e + \sigma_s \ . \tag{1}$$

For an idealized lung parenchyma which consists of the cubic alveoli with the edge length Δ as shown in Fig. 1, Vawter *et al.*[10] and Fung[5] have derived a pseudo-strain energy function W of the lung parenchyma,

$$\rho_0 W = \frac{1}{2} c \exp\left(a_1 E_x^{\ 2} + a_2 E_y^{\ 2} + 2a_4 E_x E_y\right) \\ + \text{symmetrical terms by permutation} \tag{2}$$

where ρ_0 denotes the density of parenchyma, c and $a_j(j=1,2,4)$ are material constants. Green's strain $E_i(i=x, y, z)$ is defined by

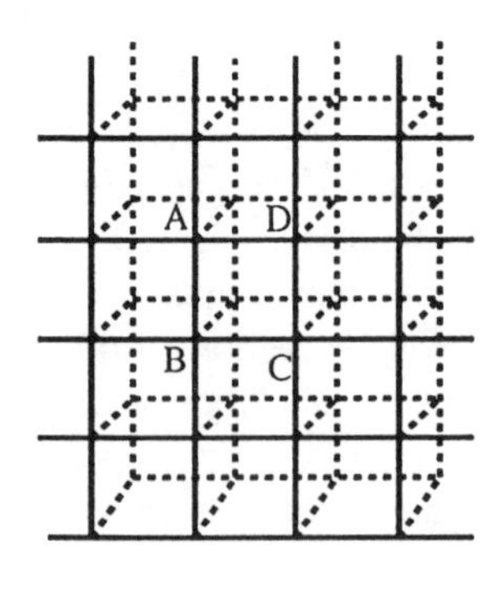

(a) Parenchyma structure

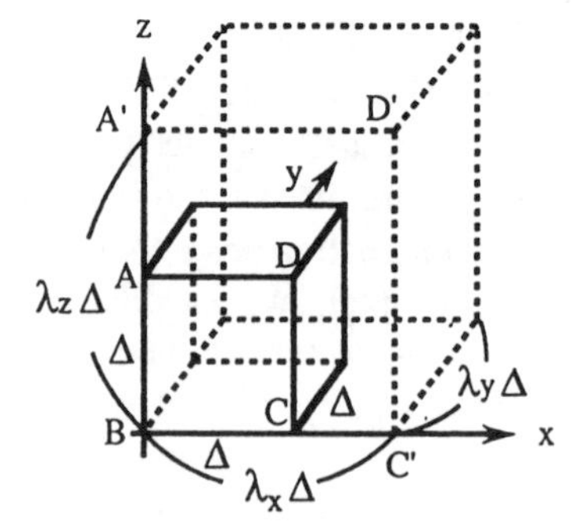

(b) Deformation of cubic alveolus

Figure 1: Idealized lung parenchyma model[5]

$$E_i = \left(\lambda_i^{\,2} - 1\right)/2 \tag{3}$$

where $\lambda_i (i=x, y, z)$ is the stretch ratio of lung tissue. The Lagrangian stress T_i is derived from the energy function as

$$T_i = \frac{\partial(\rho_0 W)}{\partial \lambda_i} \quad (i = x, y, z) \tag{4}$$

and thus the elastic stress σ_{ei} is written as

$$\sigma_{ei} = \frac{T_i}{\lambda_i^{\,2}} \quad (i = x, y, z). \tag{5}$$

Under the assumption of uniform and isotropic inflation of the lung parenchyma, the stress for each principal direction becomes identical each other,

$$\lambda = \lambda_i\ , \quad T = T_i\ , \quad \sigma_e = \sigma_{ei} \quad (i = x, y, z). \tag{6}$$

The material constants in eqn (2) are given by Zeng *et al.*[70] for the human lung and by Vawter *et al.*[71] for the dog lung.

2.1.2 Surface tension

For the idealized cubic alveolar model, the stress σ_s due to the surface tension γ is represented as

$$\sigma_s = \frac{4\gamma}{\lambda \Delta}. \tag{7}$$

It is known that the lung surfactant reduces the surface tension depending on the surface area change[14-18]. The hysteresis of the surface tension γ to the cyclic area change is expressed by the rate-type model[62]

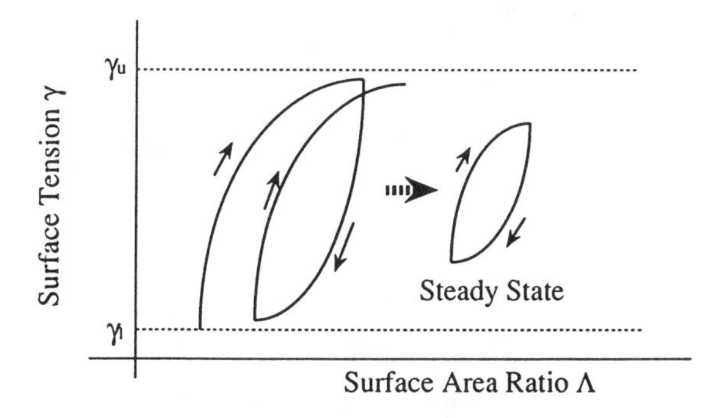

Figure 2: Surface tension model

$$\frac{d\gamma}{dt} = \begin{cases} a(\gamma_u - \gamma)\dfrac{d\Lambda}{dt} & \text{for } \dfrac{d\Lambda}{dt} \geq 0 \\ a(\gamma - \gamma_l)\dfrac{d\Lambda}{dt} & \text{for } \dfrac{d\Lambda}{dt} < 0 \end{cases} \tag{8}$$

where the coefficient a determines the degree of hysteresis, and the surface area ratio Λ is the square of the stretch ratio λ. According the previous reports,[52,72,73] the upper bound and the lower bound of the surface tension are γ_u=30dyn/cm and γ_l=2dyn/cm, respectively. This surface tension model represents the essential features obtained by several experiments[14-19]: (i) the hysteresis of surface tension to the surface area change, (ii) the nonlinear saturation of surface tension to the upper and lower bounds, and (iii) the convergence of the hysteresis loop at any range of surface area change (see Fig. 2). The last adaptive feature (iii) is only expressed by using the rate-type model not by the model fixed in the state space.[74]

2.1.3 Gas flow in airway

The inflation and deflation of the lung yield the gas flow in the airway. The airway starting from the trachea bifurcates into two smaller branches 23 times in average for the human lung, and reaches the alveoli.[29] The model equations of the gas flow in the airway are built up by assembling those for each branch at generation i from the trachea (i=0) to the alveoli (i=N). The airway branch is assumed to be an elastic cylindrical tube. Let us consider the branch at generation i with the axis x_i along it. The equation of motion of the gas flow with the pressure distribution $P_i(x_i, t)$ and the mean velocity distribution $u_i(x_i, t)$ is represented by

$$\frac{\partial u_i}{\partial t} + u_i \frac{\partial u_i}{\partial x_i} + \frac{1}{\rho_i}\frac{\partial P_i}{\partial x_i} + \lambda_f \frac{u_i^2}{4r_i} = 0 \tag{9}$$

where ρ_i is the gas density and r_i denotes the radius of the airway tube. The flow resistance is estimated by that at the steady flow,[75] and thus the coefficient of friction is expressed as $\lambda_f = 64/\mathrm{Re}$ for the laminar flow and $\lambda_f = 0.3164/\sqrt[4]{\mathrm{Re}}$ for the turbulent flow, where Re is the Reynolds number. The mass conservation of the gas flow gives the equation of continuity considering the airway deformation as

$$\frac{\partial P_i}{\partial t} + u_i \frac{\partial P_i}{\partial x_i} + P_i \frac{\partial u_i}{\partial x_i} + 2P_i \frac{d\varepsilon_{\theta i}}{dt} = 0 \tag{10}$$

which includes the equation of state under isothermal process, that is ρ_i=$k_p P_i$. Since the airway is embedded in the lung parenchyma,[25] the alveolar gas pressure P_{al} works on the airway wall at the outer surface as shown in Fig. 3. The circumferential strain $\varepsilon_{\theta i}$ of the airway wall is calculated from the transmural pressure P_i-P_{al} by using the constitutive equation and the equilibrium relation based on the finite strain theory.[76] Equations (9) and (10) define each generation i of the airway separately.

The airway bifurcation is assumed to be regular dichotomy.[29] As the outlet ($x_i=l_i$) of airway branch at generation i connects to the inlet ($x_{i+1}=0$) of branch at generation i+1, the compatibility coming from the gas continuity and the energy conservation are represented by

$$\rho_i A_i u_i \big|_{x_i=l_i} = 2\rho_{i+1} A_{i+1} u_{i+1} \big|_{x_{i+1}=0} \tag{11}$$

$$k_p \log P_i + \frac{1}{2} u_i^{\,2} \Big|_{x_i=l_i} = k_p \log P_{i+1} + \frac{1}{2}(1+\zeta) u_{i+1}^{\,2} \Big|_{x_{i+1}=0} \tag{12}$$

where A_i is the cross-sectional area of the airway branch and ζ denotes the pressure loss coefficient at the branching point. The boundary conditions of the gas flow model are given by the pressure at both edges of the airway system, that is, at the oral space ($x_0=0$) and at the alveolar space ($x_N=l_N$) as

$$P_0(0,t) = P_{or},\ P_N(l_N,t) = P_{al}\ . \tag{13}$$

2.1.4 Lung equilibrium

In the thorax space, the gas pressure in the lung P_L acts on the lung parenchyma as well as the pleural pressure P_{pl}. When the alveolar gas pressure P_{al} is the representative of the gas pressure P_L in the lung, the macroscopic stress balances the applied pressure as

$$P_{al} = P_{pl} + \sigma \tag{14}$$

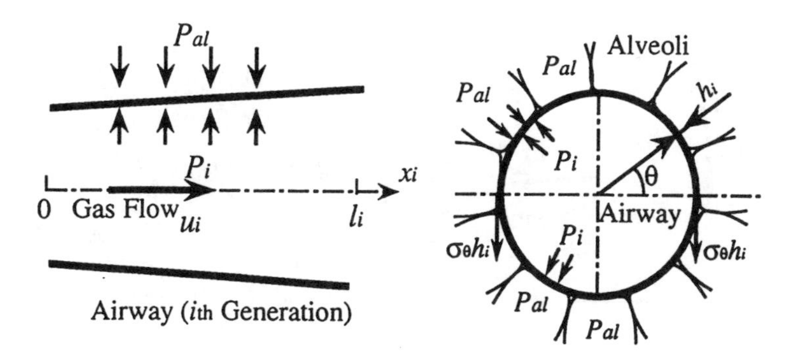

Figure 3: Airway model

where the tension of the pleura is neglected[5]. The diaphragm motion oscillates the pleural pressure P_{pl} in the normal breathing. In the case of artificial ventilation, the change of pleural pressure due to the lung volume change is approximated as

$$P_{pl} = c_1\lambda_V^{\ 2} + c_2\lambda_V + c_3 \quad (\lambda_V > 0) \tag{15}$$

where the volume ratio λ_V is defined by $\lambda_V=(V-V_{FRC})/V_{FRC}$ using the functional residual capacity V_{FRC}. Referring to the conventional observations,[3] the coefficients are set to c_1=4.0cmH_2O, c_2=16.0cmH_2O, c_3=-2.8cmH_2O, and V_{FRC}=3000cm^3 for the human lung.

The alveolar pressure P_{al} determined by the equilibrium relation of eqn (14) provides one of the boundary conditions in eqn (13) of the gas flow model. On the other hand, the alveolar volume V_{al} obtained by the gas flow model gives the stretch ratio λ of the lung tissue as

$$\lambda = \sqrt[3]{V_{al}/V_0} \tag{16}$$

for calculating the macroscopic stress σ in eqn (14), where V_0 is the normal lung volume free from the stress σ. This coupling process in breathing is illustrated in Fig. 4.

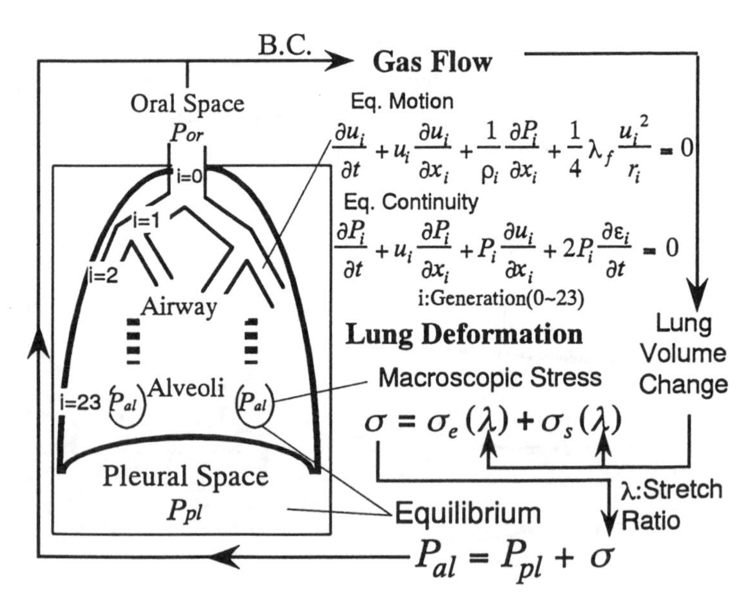

Figure 4: Breathing model

2. 2 Pulmonary circulation model

Beginning from the outlet of right ventricle, the vessel tree of pulmonary artery bifurcates successively in the lung, and reaches capillary blood vessels as shown in Fig. 5. The capillary vessels form a network structure in the alveolar wall, and converge successively to the pulmonary vein terminating at the left atrium. The vessel deformation due to lung inflation and the oppression by the gas pressure in lung affect the blood flow in the pulmonary circulation. This section describes the blood flow model in the pulmonary circulation system in order to express directly the mechanical interaction with breathing dynamics.

2.2.1 Blood flow in arterial and venous trees

The morphology of the arterial and vessel branches have been reported for the human lung[32,33] and for the cat lung.[34,35] The vessel order is numbered from the peripheral vessel. Both of the arterial and venous trees have 17 order vessels for the human lung and 11 orders for the cat lung. The blood flow in the vessel tree is expressed by assembling the tube-flow model for each vessel assumed to be an elastic cylindrical tube. The distribution of the blood pressure $P_{b(j)}(x_{b(j)}, t)$ and that of the mean velocity $u_{b(j)}(x_{b(j)}, t)$ are governed by the equation of motion

$$\frac{\partial u_{b(j)}}{\partial t}+u_{b(j)}\frac{\partial u_{b(j)}}{\partial x_{b(j)}}+\frac{1}{\rho_b}\frac{\partial P_{b(j)}}{\partial x_{b(j)}}+\frac{8\nu_{b(j)}f_{b(j)}}{r_{b(j)}^{2}}u_{b(j)}=0 \qquad (17)$$

and the equation of continuity

$$\frac{\partial r_{b(j)}}{\partial t}+u_{b(j)}\frac{\partial r_{b(j)}}{\partial x_{b(j)}}+\frac{r_{b(j)}}{2}\frac{\partial u_{b(j)}}{\partial x_{b(j)}}=0 \qquad (18)$$

where the subscript $j(0<|j|\leq N_b)$ stands for the vessel order, $f_{b(j)}$ is the geometrical friction coefficient, and ρ_b and $\nu_{b(j)}$ denote the blood density and

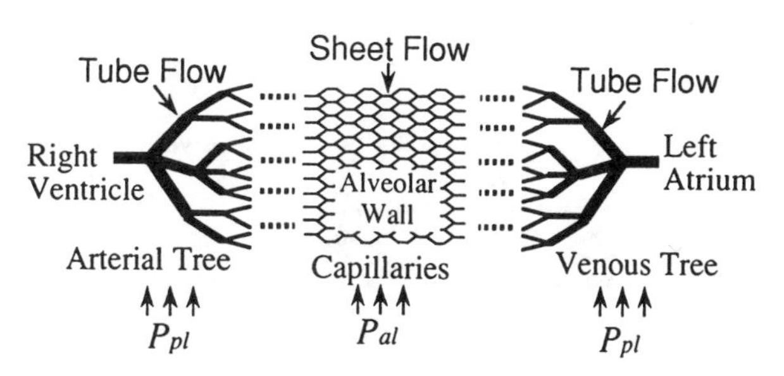

Figure 5: Pulmonary vessel structure

viscosity, respectively. The negative number of the vessel order implies the arterial vessel. According to the experimental observations,[77,78] the radius $r_{b(j)}$ of the elastic vessel varies almost linearly with the increase of transmural pressure $\Delta P_{b(j)}=P_{b(j)}-P_{ex}$ within the physiological range. Considering the visco-elasticity of the vessel wall, the radius of vessel is represented as

$$\alpha_j \beta_j \frac{\partial r_{b(j)}}{\partial t} + r_{b(j)} - r_{b0(j)}\left(1+\alpha_j \Delta P_{b(j)}\right) = 0 \tag{19}$$

where $r_{b0(j)}$ is the radius at the natural state and the external pressure P_{ex} is the gas pressure in the lung P_{al} for the small vessels and the pleural pressure P_{pl} for the large vessels near the heart. The stretch ratio of the blood vessel in the longitudinal direction due to the lung deformation is assumed to be identical to that of the lung tissue λ.

The continuity and the energy conservation at the connection between the outlet ($x_{b(j)}=l_{b(j)}$) of vessel of order j and the inlet ($x_{b(j+1)}=0$) of order j+1,

$$n_{b(j)} \pi r_{b(j)}^2 u_{b(j)} \Big|_{x_{b(j)}=l_{b(j)}} = n_{b(j+1)} \pi r_{b(j+1)}^2 u_{b(j+1)} \Big|_{x_{b(j+1)}=0} \tag{20}$$

$$P_{b(j)} + \frac{\rho_b}{2} u_{b(j)}^2 \Big|_{x_{b(j)}=l_{b(j)}} = P_{b(j+1)} + \frac{\rho_b}{2}(1+\zeta_b) u_{b(j+1)}^2 \Big|_{x_{b(j+1)}=0} \tag{21}$$

assemble the governing eqns (17) and (18) for each vessel order into the blood flow models in the arterial and venous trees, where $n_{b(j)}$ denotes the number of blood vessels of order j.

2.2.2 Capillary blood flow

The concept of sheet-flow proposed by Fung & Sobin[51] is employed for the capillary blood flow in the alveolar wall. We regard the blood flow in the capillary network as that between the alveolar membranes supported by the posts of the connecting tissue as illustrated in Fig. 6. The one dimensional blood flow in the alveolar sheet is represented by the equation of motion

$$\frac{\partial u_{b(0)}}{\partial t} + u_{b(0)} \frac{\partial u_{b(0)}}{\partial x_{b(0)}} + \frac{1}{\rho_b} \frac{\partial P_{b(0)}}{\partial x_{b(0)}} + \frac{12 \nu_{b(0)} f_b}{h_b^2} u_{b(0)} = 0 \tag{22}$$

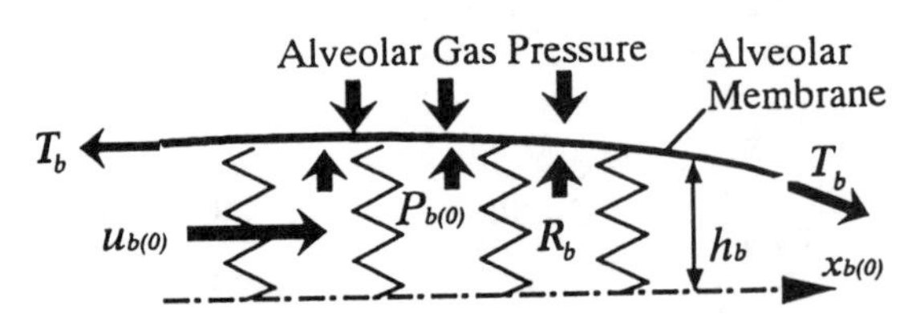

Figure 6: Sheet flow model for the pulmonary capillary blood flow (side view)[51]

and the equation of continuity

$$\frac{\partial h_b}{\partial t} + u_{b(0)} \frac{\partial h_b}{\partial x_{b(0)}} + h_b \frac{\partial u_{b(0)}}{\partial x_{b(0)}} = 0 \tag{23}$$

where subscript (0) indicates the capillary vessel. The alveolar gas pressure P_{al} applies to the alveolar membrane, and the height of alveolar sheet h_b changese depending on the transmural pressure $\Delta P_{b(0)}=P_{b(0)}-P_{al}$. Then, the equilibrium relation of the alveolar membrane is expressed as

$$\Delta P_{b(0)} + R_b - \kappa T_b = 0 \tag{24}$$

where T_b is the resultant tension of the tissue due to the lung inflation, that is, the sum of the elastic tension and the surface tension. The curvature κ of the alveolar membrane is written as

$$\kappa = \frac{-\partial^2 w_b / \partial x_{b(0)}{}^2}{\left\{1 + \left(\partial w_b / \partial x_{b(0)}\right)^2\right\}^{1.5}} \tag{25}$$

using the deflection $w_b=(h_b-h_{b0})/2$. The sheet thickness without transmural pressure is h_{b0}=3.5μm for the human lung and h_{b0}=4.2μm for the cat lung.[79] The elastic posts are assumed to be distributed uniformly like a spring system and the elastic resistance is represented as

$$R_b = k_b w_b \ \ (w_b \geq 0), \quad R_b = 0 \ \ (w_b < 0). \tag{26}$$

where the coefficient determined by the experiments[79,80] is $k_b=1.54\times10^5 \mathrm{N/m^3}$ for the human lung and $k_b=0.895\times10^5 \mathrm{N/m^3}$ for the cat lung.

The compatibility to satisfy the continuity at the connection between the outlet of arterial tree ($x_{b(-1)}=l_{b(-1)}$) and the inlet of capillary ($x_{b(0)}=0$) are

$$n_{b(-1)} \pi r_{b(-1)}{}^2 u_{b(-1)} \Big|_{x_{b(-1)}=l_{b(-1)}} = A_{b(0)} u_{b(0)} \Big|_{x_{b(0)}=0} \tag{27}$$

$$P_{b(-1)} + \frac{\rho_b}{2} u_{b(-1)}{}^2 \Big|_{x_{b(-1)}=l_{b(-1)}} = P_{b(0)} + \frac{\rho_b}{2} u_{b(0)}{}^2 \Big|_{x_{b(0)}=0} \tag{28}$$

and those between the outlet of capillary ($x_{b(0)}=l_{b(0)}$) and the inlet of venous tree ($x_{b(1)}=0$) are

$$A_{b(0)} u_{b(0)} \Big|_{x_{b(0)}=l_{b(0)}} = n_{b(1)} \pi r_{b(1)}{}^2 u_{b(1)} \Big|_{x_{b(1)}=0} \tag{29}$$

$$P_{b(0)} + \frac{\rho_b}{2} u_{b(0)}{}^2 \Big|_{x_{b(0)}=l_{b(0)}} = P_{b(1)} + \frac{\rho_b}{2} u_{b(1)}{}^2 \Big|_{x_{b(1)}=0} \tag{30}$$

where $A_{b(0)}$ is the total cross-sectional area of the capillary. For the assembled model over the pulmonary circulation system, the blood pressure or the velocity at the inlet of artery

$$P_{b(-N_b)}(0,t) = P_a(t) \text{ or } u_{b(-N_b)}(0,t) = u_a(t) \tag{31}$$

and the blood pressure at the outlet of vein

$$P_{b(N_b)}\left(l_{b(N_b)},t\right) = P_v(t) \tag{32}$$

are prescribed as the boundary conditions.

2.3 Gas exchange model

The gas exchange, which is the primal function of lung respiration, is carried out by the combination of breathing and pulmonary circulation. The former dominates the ability of ventilation between the external air and the alveolar gas, and the latter governs that of the carriage of exchanged gas by the blood flow. The gas exchange model in the respiratory system based on the previous researches which have focused on the gas transport in the airway[36-4 and the gas diffusion into the blood.[46-50]

2.3.1 Gas transport in the airway

The alveolar gas is ventilated with the external air through the airway by breathing. The airway gas primarily consists of oxygen, carbon dioxide nitrogen and water vapor. Since the vapor is almost saturated in the airway the nitrogen concentration is obtained from those of the oxygen and carbon dioxide. Thus, the ventilation problem is reduced to the gas transport problem of the oxygen and the carbon dioxide.

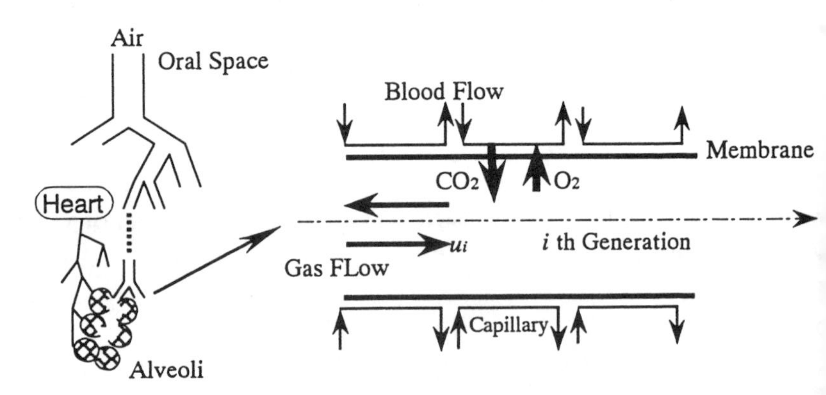

Figure 7: Gas transport and exchange in peripheral airway

The gas transport mechanism is governed by two fundamental processes: the bulk convection and the molecular diffusion. The combination of axial convection and lateral mixing by molecular diffusion results in an effective mechanism called augmented diffusion.[43] In the peripheral airway surrounded by the alveoli, the oxygen is diffused into the capillary blood and the carbon dioxide is diffused out from it (Fig. 7). Taking account of both the gas transport and the gas exchange in the airway system, the gas concentrations $C_{(g)i}(x_i, t)$ of the oxygen ($g=O_2$) and the carbon dioxide ($g=CO_2$) in each airway generation i are expressed by

$$\frac{\partial C_{(g)i}}{\partial t} + u_i \frac{\partial C_{(g)i}}{\partial x_i} - D_{eff} \frac{\partial^2 C_{(g)i}}{\partial x_i^{\,2}} + S_i D_{L(g)} \left(p_{(g)i} - \bar{p}_{b(g)} \right) = 0 \tag{33}$$

respectively, where the gas concentration $C_{(g)i}$ implies the mass fraction. The forth term indicates the diffusion of oxygen into or that of carbon dioxide out from the blood, which obeys Fick's law with diffusing capacity $D_{L(g)}$. The partial pressure $p_{(g)i}$ in the airway is converted from the gas concentration $C_{(g)i}$. As the capillary position could not be specified at the interface with the airway space, the average partial pressure $\bar{p}_{b(g)}$ of the blood gas is used. According to Weibel,[29] most of the alveoli are observed at the 20-23rd generations for the human lung, hence the alveolar area ratio to the airway volume, S_i, is set to zero for the central airway ($i<20$).

The gas velocity u_i in the convection term of eqn (33) comes from the breathing model, and the effective diffusion coefficient D_{eff} involves the augmented diffusion. Several expression have been proposed for the coefficient based on Taylor's work.[81] We refer to the theoretical analysis of the diffusion in the oscillatory pipe flow by Watson,[82] and write the coefficient as

$$D_{eff} / D_{mol} = 1 + f(\alpha_w) \cdot \mathrm{Pe}^2 \tag{34a}$$

$$f(\alpha_w) = \begin{cases} 1/192 & (\alpha_w < 3) \\ \left(\alpha_w + 3/\sqrt{2}\right) / 8\sqrt{2}\alpha_w^{\,4} & (\alpha_w \geq 3) \end{cases} \tag{34b}$$

where Pe is the Peclet number ($\mathrm{Pe}=2u_i r_i/D_{mol}$), and α_w indicates the Wormersely parameter($\alpha_w = r_i\sqrt{2\pi f/\nu}$, f:frequency). The molecular diffusion coefficient is D_{mol}=0.206cm²/s for the oxygen and D_{mol}=0.164cm²/s for the carbon dioxide.

The compatibility to satisfy the mass conservation at the branching point between branches at generation i and i+1 are written as

$$C_{(g)i}\Big|_{x_i=l_i} = C_{(g)i+1}\Big|_{x_{i+1}=0} \tag{35}$$

$$D_{eff}A_i \frac{\partial C_{(g)i}}{\partial x_i}\Bigg|_{x_i=l_i} = 2D_{eff}A_{i+1}\frac{\partial C_{(g)i+1}}{\partial x_{i+1}}\Bigg|_{x_{i+1}=0} \tag{36}$$

which assemble the model eqn (33) for each generation (i=0~N) to that for the whole airway system. The boundary condition is given at the oral space

$$C_{(g)0}(0,t) = C_{(g)or} \quad \text{for inspiration} \tag{37a}$$

$$\frac{\partial C_{(g)0}}{\partial x_0}(0,t) = 0 \quad \text{for expiration} \tag{37b}$$

and at the end of alveoli

$$D_{eff}\frac{\partial C_{(g)N}}{\partial x_N}(l_N,t) = -D_{L(g)}\left(p_{(g)N}(l_N,t) - \bar{p}_{(g)b}\right). \tag{38}$$

2.3.2 Gas carriage by blood flow

The venous blood is arterialized during passing through the alveolar capillaries. Based on the mass conservation of the blood gas in the capillary the distribution of the oxygen and carbon dioxide concentrations $C_{b(g)}(x_{b(0)}, t)$ along the capillary are respectively expressed by

$$\frac{\partial C_{b(g)}}{\partial t} + u_{b(0)}\frac{\partial C_{b(g)}}{\partial x_{b(0)}} + S_b D_{L(g)}\left(p_{b(g)} - \bar{p}_{(g)}\right) = 0 \tag{39}$$

where S_b is the ratio of the gas exchange area to the capillary volume, and $\bar{p}_{(g)}$ denotes the average partial gas pressure over the peripheral airway with the alveoli (i=20~23). The pulmonary circulation model provides the distribution of blood velocity $u_{b(0)}$ in the capillary. The gas concentration $C_{b(g)}$ is converted to the partial pressure $p_{b(g)}$ by the dissociation curve including Bohr & Haldane effects.[83] Gomez[84] has proposed the relation

$$C_{b(O2)} = 0.2S_{O2} + 3.0\times10^{-5}p_{b(O2)} \tag{40a}$$

$$C_{b(CO2)} = \left(0.149 - 0.014S_{O2}\right)p_{b(CO2)}^{\;0.35} \tag{40b}$$

where S_{O2} is the oxygen saturation of hemoglobin.

The inverse of the diffusing capacity $D_{L(g)}$ in eqns (33) and (39) implies the resistance of the diffusion. Since the oxygen combines with the hemoglobin in the blood after passing through the alveolar membrane Forster[46] has proposed the expression as

$$\frac{1}{D_{L(O2)}} = \frac{1}{D_{m(O2)}} + \frac{1}{V_c \theta_{(O2)}} \tag{41}$$

where $D_{m(O2)}$ is membrane diffusing capacity, and V_c stands for the capillary blood volume taking the gas exchange upon. Staub *et al.*[85] has measured the reaction rate $\theta_{(O2)}$ of the hemoglobin, which depends on the hemoglobin saturation S_{O2}. For the carbon dioxide diffusing capacity $D_{L(CO2)}$, the expression is also applied for convenience although the diffusion process becomes more complex.[48]

The total amount of the gas exchange calculated from the airway side is

$$F_{a(g)} = \sum_{i=20}^{N} 2^i \int_0^{l_i} A_i S_i D_{L(g)} \left(p_{i(g)} - \bar{p}_{b(g)} \right) dx_i \tag{42}$$

and that from the capillary is

$$F_{b(g)} = \int_0^{l_{b(0)}} A_{b(0)} S_b D_{L(g)} \left(\bar{p}_{(g)} - p_{b(g)} \right) dx_{b(0)} \tag{43}$$

where $l_{b(0)}$ is the capillary length. These values must be identical to the average flux

$$\bar{F}_{(g)} = A_s D_{L(g)} \left(\bar{p}_{(g)} - \bar{p}_{b(g)} \right) \tag{44}$$

where A_s stands for the total surface area for the gas exchange. The average partial pressure $\bar{p}_{(g)}$ and $\bar{p}_{b(g)}$ are determined as to satisfy $F_{a(g)} = F_{b(g)} = \bar{F}$.

3 Computational procedure

In order to simulate the lung respiration, it is necessary to discretize the mathematical models written in partial differential equations and to send them to the computational mechanics procedure. In this section, we explain the discretization by the finite element method based on the adjoint variational principle and the simulation procedure.

3.1 Adjoint variational principle

Seguchi *et al.*[76,86] extended the Lagrangian multiplier method to formulate a variational principle for the airway system. Since the detailed formulation is descried in the references,[76,79,86] this section concentrates on the pulmonary circulation model for the explanation purpose of the procedure.

For simplicity in description, we use the symbolic expression for the governing equations. That is, $f_{1(j)}=0$ stands for the equation of motion of eqn

(17) or (22), $f_{2(j)}=0$ for the equation of continuity of eqn (18) or (23), and $f_{3(j)}=0$ for the equation of vessel deformation of eqn (19) or (24), where $-N_b \leq j \leq N_b$ with the vessel order $j=0$ denoting the capillary. Similarly, $g_{1(j)}=0$ stands for the equation of connecting compatibility of continuity of eqn (20), (27) or (29), and $g_{2(j)}=0$ for that of energy conservation of eqn (21), (28) or (30), where $-N_b \leq j \leq N_b-1$. Then, Lagrangian multipliers $\phi_{k(j)}(x_{b(j)}, t)$ (k=1, 2, 3) and $\eta_{k(j)}(t)$ (k=1,2) are provided for the governing equations $f_{k(j)}=0$ and $g_{k(j)}=0$, respectively. The residual of governing equations are weighted by the multiplier and are integrated over the domain under consideration. They are summed over every vessel order giving the functional J

$$J = \sum_{j=-N_b}^{N_b} \sum_{k=1}^{3} \int_0^T \int_0^{x_{b(j)}} \phi_{k(j)}\left(x_{b(j)}, t\right) f_{k(j)} dx_{b(j)} dt + \sum_{i=-N_b}^{N_b-1} \sum_{k=1}^{2} \int_0^T \eta_{k(j)}(t) g_{k(j)} dt \,. \quad (45)$$

Then, we have the variational problem

$$\begin{array}{l}\text{Stationaly } J \text{ with respect to } u_{b(j)}, P_{b(j)}, r_{b(j)}, \phi_{k(j)}, \eta_{k(j)} \\ \text{subject to initial and boundary conditions.}\end{array} \quad (46)$$

According to the Lagrangian principle in the modern optimization theory, the first variation of the functional J after successive integration by parts leads

Table 1 Connecting compatibility of Lagrangian multipliers

Primal part	Adjoint part	
Blood flow model		
Eqns (20) and (21)	$\phi_{1(j)}\big\|_{x_{b(j)}=l_{b(j)}} = \phi_{1(j+1)}\big\|_{x_{b(j+1)}=0}$,	$\dfrac{\phi_{2(j)}}{n_{b(j)} r_{b(j)}}\bigg\|_{x_{b(j)}=l_{b(j)}} = \dfrac{\phi_{2(j+1)}}{n_{b(j+1)} r_{b(j+1)}}\bigg\|_{x_{b(j+1)}=0}$
Eqns (27) and (28)	$\phi_{1(-1)}\big\|_{x_{b(-1)}=l_{b(-1)}} = \phi_{1(0)}\big\|_{x_{b(0)}=0}$,	$\dfrac{\phi_{2(-1)}}{n_{b(-1)} r_{b(-1)}}\bigg\|_{x_{b(-1)}=l_{b(-1)}} = \dfrac{h_b \phi_{2(0)}}{A_{b(0)}}\bigg\|_{x_{b(0)}=0}$
Eqns (29) and (30)	$\phi_{1(0)}\big\|_{x_{b(0)}=l_{b(0)}} = \phi_{1(1)}\big\|_{x_{b(1)}=0}$,	$\dfrac{h_b \phi_{2(0)}}{A_{b(0)}}\bigg\|_{x_{b(0)}=l_{b(0)}} = \dfrac{\phi_{2(1)}}{n_{b(1)} r_{b(1)}}\bigg\|_{x_{b(1)}=0}$
Gas flow model		
Eqns (13) and (14)	$\psi_{1(i)}\big\|_{x_i=l_i} = \psi_{1(i+1)}\big\|_{x_{i+1}=0}$,	$2A_{i+1}\psi_{2(i)}\big\|_{x_i=l_i} = A_i \psi_{2(i+1)}\big\|_{x_{i+1}=0}$
Gas transport model*		
Eqn (35)	$2A_{i+1}\varphi_{(g)i}\big\|_{x_i=l_i} = A_i \varphi_{(g)i+1}\big\|_{x_{i+1}=0}$	

*Stationary condition of the functional J satisfies eqn (36) naturally.

the necessary condition, which includes the original set of governing equations of eqns (17)~(24) and (27)~(30) as the primal part, and the equations of the Lagrangian multipliers as the adjoint part. In this case, the variational problem is non-selfadjoint, that is, the set of equations of the adjoint part is not identical to that of the primal part, hence the physical meaning of the multipliers is not straight forward. However, the connecting compatibility of the Lagrangian multipliers $\phi_{k(j)}$ are obtained at the branching point by eliminating $\eta_{k(j)}$, which correspond to those of the primal part as listed in Table 1. Similarly, at the airway branch, we have the connecting compatibility of the Lagrangian multipliers $\psi_{1(i)}$ and $\psi_{2(i)}$ for the gas flow eqns (9) and (10), and $\varphi_{(g)i}$ for the gas transport eqn (33). The results are also summarized in Table 1, and these are used later.

3.2 Finite element discretization

The space variables in the functional J are discretized by the standard finite element method. The pulmonary circulation system is subdivided into some elements as shown in Fig. 8. In the finite element j, the space variables $\mathbf{w}_j(x_{b(j)}, t) = [P_{b(j)}(x_{b(j)}, t),\ u_{b(j)}(x_{b(j)}, t),\ r_{b(j)}(x_{b(j)}, t)]^T$ and $\boldsymbol{\phi}_j(x_{b(j)}, t) = [\phi_{1j}(x_{b(j)}, t), \phi_{2j}(x_{b(j)}, t),\ \phi_{3j}(x_{b(j)}, t)]^T$ are interpolated in terms of the element vectors $\mathbf{w}_{e(j)}=[\mathbf{w}_j(0,t)^T,\ \mathbf{w}_j(l_{b(j)},t)^T]$ and $\boldsymbol{\phi}_{e(j)}= [\boldsymbol{\phi}_j(0,t)^T,\ \boldsymbol{\phi}_j(l_{b(j)},t)^T]$ which are composed from the value of the variable at both ends as

$$\mathbf{w}_j\left(x_{b(j)},t\right) = \mathbf{N}_w{}^T\mathbf{w}_{e(j)}{}^T,\ \boldsymbol{\phi}_j\left(x_{b(j)},t\right) = \boldsymbol{\phi}_{e(j)}\mathbf{N}_\phi \tag{47}$$

where vectors $\mathbf{N}_w$ and $\mathbf{N}_\phi$ are the polynomial shape functions. The radius $r_{b(j)}$

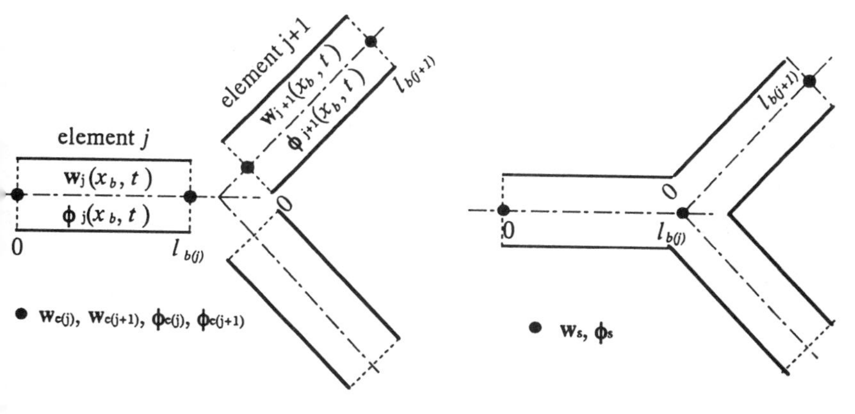

(a) Element variable (b) System variable

Figure 8: Finite elements of blood vessel

is replaced by the sheet thickness h_b at the capillary (j=0). The overall pulmonary circulation system is assembled from the elements. The element vectors $\mathbf{w}_{e(j)}$ and $\boldsymbol{\phi}_{e(j)}$ are represented by the global vectors $\mathbf{w}_s$ and $\boldsymbol{\phi}_s$ which consist of the value of independent variables at the nodal points as

$$\mathbf{w}_{e(j)}^{\mathrm{T}} = \mathbf{B}_{w(j)}^{T}\mathbf{w}_s^{T}, \quad \boldsymbol{\phi}_{e(j)} = \boldsymbol{\phi}_s\mathbf{B}_{\phi(j)} \tag{48}$$

where the matrices $\mathbf{B}_{w(j)}$ and $\mathbf{B}_{\phi(j)}$ are established from the connecting compatibility listed in Table 1 after quasi-linearization. It is noted that the matrices correspond to the Boolean in the standard finite element method but

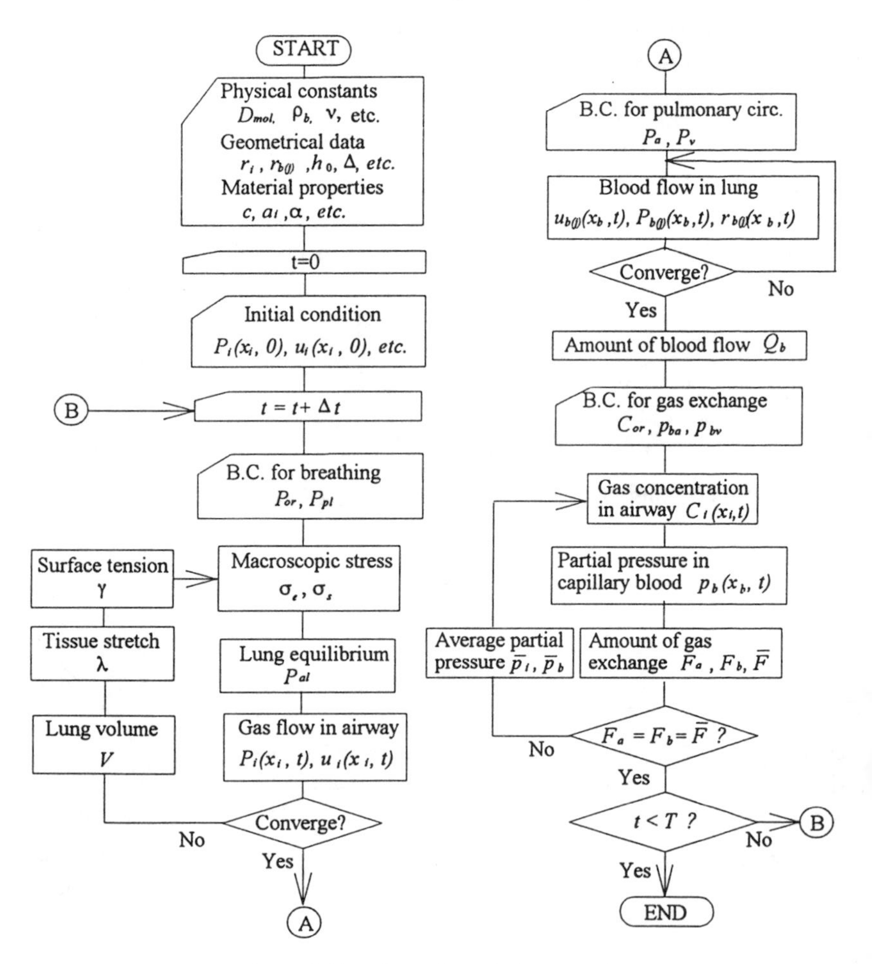

Figure 9: Flowchart of simulation procedure

their components are not limited to zero and unity. By substituting eqns (47) and (48) into eqn (45), the stationary condition of the discretized functional gives the ordinary differential equations in a matrix form,

$$\sum_{j=-N_b}^{N_b}\left[\mathbf{B}_{\phi(j)}\mathbf{K}_{1(j)}\mathbf{B}_{w(j)}{}^{\mathrm{T}}\frac{d\mathbf{w}_s{}^{\mathrm{T}}}{dt}+\mathbf{B}_{\phi(j)}\mathbf{K}_{2(j)}(\mathbf{w}_s)\mathbf{B}_{w(j)}{}^{\mathrm{T}}\mathbf{w}_s{}^{\mathrm{T}}\right]=\sum_{j=-N_b}^{N_b}\mathbf{B}_{\phi(j)}\mathbf{f}_{(j)} \tag{49}$$

where $\mathbf{K}_{1(j)}$ and $\mathbf{K}_{2(j)}$ are the elemental coefficient matrices and $\mathbf{f}_{(j)}$ denotes the elemental free-term vector.

Similar procedure leads the discretized form of the gas flow eqns (9) and (10), and the gas transport eqns (33) and (39). The obtained ordinary differential equations are solved under the appropriate boundary and initial conditions. The time derivative is dealt with the numerical integration combined with the conventional iterative technique due to the quasi-linearization for the nonlinear terms. The simulation procedure is briefly illustrated in Fig. 9.

4 Breathing dynamics and surface tension behavior

In this section, breathing dynamics are discussed with the mechanical elements. The unknown model parameters in the breathing model are identified from the system behavior observed by experiment. The inside mechanical phenomena in the lung are investigated by the simulation using the identified parameters.

4.1 Pressure-volume relation

The lung volume change of the excised lung under the controlled pleural pressure was measured by using Japanese white rabbits weighing about 3kg. The lung specimen was excised and was hung in the closed chamber shown in Fig. 10. The upper trachea was opened to the atmosphere through the flow meter. The rubber diaphragm at the bottom of the chamber was driven by the electro-hydraulic servo actuator, imitating the diaphragm motion. The chamber pressure, that works as the pleural pressure P_{pl}, was recorded simultaneously with the gas flow at the trachea. The lung volume change ΔV is calculated by the time integration of the flow.

Five rabbit lungs were examined under several conditions oscillating the pleural pressure. The upper bound of pleural pressure was fixed at -0.49 kPa, and the pressure was oscillated at several frequencies between 0.1Hz and 10Hz with three different amplitude: Case(A) 0.245kPa, (B) 0.49kPa and (C) 0.735kPa. Figure 11 shows examples of the obtained P_{pl}-ΔV relations. In

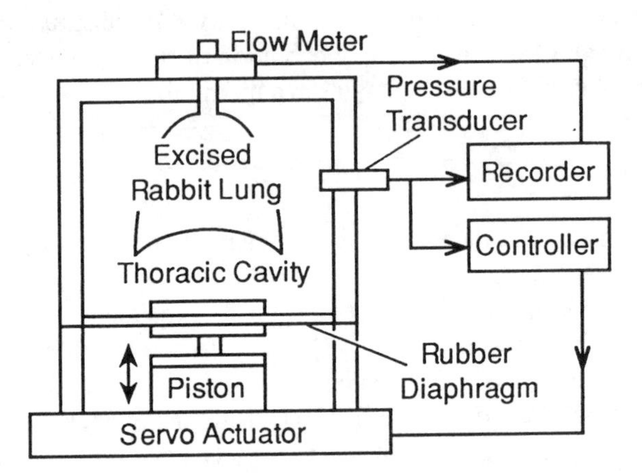

Figure 10: Illustration of experimental apparatus for P_{pl}-ΔV test of excised rabbit lung

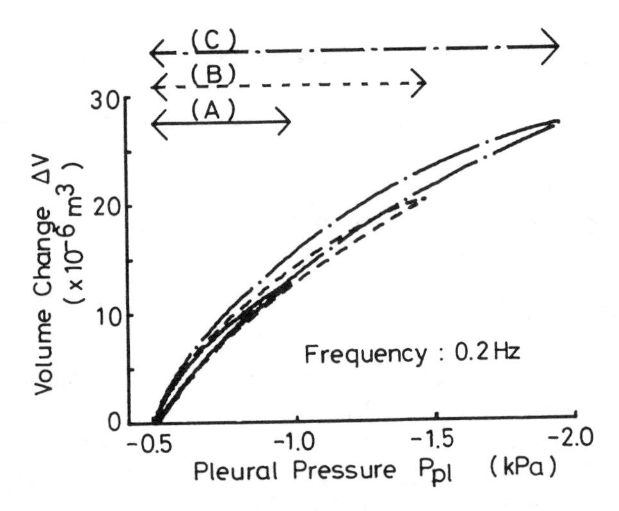

Figure 11: Experimental observation of P_{pl}-ΔV relation

order to characterize the hysteresis of the P_{pl}-ΔV relations, the tidal volume V_T and the hysteresis area H are calculated. Table 2 gives these characteristic values at the frequencies of 0.2Hz and 5Hz.

4.2 Tissue elasticity

Nonlinear material properties of the lung tissue is represented by the constitutive equation based on the pseudo-energy function described in

Table 2 Tidal volume V_T and hysteresis area H of P_{pl}-ΔV relation

Specimen	Tidal Volume V_T (cm³)			Hysteresis area H (μJ)		
	Case(A)	Case(B)	Case(C)	Case(A)	Case(B)	Case(C)
Frequency:0.2Hz						
I	12.7	20.3	27.3	0.41	1.25	2.64
II	12.4	18.3	23.6	0.49	1.20	2.66
III	14.7	22.3	29.2	0.48	1.43	2.91
IV	18.5	27.5	34.9	0.73	1.87	3.44
V	20.1	29.3	37.1	1.28	3.79	6.62
Frequency:5Hz						
I	8.7	13.9	17.8	2.03	6.15	11.8
II	8.2	12.6	16.1	1.81	4.77	8.99
III	9.4	14.4	18.1	2.31	6.26	11.6
IV	10.0	15.8	19.6	2.70	7.82	14.0
V	10.8	16.1	19.5	3.28	8.43	14.2

Case(A):P_{pl}=-0.49~-0.98kPa, Case(B):P_{pl}=-0.49~-1.47kPa, Case(C):P_{pl}=-0.49~-1.96kPa

Section 2.1.1. The material constants were experimentally determined by the biaxial tension test following the P_{pl}-ΔV test in the above. The testing apparatus shown in Fig. 12 was designed by referring to the experiment by Vawter *et al.*[71] The test piece with dimensions of about 20×20×3 mm was immersed in a saline bath in order to keep it free from the surface tension. Uniform tensile force was loaded to the test piece at each edge through 6 threads. The small target marks dotted on the specimen by Indian ink were monitored by the CCD camera system, and the tissue deformation was calculated from the movement of the target marks. Quasi-static tension was applied to the specimen in a cyclic manner as that the tension in the x direction was equalized to that in the y direction. Figure 13 shows the relationship between the stretch ratio λ_x and the Lagrangian stress T_x in the x direction, and the results are almost identical to that in the y direction.

By using the constitutive eqn (4) under the plane stress condition, T_z=0, the theoretical Lagrangian stresses T^{th}_x and T^{th}_y are calculated for the stretch ratio λ_x and λ_y sampled in the experiment. They are compared with the experimental values T^{ex}_x and T^{ex}_y, and the material constants are determined so as to minimize the sum of the squared error

$$R = \sum_{m=1}^{M} \left\{ \left(T^{ex}_{x,m} - T^{th}_{x,m} \right)^2 - \left(T^{ex}_{y,m} - T^{th}_{y,m} \right)^2 \right\} \tag{50}$$

where subscript m denotes the sample number along the loading. The identified material constants for the individual specimens are listed in Table 3.

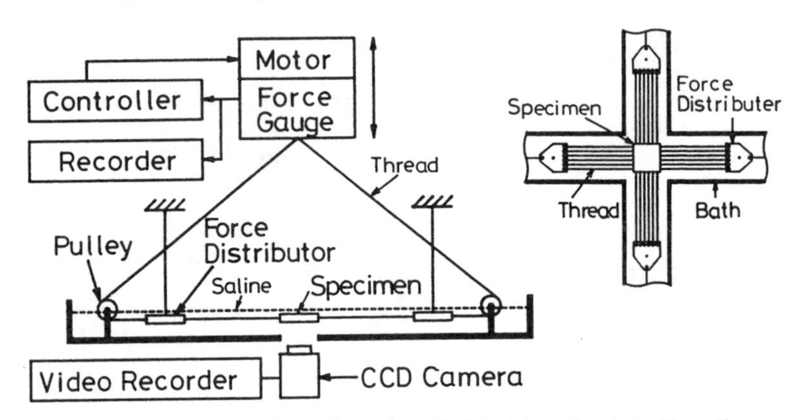

Figure 12: Illustration of experimental apparatus for biaxial tension test of lung tissue

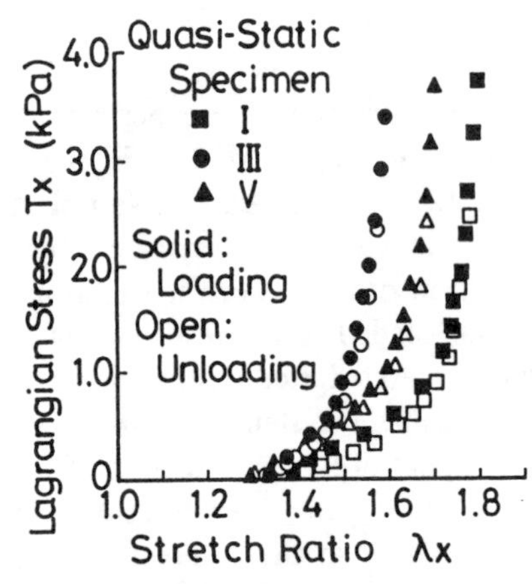

Figure 13: Elastic properties of lung tissue (rabbit)

Table 3 Material constants of lung tissue (rabbit)

Specimen	c (Pa)	$a_1 = a_2$	a_3
I	19.1	1.48	0.117
II	22.1	1.75	0.411
III	39.7	1.90	0.509
IV	106	0.824	0.152
V	62.5	1.35	0.316

4.3 Surface tension characteristics

The surface tension at alveoli is substantial to the breathing dynamics. Bachofen *et al.*[14] attempted to evaluate the surface tension from the difference of the pressure-volume relations between the saline-filled lung and the air-filled lung. A pulsating bubble technique was also developed to measure it *in vitro* by Enhorning[15]. However, the accuracy of the measurement and the difference in environment are bottle-neck to make it clear.[52] The purpose of this section is to examine the surface tension characteristics in the model-based context.

4.3.1 Identification procedure

By using the material constants listed in Table 3, the breathing dynamics of each rabbit lung is simulated. The edge length of cubic alveolus is assumed to be Δ=100μm, and the airway size of the human lung[29] is scaled down to the one third for the rabbit lung. The boundary conditions are set as corresponding to the previous P_{pl}-ΔV test, that is, the oral pressure P_{or} is maintained at atmospheric pressure, and the pleural pressure P_{pl} is oscillated with the frequency and amplitude as those used in the measurements.

The minimum error is the criterion in identifying the model parameters: the lung volume at the natural state V_0, the hysteresis coefficient a of the surface tension, and the pressure loss coefficient ζ at the branching points in the airway. The errors for the tidal volume V_T and the hysteresis area H of the P_{pl}-ΔV relations are defined as

$$R_{V(k)} = \left(V_{T(k)}^{e} - V_{T(k)}^{s}\right) / V_{T(k)}^{e} \tag{51a}$$

$$R_{H(k)} = \left(H_{(k)}^{e} - H_{(k)}^{s}\right) / H_{(k)}^{e} \tag{51b}$$

where the superscripts e and s denote the experiment and the simulation, respectively, and the subscript (k) distinguishes the pressure amplitude of Case (A), (B) and (C). Then the model parameters are identified by minimizing the resultant error

$$R_{VH} = \sum_{k} \left(R_{V(k)}{}^{2} + R_{H(k)}{}^{2}\right). \tag{52}$$

4.3.2 Surface tension based on identified parameter

The model parameters a and V_0 are identified using the P_{pl}-ΔV relations at the frequency of 0.2Hz. The pressure loss coefficient is assumed to be ζ=1.0 because it influences little the result for such a slow oscillation rate. Figures 14(a) and (b) show the case-wise resultant error $R_{VH(k)}$ against the parameters

a and V_0 in the vicinity of the identified value. No significant difference is found among the values which minimize the error $R_{VH(k)}$ for each case of pressure amplitude, although the model parameters have a significant sensitivity to the error. These results indicate that the identification process gives a meaningful result and the identified parameter is applicable to every pressure amplitude.

The identified parameters for each specimen are listed in Table 4 along with the errors in tidal volume $R_{V(k)}$ and hysteresis area $R_{H(k)}$. The errors in tidal volume and hysteresis are kept within a reasonable level. It seems that the identified values vary specimen by specimen due to the individuality. The simulated P_{pl}-ΔV relations based on the identified parameters are compared with the experimental results in Fig. 15. It is confirmed that the simulated curves coincide with the experimental one.

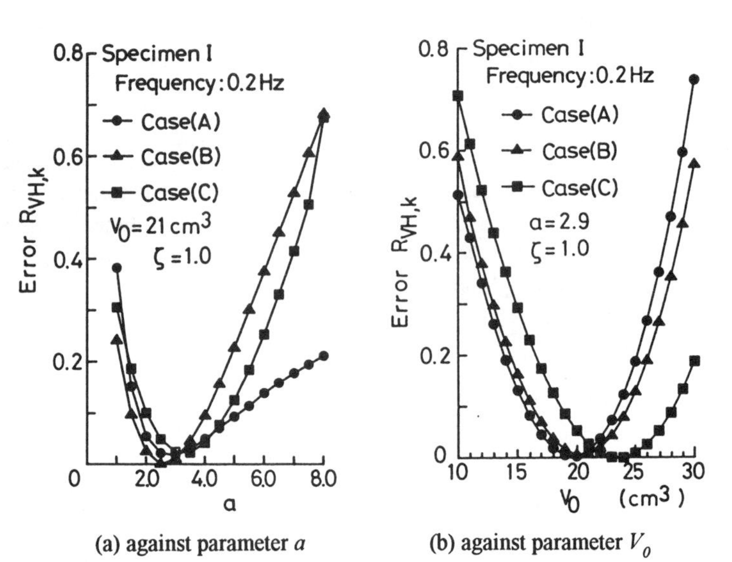

(a) against parameter a (b) against parameter V_0

Figure 14: Case-wise resultant error $R_{VH(k)}$ in the vicinity of identified parameter

Table 4 Identified natural lung volume V_0 and hysteresis coefficient a

Specimen	V_0 (cm^3)	a	Error in tidal volume $R_{V(k)}$ (%) Case(A)	Case(B)	Case(C)	Error in hysteresis area $R_{H(k)}$ (%) Case(A)	Case(B)	Case(C)
I	21	2.9	2.8	-2.6	-10.5	11.3	7.6	-12.8
II	22	3.1	-2.6	-1.4	-6.9	18.5	6.4	-20.5
III	26	2.7	2.8	-0.1	-8.7	13.3	12.1	-12.9
IV	18	1.7	1.4	0.1	-4.0	3.4	5.4	-8.5
V	37	3.9	-3.8	9.3	9.0	-13.1	3.3	3.2

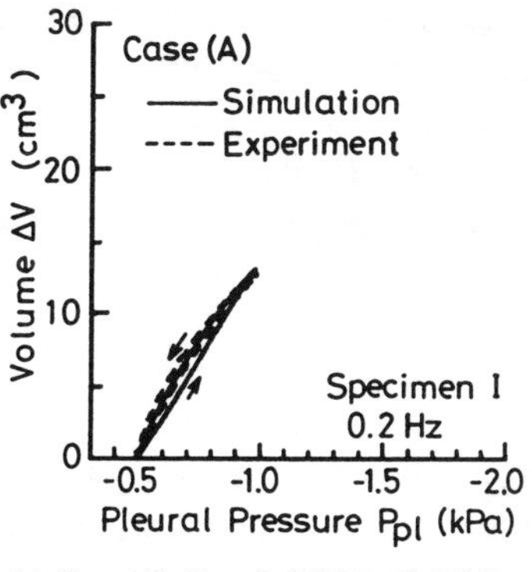

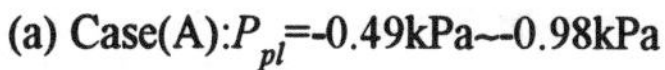
(a) Case(A):P_{pl}=-0.49kPa~-0.98kPa

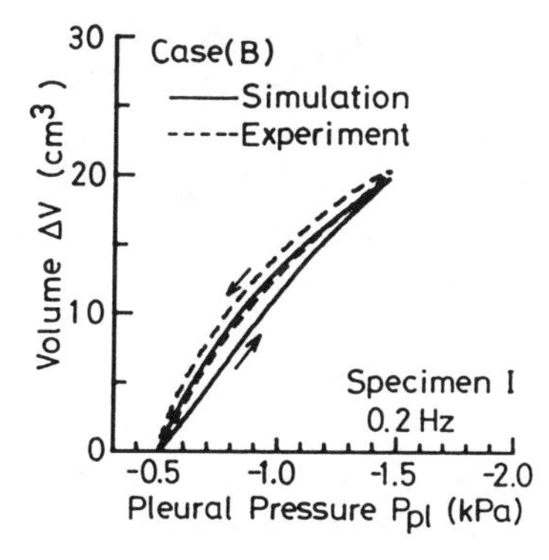

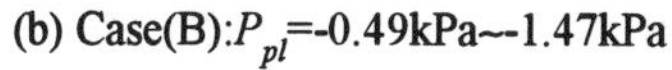
(b) Case(B):P_{pl}=-0.49kPa~-1.47kPa

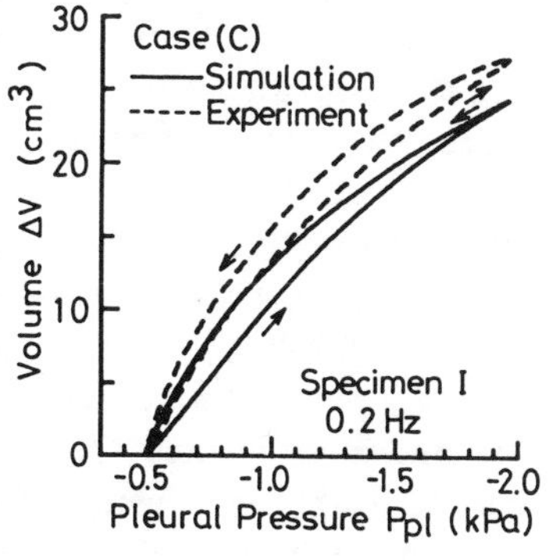

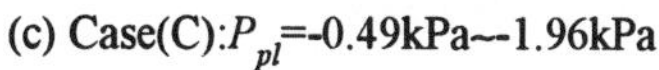
(c) Case(C):P_{pl}=-0.49kPa~-1.96kPa

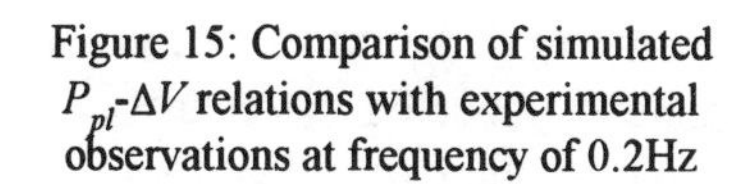
Figure 15: Comparison of simulated P_{pl}-ΔV relations with experimental observations at frequency of 0.2Hz

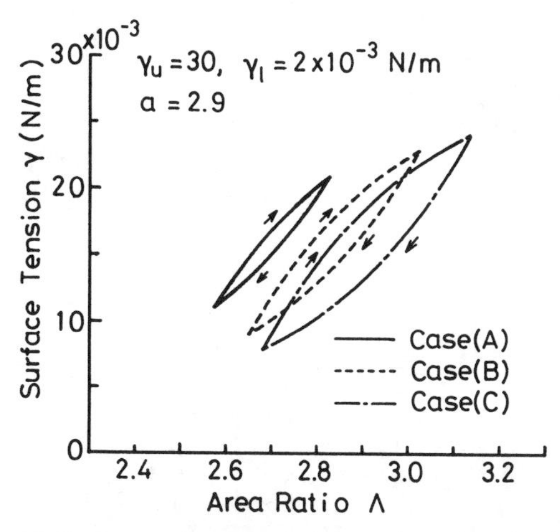

Figure 16: Surface tension characteristics estimated by simulation

Figure 16 shows the relation between the surface tension γ and the area ratio Λ, attained by the simulation using the identified parameters. The applied rate-type model of eqn (8) may restrict the surface tension behavior rather simple, but the difference in P_{pl}-ΔV relation between the simulation and the experiment is small reasonably. This finding supports the validity of the prediction for the surface tension behavior by the simulation.

4.4 Frequency response

The gas flow in the airway dominates the dynamic behavior of the lung. The pressure loss coefficient ζ is identified by the same way described in Section 4.3, using the P_{pl}-ΔV relations in the dynamic condition at the frequency of 5Hz. The coefficient ζ is assumed to be same value at any branching points in the airway system, and the model parameters a and V_0 are fixed to those in Table 4. The identified coefficients and the comparison with experimental results are shown in Table 5 and Fig. 17(c), respectively. It is confirmed that the deviation between the simulated results and the experimental observations is not significant for any case. The pressure distribution along the airway is demonstrated in Fig. 18. It is shown that the pressure drop is primarily occurred in the airway at generation of the first half.

We applied the identified parameters to simulate the P_{pl}-ΔV relation at several frequencies different from that used for the identification. As the results, the simulation shows a good agreement with the experimental observations as illustrated in Figs. 17(a), (b) and (d). Figure 19 shows the frequency response of the tidal volume. The simulation gives the sufficient prediction for the frequency response at any frequency and amplitude of the pressure oscillation. These results represent the capabilities of the simulation to analyze the system characteristics of breathing dynamics, and demonstrate that the simulation based on the identified model parameters provides the unmeasurable behavior in the lung such as the surface tension characteristics and the gas pressure distribution as shown in Figs. 16 and 18, respectively.

Table 5 Identified pressure loss coefficient ζ and resultant error $R_{V(k)}$ and $R_{H(k)}$

Specimen	ζ	Error in tidal volume $R_{V(k)}$ (%)			Error in hysteresis area $R_{H(k)}$(%)		
		Case(A)	Case(B)	Case(C)	Case(A)	Case(B)	Case(C)
I	1.5	2.3	-6.3	-10.7	21.8	14.5	2.8
II	3.0	-14.9	-18.7	-22.1	16.3	24.9	18.8
III	2.0	-9.5	-14.2	-16.9	14.6	17.7	9.8
IV	1.5	-3.7	-8.9	-9.9	12.8	13.9	12.3
V	1.5	-10.4	-10.0	-36.3	-4.7	10.4	46.0

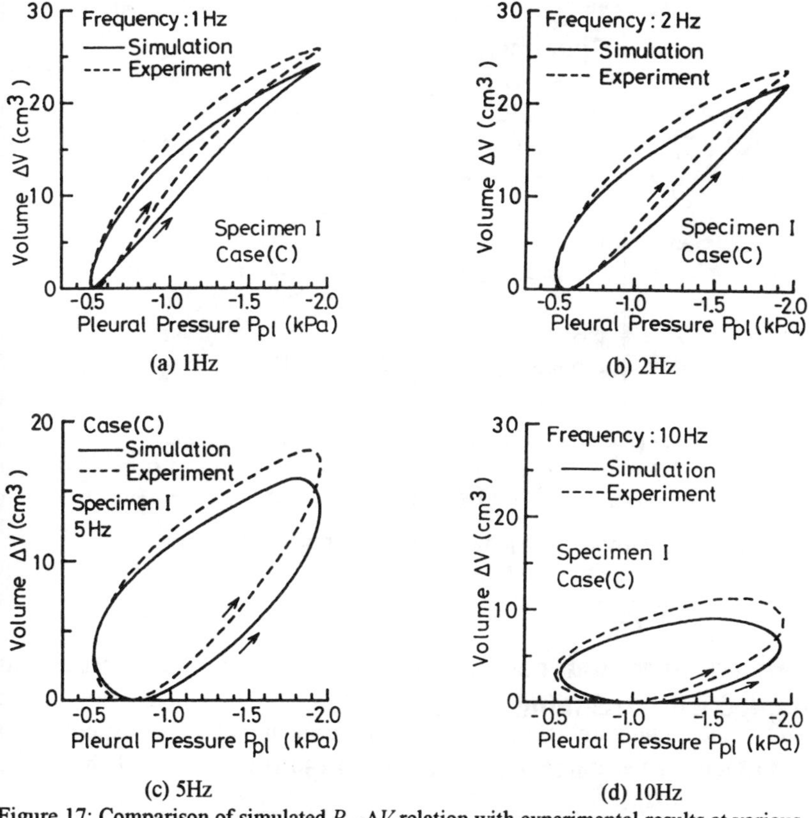

Figure 17: Comparison of simulated P_{pl}-ΔV relation with experimental results at various frequencies (Case(C): P_{pl}=-0.49kPa~-1.96kPa)

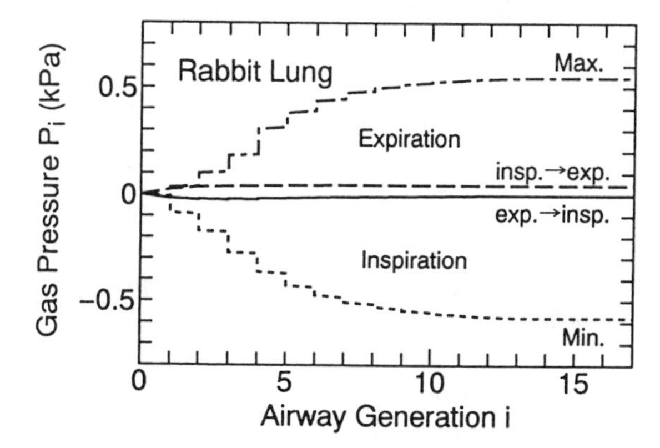

Figure 18: Pressure distribution along airway

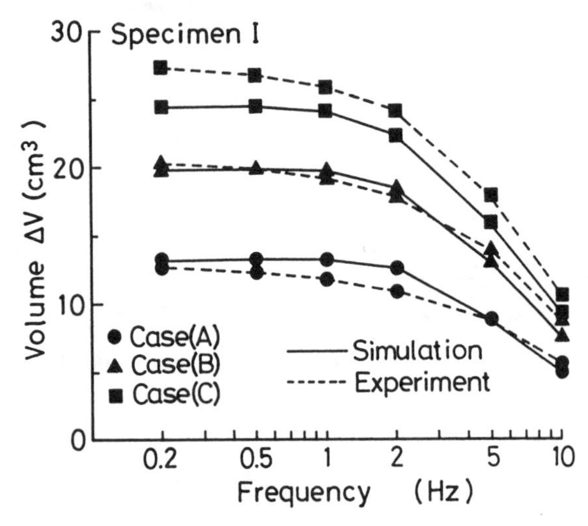

Figure 19: Frequency response of breathing dynamics

5 Blood flow and gas exchange in pulmonary circulation coupled with breathing and ventilation

5.1 Influence of gas pressure to pulmonary blood flow

The gas pressure in the lung oppresses the blood vessels and affects the blood flow in the pulmonary circulation. This section examines the influence of it at steady blood flow by experiment and simulation.

5.1.1 Experimental observation

The experimental setup is illustrated in Fig. 20. Four Japanese white rabbits weighing about 3kg were sacrificed by an anesthetic following the heparin treatment. The gas pressure P_L in the lung was regulated to keep a positive constant connecting the trachea to air-supply. The chest wall was opened and the elastic tubes were inserted into the pulmonary artery and vein from the right ventricle and left atrium, respectively. Keeping the hydrostatic pressure P_t of the reservoir tank a constant value, P_t=2.45kPa, the working fluid run into the artery and the volumetric flow Q_b coming out of the vein was measured. The working fluid was the blood diluted with the saline to about 20% to prevent the leak into the alveolar space. The relation between the volumetric blood flow Q_b and the lung gas pressure P_L is shown by solid marks in Fig. 21. The blood flow obviously decreases with the raise of the gas pressure in the lung. The decrease tends to remarkable over the gas pressure P_L=1.96kPa.

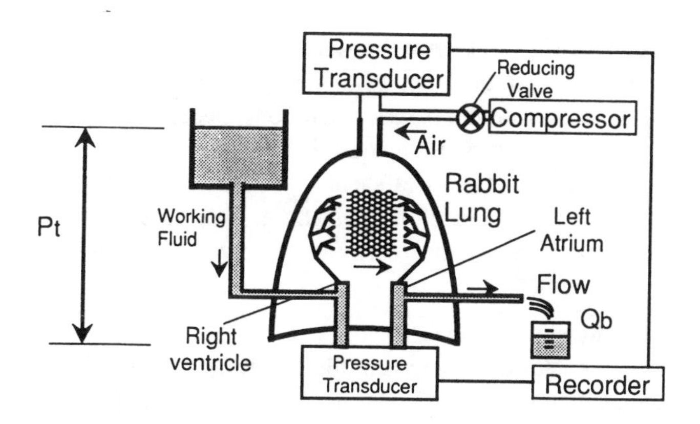

Figure 20: Illustration of experimental apparatus for pulmonary blood flow measurement

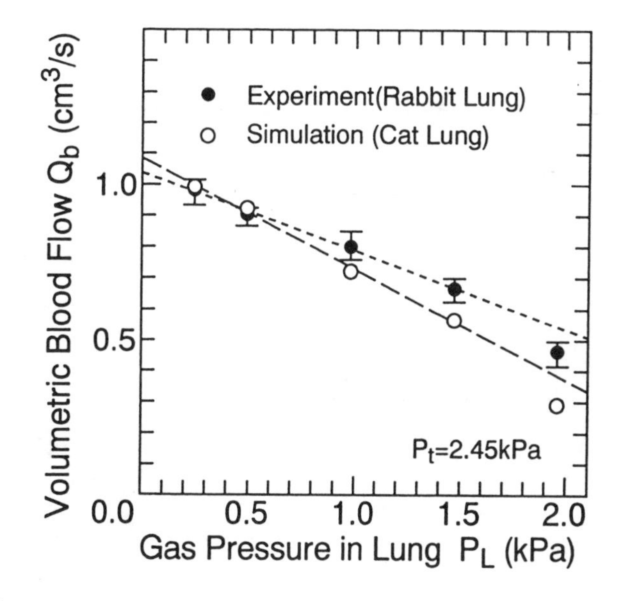

Figure 21: Pulmonary blood flow subject to static gas pressure in lung

5.1.2 Blood flow properties by simulation

The above experiment was also simulated using the pulmonary circulation model. In this case, the coupling with breathing is considered by prescribing the alveolar gas pressure P_{al} as the inhaled gas pressure P_L and the stretch ratio λ of the lung tissue corresponding to the inflation state. The geometry and compliance of the blood vessel are borrowed from those of cat's lung.[60] The blood pressures at the inlet of artery (j=-11) and the outlet of vein (j=11) are given as the boundary conditions by taking account of the pressure loss in the elastic tube used in the experiment. By using the adequate geometrical friction coefficient $f_{b(j)}$=2.2 in eqn (17), the volumetric blood flow and its dependency on the static gas pressure P_L show the good correspondence to the experimental result as shown by the open marks in Fig 21.

5.1.3 Analysis by simulation

The simulation enables us to interpret the experimental result in terms of the mechanical phenomena in the lung. Figure 22 shows the distribution of the alveolar sheet thickness, that is, the height of capillary network from the entrance to the exit of alveolar sheet. The gas pressure P_L in the lung oppresses the alveolar sheet, and the collapse of the alveolar capillaries appears at the down stream in the sheet flow when the gas pressure is P_L=1.96kPa. This phenomenon corresponds to the following result which is observed by both simulation and experiment. The volumetric blood flow Q_b decreases almost linearly below the gas pressure up to P_L=1.47kPa as was

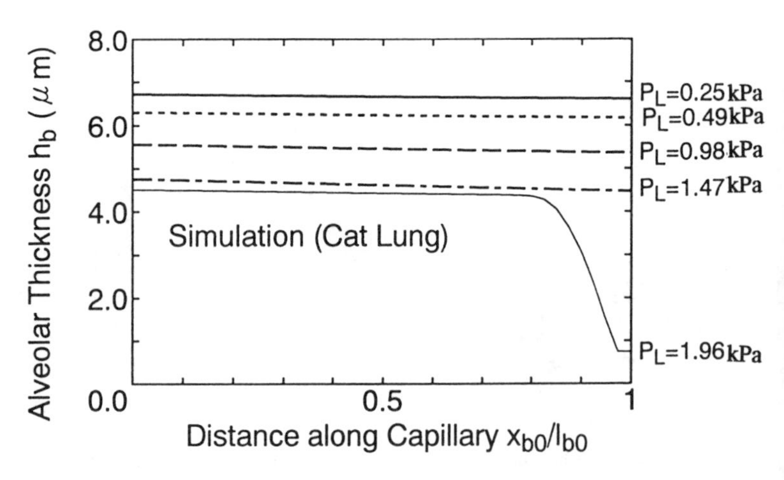

Figure 22: Alveolar sheet thickness along capillary

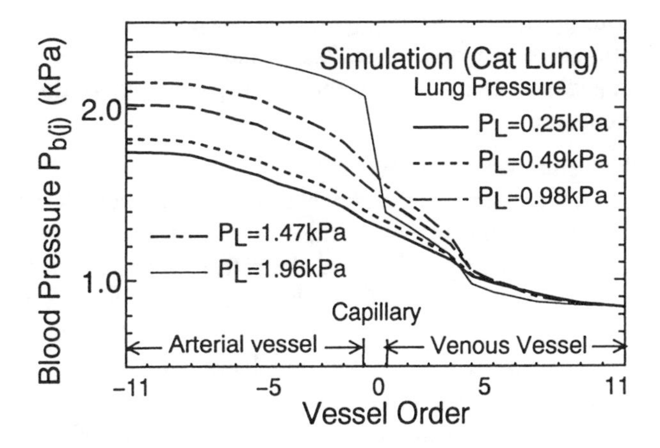

Figure 23: Blood pressure distribution in pulmonary circulation

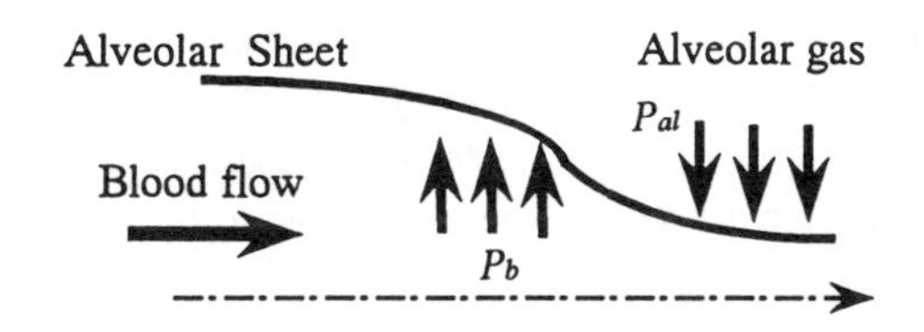

Figure 24: Alveolar sheet supported by elevated blood pressure

shown by the dotted line in Fig. 21. However the decrease of Q_b is accelerated at gas pressure around P_L=1.96kPa.

Figure 23 shows the distribution of the blood pressure along the pulmonary vessels. The abscissa indicates the vessel order. As to the increase of gas pressure P_L in the lung, the alveolar thickness h_b decreases (Fig. 22) and the blood pressure $P_{b(j)}$ increase in the upstream region of the arterial vessels (Fig. 23). In the case of relative high gas pressure of P_L=1.96kPa, the collapsing of the capillary increases the flow resistance significantly resulting in high blood pressure in the arterial vessels. It is also found that the blood pressure coming from the upper stream supports the alveolar sheet and that the alveolar capillary does not suffer from the complete collapse. This mechanism is schematically illustrated in Fig. 24.

5.2 Pulmonary blood flow in normal breathing

By using the coupled model of breathing and pulmonary circulation, the blood flow during normal breathing is examined by simulation. The simulation estimates the blood vessel compliance of the human lung and shows the impedance characteristics of the pulsatile blood flow in the pulmonary circulation.

5.2.1 Boundary condition at artery

When the arterial pressure oscillates as a sinusoidal wave form in the simulation, the inverse flow appears at the artery due to the vessel elasticity Since the pulmonary valve prevents the blood from returning to the right ventricle, the boundary condition of the blood flow is set as follows. The pressure boundary condition of eqn (31) is activated when the cardiac output pressure $P_a(t)$ becomes larger than the blood pressure $P_{b(-Nb)}(0, t)$ at the entrance of artery. And, when the blood velocity $u_{b(-Nb)}(0, t)$ at the entrance of artery decreased and reaches zero as to the decrease of the cardiac pressure $P_a(t)$, the velocity boundary condition of eqn (31) is activated with $u_a(t)=0$ for the valve closed. The cardiac output pressure at the right ventricle is given by a intermittent wave pattern;

$$P_a(t) = \begin{cases} \dfrac{P_{a\max} - P_{a\min}}{2}\left(1 - \cos\dfrac{4\pi\tau}{t_b}\right) + P_{a\min} & (0 \le \tau < \dfrac{t_b}{2}) \\ P_{a\min} & (\dfrac{t_b}{2} \le \tau < t_b) \end{cases} \tag{53}$$

where P_{amax} and P_{amin} are the upper and lower bounds of cardiac output pressure, respectively, τ denotes the elapsed time from the constriction of heart, and t_b is the output interval.

5.2.2 Vessel compliance of human lung

The pulmonary circulation in normal breathing is simulated based on the geometrical data of the airway reported by Weibel[29] and those of the blood vessel by Singhal *et al.*[32] and Horsfield & Gordon.[33] For the breathing model the oral pressure P_{or} is set to the atmospheric pressure and the pleural pressure P_{pl} oscillates between -0.49kPa and -0.88kPa at the frequency of 0.2Hz. For the pulmonary circulation model, the blood pressure $P_v(t)$ at the exit of the vein is kept constant 0.54kPa, the upper and lower bounds of the cardiac output pressure are P_{amax}=3.88kPa and P_{amin}=0kPa, respectively, and the interval is t_b =1sec.

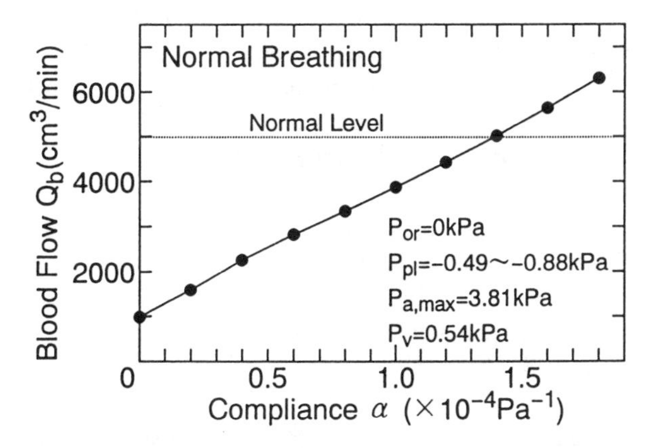

Figure 25: Relationship between volumetric blood flow and vessel compliance

The amount of blood flow in the lung depends on the vessel compliance as simulated in Fig. 25 where all the vessel compliance of the arterial and venous trees is assumed same. As the pulmonary blood flow is approximately 5000cm^3/min in the physiological state, the vessel compliance α is estimated to be 1.4×10^{-4}Pa^{-1}. This is larger than 0.6×10^{-4}Pa^{-1} measured by Greenfield & Griggs[87] using the patient lungs. However, according to the detailed measurements using the cat lungs,[77,78] the compliance of the vessels varies widely between 0.71×10^{-4}Pa^{-1} and 2.66×10^{-4}Pa^{-1} place by place. Thus, the estimated compliance falls in the acceptable range.

5.2.3 Input impedance

The pulmonary vascular input impedance is defined as the ratio between the amplitude of pulsatile blood pressure and that of the flow at the pulmonary artery, and it represents a characteristics of the blood flow dynamics. Since Caro & McDonald[88] measured it using the animal lungs in 1961, a number of experiments have reported that the impedance modulus depends on the frequency of the flow oscillation by the pulse wave reflection.[27] We investigate the impedance characteristics by the simulation based on the pulmonary circulation model coupled with breathing dynamics.

Figure 26 shows the pulsatile blood pressure and flow at the artery obtained by the simulation of normal breathing. The heart throbs five times during single respiratory cycle for 5 seconds. The continuous pulsatile flow is observed at the outlet of artery, although the blood flow at the inlet of artery is intermitted due to the pulmonary valve close. The simulated wave form is qualitatively close to the experimental observation[89] except the relative high blood pressure at diastole. This may be the reason why the simulation model disregards the active deformation of the blood vessel.

The pulse wave of the blood pressure and flow in Fig. 26 is analyzed by discrete Fourier transform and the input impedance is calculated for the harmonic oscillation. The obtained impedance modulus and the phase lag are shown in Figs. 27(a) and (b). The impedance becomes a local minimum at 3Hz, around which the phase lag becomes positive, that is, the pressure oscillation becomes to lead on the flow oscillation. The local maximum is occurred at 7Hz almost double of the minimum. The average impedance Z_0 over the first impedance minimum is 17dyn·s/cm^5. The fluctuation of the input impedance to the frequency and the characterized values are observed by the experiments.[27] The input impedance of the cat lung is also examined by the simulation using a complete set of the vessels properties,[60] and the results are compiled in Table 6. It is confirmed that the simulated results represent the difference of the impedance characteristics between the human and cat lungs corresponding to the experimental results.[27,89,90]

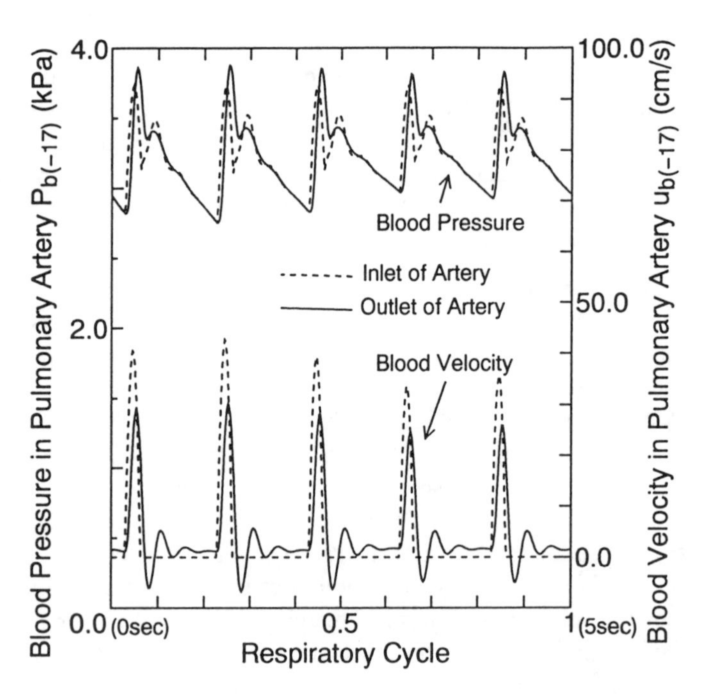

Figure 26: Pulsatile blood pressure and velocity at pulmonary artery

Table 6 Frequency of first impedance minimum f_{min} and average impedance Z_0

	f_{min}(Hz)		Z_0 (dyn·s/cm^5)	
	Human	Cat	Human	Cat
Experiment[27]	2-5	6-8	20-200	1000
Simulation	3	8	17	537

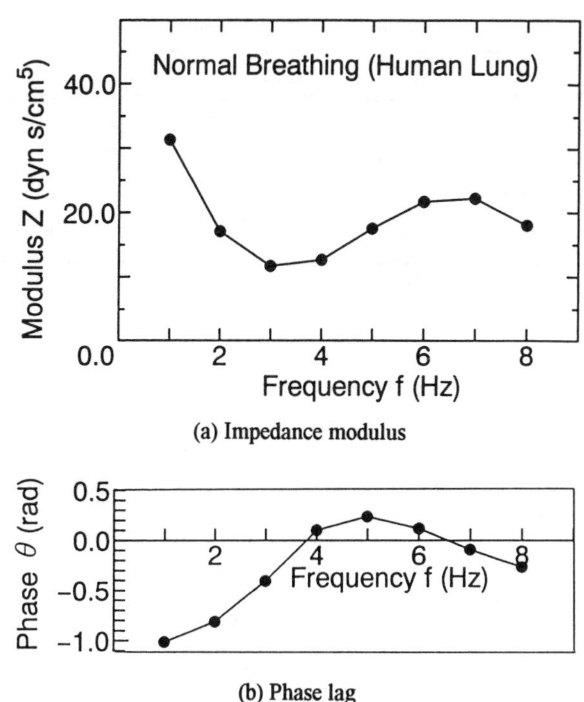

(a) Impedance modulus

(b) Phase lag

Figure 27: Pulmonary vascular input impedance

5.3 Respiratory function evaluated by simulation

The breathing dynamics influence not only the pulmonary blood flow but also the ventilation function in the airway. In the case of artificial ventilation, the pressure condition at the oral space should be determined taking account of both effects.[91] In this section, the total function of the lung respiratory system is evaluated by the simulation considering the coupled behavior among breathing, ventilation and pulmonary circulation.

5.3.1 Gas exchange in normal respiration

The simulation of the normal respiration using the model parameters listed in Table 7 shows the reasonable results compared with the physiological state. This validates the total respiratory model including gas exchange behavior. Figures 28(a) and (b) represent the oxygen and carbon dioxide distribution in the airway during the inspiration of normal respiration. It is straight forward

Table 7 Boundary condition for normal respiration and simulated results

Boundary condition			
Pleural pressure P_{pl}	-0.49~-0.88kPa	Respiratory cycle	12cyc/min
Oral pressure P_{or}	0kPa	C_{O2} of inhaled gas	23.2%
Blood pressure P_v	0.54kPa	C_{CO2} of inhaled gas	0%
Cardiac pressure P_{amax}	3.81kPa	p_{O2} at artery	40mmHg
Cardiac cycle	60cyc/min	p_{CO2} at artery	46mmHg
Simulated results			
Tidal volume V_T	423cm^3	Blood flow Q_b	5027cm^3/min
Lung volume V	3018~3441cm^3	p_{O2} at vein	97.2~102mmHg
O_2 intake	255cm^3/min	p_{CO2} at vein	39.6~41.6mmHg
CO_2 exhaust	178cm^3/min		

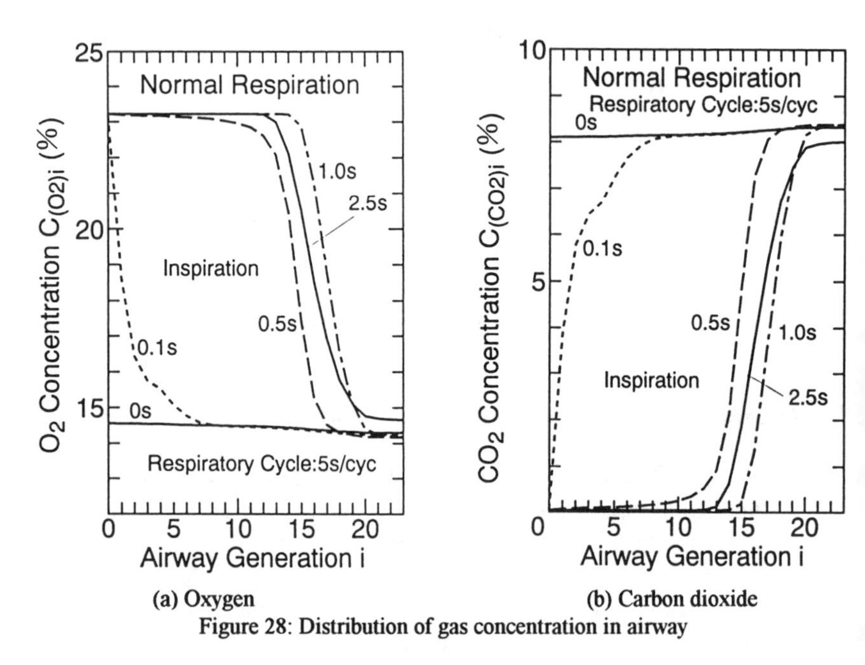

(a) Oxygen (b) Carbon dioxide

Figure 28: Distribution of gas concentration in airway

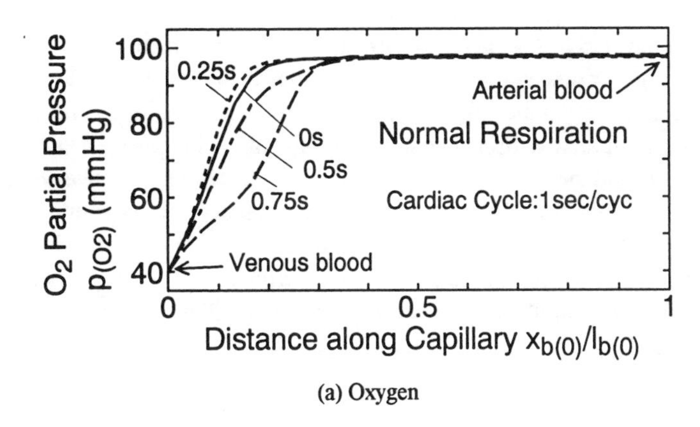

(a) Oxygen

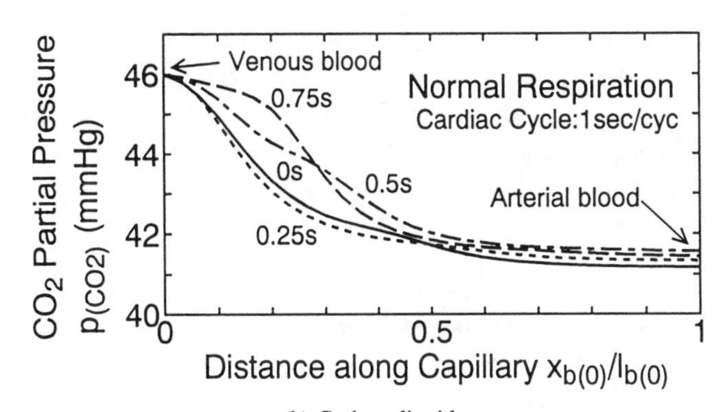

(b) Carbon dioxide

Figure 29: Distribution of partial gas pressure along capillary

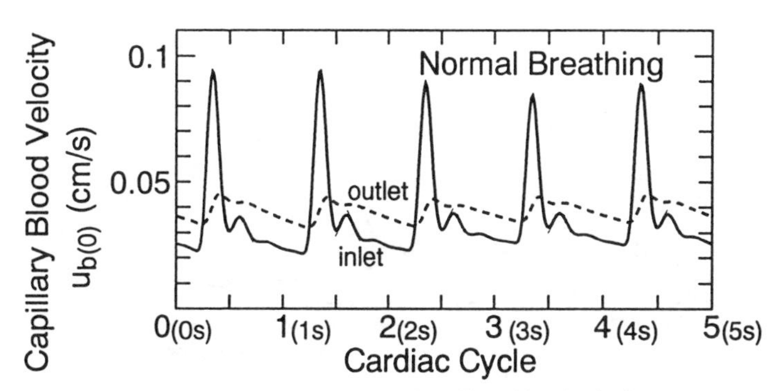

Figure 30: Time variation of capillary blood velocity

to understand from the gradient of the gas concentration that the gas transport is governed mainly by the bulk flow in the airway branches with smaller generation number up to about 15th and additionally by the diffusion at the peripheral airway branches including alveoli. The gas concentration at the alveoli is balanced with that at the capillary blood. Figure 29 is the partial pressure of the blood gas along the capillary. It is found that the pulsatile blood velocity shown in Fig. 30 influences the distribution of the partial gas pressure. As this distribution determines the average partial gas pressure $\bar{p}_{b(g)}$ of capillary blood, the resultant amount of gas exchange is influenced by the blood velocity at the capillary as well as the volumetric blood flow.

5.3.2 Pulmonary blood flow in artificial ventilation

The pulmonary blood flow in artificial ventilation is evaluated by the simulation. The oral pressure P_{or} oscillates between 0kPa and 0.98kPa at the frequency of 0.2Hz. In this case, the alveolar gas pressure P_{al} and the pleural pressure P_{pl} change as shown in Fig. 31 depending on the inhaled gas pressure P_{or} and the lung volume. Figures 32(a) and (b) show the average blood flow per one cardiac cycle at the pulmonary artery and capillary for single respiratory cycle during the artificial ventilation and normal breathing in the pervious section. In the case of normal breathing, the blood flow at the artery fluctuates by the change of pleural pressure (Fig. 31), but the capillary blood flow is maintained almost constant regardless the cardiac cycle number. On the other hand, the change of pleural pressure P_{pl} and the elevated alveolar gas pressure P_{al} in artificial ventilation varies not only the blood flow at the artery but also that at the capillary. Subsequently, the volumetric blood flow for single respiratory cycle reduces to 94.3 % of that at normal breathing.

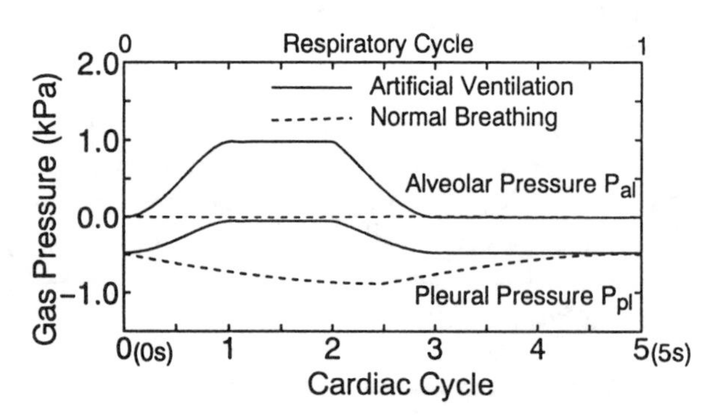

Figure 31: Change of pleural pressure and alveolar pressure

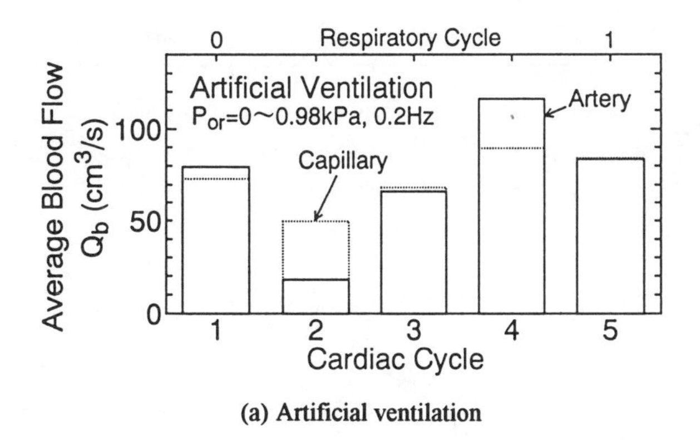

(a) Artificial ventilation

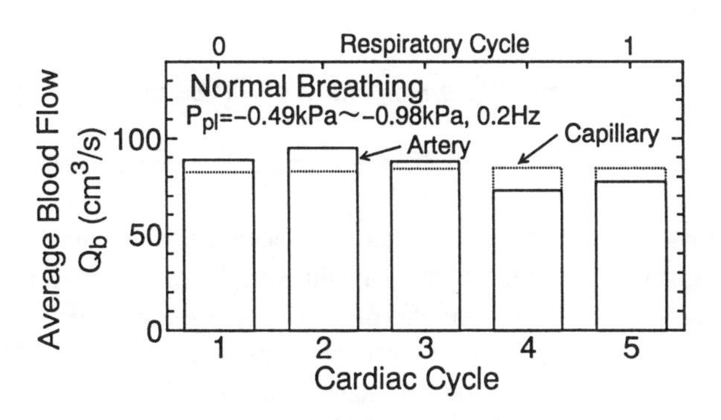

(b) Normal breathing

Figure 32: Average blood flow per cardiac cycle

5.3.3 Attempt to optimal artificial ventilation

Let consider a stiffened lung as an example of abnormal case, and evaluate the total respiratory function during the artificial ventilation by simulation. The material constant c in eqn (2) is assumed to be three time as large as that of normal lung. The raise of the pressure amplitude of the inhaled air increases the tidal volume V_T and decreases the amount of blood flow Q_b as shown in Fig. 33. The former encourages the ventilation function in the airway but the latter discourages the circulation function. This trade-off effect to the gas exchange determines the amount of the oxygen intake and the carbon dioxide exhaust as demonstrated in Fig. 33. When the pressure amplitude is smaller than 0.8kPa, the ventilation function is not sufficient, and then the tidal volume dominates the ability of the gas exchange. For the larger amplitude, however, the oxygen intake is mainly subject to the amount

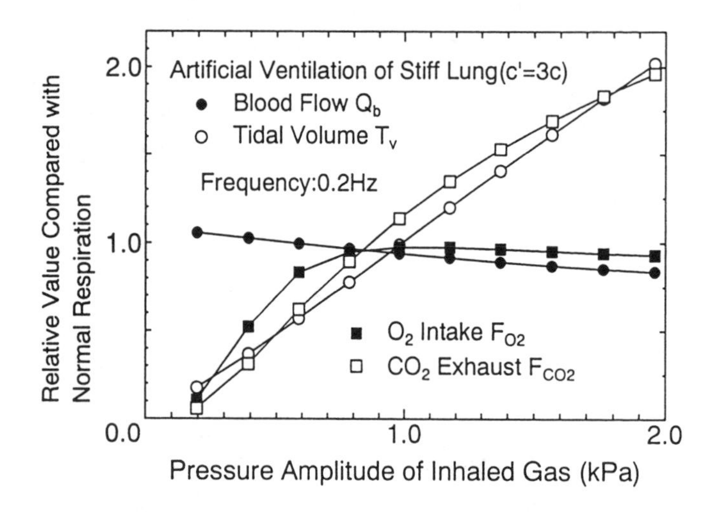

Figure 33: Total respiratory function in artificial ventilation of stiff lung

of blood flow rather than the tidal volume, because the oxygen content in the blood is saturated. On the other hand, the ventilation function continues to dominate the exhaust of carbon dioxide for any pressure amplitude. These quantitative evaluations assisted by clinical experience will be effective to operate the artificial ventilation optimally. In this stiff lung, it is suggested that the pressure amplitude around 0.8kPa is an appropriate condition to maintain the gas exchange in physiological state judging from both the ventilation and circulation functions.

6 Conclusions

In this chapter, we discussed a computational biomechanics model of the lung respiratory system. The model consists of the submodels for fundamental mechanical phenomena in the lung. The breathing model is made up from the gas flow in the airway and the tissue deformation considering the surface tension activity. The blood flow model for the overall pulmonary circulation takes account of the influence of gas pressure in the lung due to breathing. The gas exchange in the lung is modeled as the diffusion phenomena which is coupled with the ventilation of airway gas by breathing and the blood flow in the capillary by pulmonary circulation. This direct physical model represents the mechanical interaction explicitly, and is effective in expressing the complex mechanical structure in the respiratory system compared with the traditional equivalent model such as electric circuits. In order to solve the model equations numerically, we employed the finite element technique based on the adjoint variational principle. The Lagrangian functional is formulated by Lagrangian multiplier method and is discretized by the finite elements approximation taking account of the connecting compatibility between the elements which is derived from the variational consideration.

The simulation approach by computational mechanics procedure was applied to investigate the mechanical phenomena in the lung. The surface tension activity was estimated through the model parameter identification from the pressure and volume relations of the lung. The pulmonary blood flow affected by the gas pressure in the lung was examined by the experiments whose results are interpreted by the simulation. The total respiratory function of the artificial ventilation was evaluated by the simulation from both the ventilation function by breathing and the circulation function by the blood flow. Through case studies compared with experimental observations, it was demonstrated that detailed simulation based on computational mechanics procedures enable the unification of the mechanical phenomena at the subsystems and elements with total system behavior. It also becomes possible to identify and visualize the internal phenomena which could not be directly observed by experiments. The concept of analysis by synthesis in the simulation procedure allows the investigation of the complex system such as the lung respiration.

Since the respiratory model was built up from the fundamental mechanics, it is possible to take the difference of geometrical size and material properties into the model. The simulation based on the personal properties allows us to compare the experimental data of different species and to cope with the difference due to individuality. Such a personalized insight into the simulation is a key to evolve the computational biomechanics findings toward those in the clinical field.

Acknowledgment

The authors wish to thank Prof. H. Nakamura and Prof. T. Horikawa of Ryukoku University and Dr. H. Togari of Nagoya City University for their encouragement. This work was supported in part by Grant-in-aid for Scientific Research on Priority Areas (No.06213235) and Grant-in-aid for Encouragement of Young Scientists (No.06780738) from the Ministry of Education, Science & Culture of Japan, and the research grant from Graduate School of Ryukoku University.

References

[1]Fung, Y.C., *Biomechanics*, Springer-Verlag, New York, 1981.
[2]*Bioengineering Aspects of Lung Biology*, ed. West, J.B., Marcel Dekker, N.Y., 1977.
[3]Fishman, A.P., *Assessment of Pulmonary Function*, McGraw-Hill, New York, 1980.
[4]Fung, Y.C., A Theory of elasticity of the lung, *Trans. ASME, J. Biomechanical Eng.*, 1974, **41**, 8-14.
[5]Fung, Y.C., Stress, deformation, and atelectasis of the lung, *Circulation Research*, 1975, **37**, 481-496.
[6]Karakaplan, A.D., Bieniek, M.P. & Skalak, R., A mathematical model of lung parenchyma, *Trans. ASME, J. Biomechanical Eng.*, 1980, **102**, 124-136.
[7]Wilson, T. A., Relations among recoil pressure, surface area, and surface tension in the lung, *J. Appl. Physiol.: Respirat. Environ. Exercise Physiol.*, 1981, **50**(5), 921-926.
[8]Kimmel, E., Kamm, R.D. & Shapiro, A.H., A cellular model of lung elasticity, *Trans. ASME, J. Biomechanical Eng.*, 1987, **109**, 126-131.
[9]Kimmel, E. & Budiansky, B., Surface tension and the dodecahedron model for lung elasticity, *Trans. ASME, J. Biomechanical Eng.*, 1990, **112**, 160-167.
[10]Vawter, D.L., Fung, Y.C. & West, J.B., Constitutive equation of lung tissue elasticity, *Trans. ASME, J. Biomechanical Eng.*, 1979, **101**, 38-45.
[11]Lanir, Y., Constitutive equations for the Lung tissue, *Trans. ASME, J. Biomechanical Eng.*, 1983, **105**, 374-380.
[12]Stamenovic, D. & Wilson, T.A., A strain energy function for lung parenchyma, *Trans. ASME, J. Biomechanical Eng.*, 1985, **107**, 81-86.
[13]Budiansky, B. & Kimmel, E., Elastic moduli of lungs, *J. Applied Mechanics*, 1987, **54**, 351-358.
[14]Bachofen, H., Hildebrandt, J. & Bachofen, M., Pressure-volume curves of air- and liquid-filled excised lung: surface tension in situ, *J. Appl. Physiol.*, 1970, **29**(4), 422-431.
[15]Enhorning, G., Pulsating bubble technique for evaluating pulmonary surfactant, *J. Appl. Physiol.: Respirat. Environ. Exercise Physiol.*, 1977, **43**(2), 198-203.
[16]Schürch, S., Goerke, J. & Clements, J. A., Direct determination of volume- and time-dependence of alveolar surface tension in excised lungs, *Proc. Natl. Acad. Sci. USA*, 1978, **75**(7), 3417-3421.
[17]Wilson, T.A., Surface tension-surface area curves calculated from pressure-volume loops, *J. Appl. Physiol.: Respirat. Environ. Exercise Physiol.*, 1982, **53**(6), 1512-1520.
[18]Bachofen, H., Schürch, S. & Urbineli, M., Surfactant and alveolar micromechanics, *Basic Research of Lung Surfactant*, eds. Wichert, P. & Muller, B., Prog. Respir. Research., Basel Karger, 1990, **25**, 158-167.
[19]Clements, J.A. & Tierney, D.F., Alveolar instability associated with altered surface tension, *Handbook of Physiology*, eds. Fenn, W.O. & Rahn, H., Am. Physiological Society, Washington, D.C., 1965, **2**(3), 1565-1583.
[20]Fung, Y. C., Does the surface tension make the lung inherently unstable?, *Circulation Research*, 1975, **37**, 497-502.
[21]Stamenovic, D. & Wilson, T.A., Parenchymal stability, *J. Appl. Physiol.*, 1992, **73**(2), 596-602.
[22]Olson, D.E., Dart, G.A. & Filley, G.F., Pressure drop and fluid flow regime of air inspired into the human lung, *J. Appl. Physiol.*, 1970, **28**(4), 482-410.
[23]Pedley, T.J., Schroter, R.C. & Sudlow, M.F., Flow and pressure drop in systems of

repeatedly branching tubes, *J. Fluid Mechanics*, 1971, **46**(2), 365-383.
[24]Jaffrin, M.Y. & Kesic, P., Airway resistance: a fluid mechanical approach, *J. Appl. Physiol.*, 1974, **36**(3), 354-361.
[25]Elad, D., Kamm, R. D. & Shapiro, A. H., Tube law for the intrapulmonary airway, *J. Appl. Physiol.*, 1988, **65**(1), 7-13.
[26]Olson, L. E., Mechanical properties of small airways in excised pony lungs, *J. Appl. Physiol.*, 1992, **73**(2), 522-529.
[27]Piene, H., Pulmonary arterial impedance and right ventricular function, *Physiological Reviews*, 1986, **66**(3), 606-652.
[28]Yen, R.T. & Fung, Y.C., Model experiments on apparent blood viscosity and hematocrit in pulmonary alveoli, *J. Appl. Physiol.*, 1973, **35**, 510-517.
[29]Weibel, E.R., *Morphometry of the Human Lung*, Springer-Verlag, New York, 1963.
[30]Horsfield, K. & Cumming, G., Morphology of the bronchial tree in man, *J. Appl. Physiol.*, 1968, **24**, 373-383.
[31]Horsfield, K., Dart, G., Olson, D.E., Filley, G.F. & Cumming, G., Models of the human bronchial tree, *J. Appl. Physiol.*, 1971, **31**, 207-217.
[32]Singhal, S., Henderson, R., Horsfield, K., Harding, K. & Cumming, G., Morphometry of the human pulmonary arterial tree, *Circulation Research*, 1973, **33**, 190-197.
[33]Horsfield, K. & Gordon, W. I., Morphometry of pulmonary veins in man, *Lung*, 1981, **159**, 211-218.
[34]Yen, R.T., Zhuang, F.Y., Fung, Y.C., Ho, H.H., Tremer, H. & Sobin, S.S., Morphometry of cat pulmonary venous tree, *J. Appl. Physiol.: Respirat. Environ. Exercise Physiol.*, 1983, **55**(1), 236-242.
[35]Yen, R.T., Zhuang, F.Y., Fung, Y.C., Ho, H.H., Tremer, H. & Sobin, S.S., Morphometry of cat's pulmonary arterial tree, *Trans. ASME, J. Biomech. Eng.*, 1984, **106**, 131-136
[36]Paiva, M., Gas transport in the human lung, *J. Appl. Physiol..*, 1973, **35**(3), 401-410.
[37]Jebria, A.B., A mechanical approach to the longitudinal dispersion of gas flowing in human airways, *J. Biomechanics*, 1985, **18**(5), 399-405.
[38]Paiva, M. & Engel, L. A., Theoretical studies of gas mixing and ventilation distribution in the lung, *Physiological Reviews*, 1987, **67**(3), 750-796.
[39]Scherer, P.W., Shendalman, L.H., Greene, N.M. & Bouhuys,A., Measurement of axial diffusivities in a model of the bronchial airways, *J. Appl. Physiol..*, 1975, **38**(4), 719-723.
[40]Ultman, J.S., Gas mixing in the pulmonary airways, *Annals of Biomedical Eng.*, 1981, **9**, 513-527.
[41]Tarbell, J.M., Ultman, J.S. & Durlofsky, L., Oscillatory convective dispersion in a branching tube network, *Trans. ASME, J. Biomechanical Eng.*, 1982, **104**, 338-342.
[42]Kamm, R.D., Collins, J., Whang, J., Slutsky, A.S. & Greiner, M., Gas transport during oscillatory flow in a network of branching tubes, *Trans. ASME, J. Biomechanical Eng.*, 1984, **106**, 315-320.
[43]Fredberg, J.J., Augmented diffusion in the airways can support pulmonary gas exchange, *J. Appl. Physiol.: Respirat. Environ. Exercise Pysiol.*, 1980, **48**(3), 710-716.
[44]Drazen, J.M., Kamm, R.D. & Slutsky, A.S., High-frequency ventilation, *Physiological Reviews*, 1984, **64**(2), 505-543.
[45]Chang, H.K., Mechanisms of gas transport during ventilation by high-frequency oscillation, *J. Appl. Physiol.: Respirat. Environ. Exercise Physiol.*, 1984, **56**(3), 553-563.
[46]Forster, R.E., Exchange of gases between alveolar air and pulmonary capillary blood:

Pulmonary diffusing capacity, *Physiological Reviews*, 1957, **37**(4), 391-542.
[47]Milhorn, H.T. & Pulley, P.E., A theoretical study of pulmonary capillary gas exchange and venous admixture, *Biophysical J.*, 1968, **8**(3), 337-357.
[48]Hill, E.P., Power, G.G. & Longo, L.D., A mathematical model of carbon dioxide transfer in the placenta and its interaction with oxygen, *Am. J. Physiology*, 1973, **224**(2), 283-299.
[49]Wagner, P.D., Diffusion and chemical reaction in pulmonary gas exchange, *Physiological Reviews*, 1977, **57**(2), 257-312.
[50]Singh, M.P., Khetarpal, K. & Sharan, M., A theoretical model for studying the rate oxygenation of blood in pulmonary capillaries, *J. Mathematical Biology*, 1980, **9**, 305-330.
[51]Fung, Y.C. & Sobin, S.S., Theory of sheet flow in lung alveoli, *J. Appl. Physiol.*, 1969, **26**(4), 427-488.
[52]Notter, R.H. & Finkelstein, J.N., Pulmonary surfactant: an interdisciplinary approach, *J. Appl. Physiol.: Respirat. Environ. Exercise Physiol.*, 1984, **57**(6), 1613-1624.
[53]Pollack, G.H. & Reddy, R.V., Input impedance, wave travel, and reflections in the human pulmonary arterial tree: Studies using an electrical analog, *IEEE Trans. Biomedical Eng.*, 1968, **BME-15**(3), 151-164
[54]Piene, H., Influence of vessel distention and myogenic tone on pulmonary arterial input impedance: A study using a computer model of rabbit lung, *Acta physiol. scand.*, 1976, **98**, 54-66.
[55]Hill, E.P., Power, G.G. & Longo, L.D., Mathematical simulation of pulmonary O_2 and CO_2 exchange, *Am. J. Physiology*, 1973, **224**(4), 904-917.
[56]Fredberg, J.J. & Horning, A., Mechanical response of the lungs at high frequencies, *Trans. ASME, J. Biomechanical Engineering*, 1978, **100**, 57-65.
[57]Shore, D.L. & Tabrizi, M., Acoustic modeling of lung dynamics using bond graphs, *Trans. ASME, J. Biomech. Eng.*, 1983, **105**, 84-91.
[58]Hoshimiya, N., Tanaka, M. & Matsuo, T., Finite-length collapsible-tube of the check-valve in expiration, *Japanese J. Medical Electronics and Biological Eng.*, 1974, **12**(6), 41-48.
[59]Schmid-Schoenbein, G. W. & Fung, Y.C., Forced perturbation of respiratory system, *Annals of Biomedical Eng.*, 1978, **6**, 194-211 & 367-398.
[60]Zhuang, F.Y., Fung, Y.C. & Yen, R.T., Analysis of blood flow in cat's lung with detailed anatomical and elasticity data, *J. Appl. Physiol.: Respirat. Environ. Exercise Physiol.*, 1983, **55**(4), 1341-1348.
[61]Wada, S., Seguchi, Y., Tanaka, M. & Fung, Y.C., Simulation of respiratory dynamics considering breathing and ventilation, *1988 Advances in Bioengineering*, ed. Miller, G.R., ASME, New York, 1988, **BED-8**, 135-138.
[62]Seguchi, Y., Wada, S., Fung, Y.C. & Tanaka, M., Simulation of combined breathing and ventilation of a lung, *Progress and New Direction of Biomechanics*, eds. Fung, Y.C., Hayashi, K. & Seguchi, Y., Mita Press, Tokyo, 1989, 171-181.
[63]Wada, S., Seguchi, Y. & Tanaka, M., Simulation of breathing and ventilation combined with body circulation, *1990 Advances in Bioengineering*, ed. Goldstein, S.A., ASME, New York, 1990, **BED-17**, 67-70.
[64]Wada, S., Seguchi, Y. & Tanaka, M., Breathing-ventilation model, and simulation of high-frequency ventilation, *JSME International J.*, 1991, **I-34**(1), 98-105.
[65]Seguchi, Y., Wada, S., Tanaka, M. & Togari, H., Model, simulation and application on total respiration in terms of computational biomechanics, *Computational Mechanics*

'91, ICES Publications, Atlanta, 1991, 924-929.
[66]Wada, S., Seguchi, Y., Tanaka, M., Matsuda, M., Adachi, T. & Fujigaki, M., Parameter identification for respiratory dynamics by personalized simulation and experiment, *JSME International J.*, 1991, **I-35**(2), 170-178.
[67]Wada, S., Tanaka, M., Horikawa, T. & Nakamura, H., Computational respiratory mechanics approach to optimal artificial ventilation, *Computational Biomedicine*, eds. Held, K.D., Brebbia, C.A., Ciskowski, R.D. & Power, H., Computational Mechanics Publications, Southampton, 1993, 133-140.
[68]Rappitsch, G., Perktold, K. & Guggenberger, W., Numerical analysis of intramural stresses and blood flow in arterial bifurcation models, *Computational Biomedicine*, eds. Held, K.D., Brebbia, C.A., Ciskowski, R.D. & Power, H., Computational Mechanics Publications, Southampton, 1993, 149-156.
[69]Taylor, T. W. & Yamaguchi, T., Three-dimensional simulation of blood flow in an abdominal aortic aneurysm: Steady and unsteady flow cases, *Trans. ASME, J. Biomech. Eng.*, 1994, 116, 89-97.
[70]Zeng, Y.J., Yager, .D. & Fung, Y.C., Measurement of the mechanical properties of the human lung tissue, *Trans. ASME, J. Biomechanical Eng.*, 1987, **109**, 169-174.
[71]Vawter, D.L., Fung, Y.C. & West, J.B., Elasticity of excised dog lung parenchyma, *J. Appl. Physiol.: Respirat. Environ. Exercise Physiol.*, 1978, **45**(2), 261-269.
[72]Schürch, S., Surface tension at low lung volumes: dependence on time and alveolar size, *Respiration Physiology*, 1982, **48**, 339-355.
[73]Smith, J. C. & Stamenovic, D., Surface forces in lung: I. Alveolar surface tension-lung volume relationships, *J. Appl. Physiol.*, 1986, **60**(4), 1341-1350.
[74]Vawter, D. L. & Shields, W. H., Deformation of the lung: The role of interfacial forces, *Finite Elements in Biomechanics*, eds. Gallagher, R.H., Simon, B.R., Johnson, P.C. & Gross, J.F., John Wiley & Sons, Ltd, 1982, 84-110.
[75]Streeter, V.L. & Wylie, E.B., *Hydraulic Transients*, McGraw-Hill, New York, 1967.
[76]Seguchi, Y., Fung, Y.C. & Ishida, T., Respiratory dynamics-computer simulation, *In Frontiers in Biomechanics*, Eds. Schmid-Schönbein, G.W., Woo, S.L.Y. & Zweifach, B.M., Springer-Verlag, New York, 1986, 377-391.
[77]Yen, R.T., Fung, Y.C. & Bingham, N., Elasticity of small pulmonary arteries in the cat, *Trans. ASME, J. Biomechanical Eng.*, 1980, **102**, 170-177.
[78]Yen, R.T. & Foppiano, L., Elasticity of small pulmonary veins in the cat, *Trans. ASME, J. Biomechanical Eng.*, 1981, **103**, 38-42.
[79]Fung, Y.C., *Biomechanics: Motion, Flow, Stress, and Growth*, Springer-Verlag, New York, 1990.
[80]Fung, Y.C. & Sobin, S.S., Pulmonary alveolar blood flow, *Bioengineering Aspects of Lung Biology*, ed. West, J.B., Marcel Dekker, New York, 1977, 267-359.
[81]Taylor, G.I., Dispersion of soluble matter in solvent flowing slowly through a tube, *Proc. Royal Society of London*, 1953, **A219**, 186-203.
[82]Watson, E.J., Diffusion in oscillatory pipe flow, *J. Fluid Mech.*, 1983, **133**, 233-244.
[83]Hlastala, M.P., Significance of the Bohr and Haldane effects in the pulmonary capillary, *Respiration Physiology*, 1973, **17**, 81-92.
[84]Gomez, D.M., Considerations of oxygen-hemoglobin equilibrium in the physiological state, *Am. J. Physiol.*, 1961, **200**(1), 135-142.
[85]Staub, N.C., Bishop, J.M. & Forster, R.E., Importance of diffusion and chemical reactions in O_2 uptake in the lung, *J. Appl. Physiol.*, 1962, **17**, 21-27.

[86]Seguchi, Y., Fung, Y.C. & Maki, H., Computer simulation of dynamics of fluid-gas-tissue systems with a discretization procedure and its application to respiration dynamics, *In Biomechanics in China, Japan, and USA*, Eds. Fung, Y.C., Fukada, E. & Wang, J.J., Chinese Science Press, Beijing, 1984, 224-239.
[87]Greenfield, J.C. & Griggs, D.M., Relation between pressure and diameter in main pulmonary artery of man, *J. Appl. Physiol.*, 1963, **18**(3), 557-559.
[88]Caro, C.G. & McDonald, D.A., The relation of pulsatile pressure and flow in the pulmonary vascular bed, *J. Physiol.*, 1961, **157**, 426-453.
[89]Milnor, W.R., Conti, C.R., Lewis, K.B. & O'Rourke, M.F., Pulmonary arterial pulse wave velocity and impedance in man, *Circulation Research*, 1969, **25**(6), 637-649.
[90]Piene, H., The influence of pulmonary blood flow rate on vascular input impedance and hydraulic power in the sympathetically and noradrenaline stimulated cat lung, *Acta Physiol.*, 1976, **98**, 44-53
[91]Pinsky, M.R., The effects of mechanical ventilation on the cardiovascular system, *Critical Care Clinics*, 1990, **6**(3), 663-678.

Chapter 9

Micropolar fluid model for the brain fluid dynamics

H. Power*

Wessex Institute of Technology, University of Portsmouth, Ashurst Lodge, Ashurst, Southampton, SO40 7AA, UK

Abstract

The problem of determining the low Reynolds number, R_e, flow of the cerebrospinal fluid through the subarachnoid space passing around the brain and the spinal cord is formulated exactly as a system of linear Fredholm integral equations of the second kind. The formulation is based upon the assumption that the fluid belongs to the type described by the non–Newtonian theory of micropolar fluid. The resulting system of integral equation possesses a unique continuous solution, and it shows the different mechanisms that the brain have to control the flow in the subarachnoid space.

1 Introdution

In the central nervous system, there is a tissue fluid known as cerebrospinal fluid. This fluid is a crystal–clear, colorless, almost protein free solution which looks like water and is found in motion at the interior (ventricular system) and the exterior (subarachnoid space) of the brain and spinal cord. It also serves as a cushion between the central nervous system and the surrounding bones. The volume of cerebrospinal fluid in the average adult is estimated to be about 135 ml (75 to 150 ml), of

*On leave from Instituto de Mecanica de Fluidos, Universidad Central de Venezuela.

which roughly 80 ml is in the ventricles and 55 ml is in the subarachnoid space. Daily production of the fluid is roughly estimated at about 300 ml, giving a very low flow discharge. The 1400 g brain has a net weight of about 50 to 100 g while suspended in the cerebrospinal fluid, the specific gravities of the brain and the cerebrospinal fluid are 1.040 and 1.007, respectively. The cerebrospinal fluid is continuously formed at the choroid plexus in the ventricles, inside the brain, as a secretion produced out of blood coming from the pia vascular system.

The subarachnoid space is the interval between the arachnoid matter and pia matter and therefore is present where these meninges envelop the brain and spinal cord, which have a characteristic thickness of the order of one millimeter. This space is filled with moving cerebrospinal fluid and contains large blood vessels of the brain (see Fig. 1).

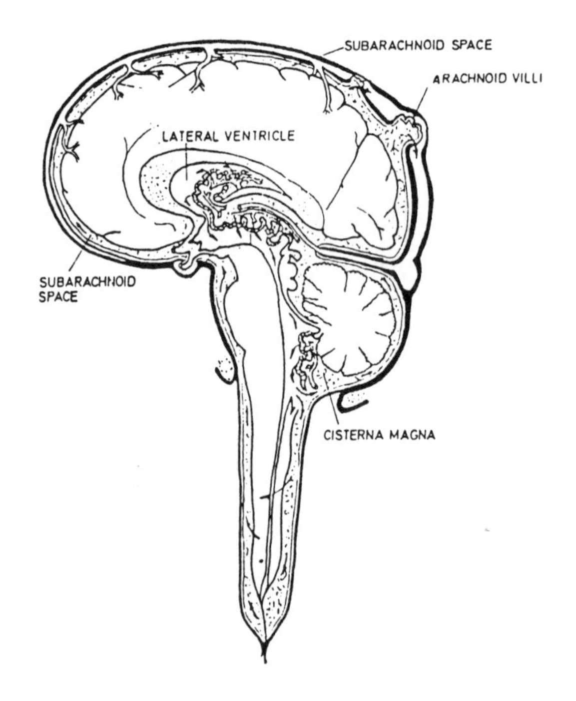

Figure 1: Illustration of the circulation of the cerebrospinal fluid

The cerebrospinal fluid leaves the fourth ventricle by way of certain foramina (*i.e.* three channels leading from the fourth ventricle to the cisterna magna, a large subarachnoid space over the medulla and below the posterior border of the cerebellum), into the subarachnoid space, where after passing out over the surface of the brain, moving in a very slow manner with characteristic Reynolds number much smaller than one, it enters the blood in the dural sinuses located at the dura matter. The drainage out of the subarachnoid space is considered through the arachnoid villi, which are essentially herniations of the arachnoid into the dura, and through these into the large blood sinuses enclosed in the dura (for more details on the neurobiologic aspect see Noback & Demarest[11]).

The main problems that have concerned physiologists in this field have been (see Davson[2])

a) the nature of the fluid and the mechanism of its formation,
b) the path follows during the circulation,
c) the mechanism of absorption of the fluid,
d) the relationship of the cerebrospinal fluid to the blood, and
e) the relationship of the cerebrospinal fluid to the nervous tissue.

Under normal conditions, the flow of the cerebrospinal fluid has to be such that the total force exerted by the fluid upon the brain be equal to the submerged weight of the brain, to keep the brain in its equilibrium position inside the cranial cavity. Power & Miranda[14] presented a mathematical model for the flow of the cerebrospinal fluid at the subarachnoid space based on an integral equation formulation for the Stokes' equation of a Newtonian fluid enclosed by two arbitrarily surfaces.

Theories considering the mechanical behavior of a fluid which possesses substructures, such as dilute polymer liquids, liquid crystals, blood, human fluids etc., for which the classical Navier–Stokes theory is inadequate, have been developed considerable. Classical continuum mechanics is based on the idea that all material bodies possess continuous mass densities, and that constitutive equations are for every part of the body, regardless of its size. However, this postulate is doubtful in the treatment of a fluid possessesing substructures. As the size of a material volume element approaches zero, it is found that, past a certain volume, the mass density begins to show a dependence on its volume and the continuity assumption for mass density is no longer applicable. Since the macroscopic limitation of the material volume element exists, the stress is no longer symmetric, and hence the arise of distributed couples per unit area across internal surfaces, i.e. couple stresses.

The inadequacy of the classical continuum approach to describe the mechanics of complex fluids has led to develop the theories of microcontinua in which continuous media possesses not only mass and velocity but also a substructure. One such theory of fluids is the specialized theory of Micropolar fluids advanced by Eringen[1]. Basically, these fluids can support couple stresses and body couples and exhibit micro–rotational effects. Fluid points contained in a small volume element, in addition to their usual rigid body motion, now posses the ability to rotate about the centroid of the volume element, in an average sense, described by an independent kinematic gyration vector $\vec{\nu}$.

In this chapter we present a mathematical model for the flow of the cerebrospinal fluid based on an integral equation formulation for the low Reynolds number approximation of micropolar fluid theory. The model has the possibility to show the different mechanisms that the brain has to control the flow in the subarachnoid space.

Integral equation technique offers an attractive method to be used as a basis for the numerical solution of a wide variety of problems concerning the slow motion of Micropolar fluid. Integral representations, analogous to those employed in potential theory, exist for Stokes flow of a Newtonian fluid and their use can be traced back to the work of Lorentz[10]; Odqvist[12] realized that the terms of the integral representation could be inspected separately and used to created Stokes solutions. He called these terms the single and double layer, and established its properties for closed surfaces. Odqvist proved that if we attempted to express the flow field in an unbounded domain outside a solid particle, solely in terms of a double–layer surface potential, the resulting Fredholm integral equation of the second kind, has solutions, if and only if, its nonhomogeneous term is orthogonal to each of the six eigenfunctions of the adjoint integral operator, which is not, generally, satisfied. In the general case, the flow field cannot be represented in terms of double–layer alone, but rather must be written as the sum of a double layer potential with unknown density, plus a linear combination of six single–layer potentials with the former eigenfunctions as densities. On the other hand, for the interior problem, where beside the non–slip boundary condition, the given boundary velocity has to satisfy a non–flux condition so the resuting interior velocity field is a solenoidal vector field, the general flow field can be given in terms of a double layer potential alone, since it satisfies the non–flux condition requirements (for more details about this two problems see Ladyzhenskaya[9] book, chapter 3). Although Odqvist's work was published as early as 1930, it has received little attention to date. In fact, his work was partly recaptured by Ladyzhenskaya[9], and sometimes his results have been attributed to her.

As observed by Youngren & Acrivos[18], Odqvist's and Ladyzhenskaya's approach for the exterior problem is not constructive, since the eigenfunctions are not explicitly known. One way around this difficulty, proposed by these authors, was the use of a first kind Fredholm integral equation formulation derived from an integral representation formulae of Green's type for the exterior flow field. As it is known, Fredholm integral equations of the first kind generally give rise to unstable numerical schemes based upon discretization of the surface integrals involved (see Goldberg[3] page 36), the instability manifesting itself in the ill-conditioning of the matrix approximation of the kernel. Nonetheless, it is possible to apply the discretization method if only low-order accuracy is desired and the system of linear equations to be solved is not too ill-conditioned. On the other hand, solving an equation of the second kind is a well-posed problem.

Power & Miranda[13] explained how integral equations of the second kind can be obtained for general unbounded three dimensional Newtonian Stokes flows around a particle, without recourse of eigenfunctions. They observed that although the double layer representation, that originates a second kind integral equation, can represent only those flow fields corresponding to a force and torque free surface, the representation may be completed by adding terms that give arbitrary total force and torque in suitable linear combinations, precisely a Stokeslet and a Rotlet, which are singular solutions of the Stokes' equation, located in the interior of the particle, and with strengths depending linearly upon the unknown double–layer density. Karrila & Kim[5] and Karrila, Fuentes & Kim[6] give an elegant mathematical interpretation of Power & Miranda's method. They observed that this method relates to Wielandt's deflation, by removing the end points of the spectrum of the integral operator of an integral equation of the second kind coming from a double layer representation without any completion. Such modification will allow a direct iterative solution, making the method suitable for a numerical scheme. Karrila & Kim called Power & Miranda's method the *Completed Double Layer Boundary Integral Equation Method*, since it involves the idea of completing the deficient range of the double layer operator. Power & Miranda[14] extended the completed double layer method to solve the problem of Newtonian Stoke's flow between two arbitrarily closed surfaces S and S_0. In this case the fluid moves exteriors to the contour S but it is contained by the contour S_0, therefore it appears as an exterior flow, looking it from the contour S, but as an interior one from contour S_0. The completed double layer formulation for this problem is not a trivial extension of the previous one, since the completeness procedure of the deficient range of the double layer potential due to the existence of an exterior container requires special care (for more detail see Kim & Karrila[7] book). This problem is physically more relevant than the previous two, because in nature most fluid bodies are contained by one or more contours.

In a series of papers published by Ramkissoon since 1980 (see for instances Ramkissoon[16], (1984)), he extended Odqvist's and Ladyzhenskaya's integral equation theory for a Newtonian fluid, to non–Newtonian Micropolar fluids, dealing with the cases of interior and exterior problem. As Odqvist's and Ladyzhenskaya's approach for the exterior problem, Ramkissoon's approach for this problem is not constructive, since it uses the corresponding eigenfunctions. Power & Ramkissoon[15] extended the completed double layer method for exterior Newtonian fluid to the case of exterior Micropolar fluid, in the next section we will describe the completed method for the problem of determining the slow flow motion of a Micropolar fluid exterior to a closed surface but contained by an exterior contour, that encloses it.

2 Statement of the problem

Let us consider a Micropolar fluid moving exterior to a closed surface S with a surrounding container S_0.

The equations of motion characterizing the slow steady flow of an incompressible Micropolar fluid (Stokes flow) are given by

$$\frac{\partial \sigma_{ij}(x)}{\partial x_j} = 0 \tag{1-a}$$

$$\frac{\partial m_{ij}(x)}{\partial x_j} + \varepsilon_{ijk}\sigma_{jk} = 0 \tag{1-b}$$

$$\frac{\partial u_i(x)}{\partial x_i} = 0 \tag{1-c}$$

while the linear constitutive laws take the form

$$\sigma_{ij}(\vec{u},\vec{\nu}) = -p\delta_{ij} + \frac{1}{2}(2\mu+\chi)\Big(\frac{\partial u_i}{\partial x_j} + \frac{\partial u_j}{\partial x_i}\Big) + \chi\varepsilon_{ijk}(w_k - \nu_k) \tag{2-a}$$

$$m_{ij}(\nu) = \alpha\frac{\partial \nu_k}{\partial x_k}\delta_{ij} + \beta\frac{\partial \nu_i}{\partial x_j} + \gamma\frac{\partial \nu_j}{\partial x_i} \tag{2-b}$$

In these equations σ_{ij} are the components of the Cauchy tensor, m_{ij} are the components of the couple stress tensor, $\vec{u}$ is the velocity vector, $\vec{\nu}$

is the microrotation vector, p is the pressure, ε_{ijk} is the alternative tensor, δ_{ij} is the Kronecker's delta, $(\alpha, \beta, \gamma, \mu, \chi)$ are constants characteristic of the particular fluid under consideration and

$$w_i = \frac{1}{2}\varepsilon_{ijk}\frac{\partial u_k}{\partial x_j} \tag{2-c}$$

is the vorticity vector.

In this work, we will present a constructive way of finding the solution of the system (1), (2), in the domain Ω, the bounded fluid domain with S as the interior contour and S_0 as the exterior one, subject to the following non–slip boundary conditions:

$$\begin{aligned} u_i(\xi) &= a_i^0(\xi) \\ \nu_i(\xi) &= b_i^0(\xi) \end{aligned} \qquad \text{for every } \xi \in S_0 \tag{3 – a}$$

and

$$\begin{aligned} u_i(\xi) &= a_i^1(\xi) \\ \nu_i(\xi) &= b_i^1(\xi) \end{aligned} \qquad \text{for every } \xi \in S \tag{3 – b}$$

where the given continuous velocity field a^l, for $l = 0, 1$, cannot be arbitrary, since $\vec{u}$ has to be a solenoidal vector field in Ω, then

$$\int_\Omega \frac{\partial u_j}{\partial x_j} dx = \int_{S_0} u_j n_j dS - \int_S u_j n_j dS = 0, \tag{4}$$

here $\vec{n}$ is the unit normal vector outwardly directed at points of S_0 and S, so that a necessary condition for our problem to have a solution in Ω is that

$$\int_{S_0} a_j^0 n_j dS = \int_S a_j^1 n_j dS. \tag{5}$$

The above boundary conditions, together with the solenoidal compatibility condition, and assuming that the datum $\vec{a}^l(\xi)$ and $\vec{b}^l(\xi)$, for $l = 0, 1$, are continuous on $S_0 \cup S$, and S_k, for $k = 0, 1$, are Lyapunov surfaces, suffices to guarantee the uniqueness of the fields $\vec{u}(x)$, $\vec{\nu}(x)$ and the corresponding pressure gradient solution of the system of partial differential eqns (1)-(2).

The analogue of Green's second identity corresponding to this problem, energy relation, is given by (see Ramkissoon[16]):

$$\int_\Omega H \; dx = \int_{S_0 \cup S} \tau_i^{\dagger} U_i dS_y, \tag{6-a}$$

where

$$H = (2\mu+\chi)\left(\frac{\partial u_i}{\partial x_j}+\frac{\partial u_j}{\partial x_i}\right)\left(\frac{\partial u_j}{\partial x_i}+\frac{\partial u_i}{\partial x_j}\right)+2\chi(w_j-\nu_j)(w_j-\nu_j)+ \left(\alpha\frac{\partial \nu_j}{\partial x_j}\frac{\partial \nu_i}{\partial x_i}+\beta\frac{\partial \nu_j}{\partial x_i}\frac{\partial \nu_i}{\partial x_j}+\gamma\frac{\partial \nu_i}{\partial x_j}\frac{\partial \nu_j}{\partial x_i}\right) \tag{6-b}$$

and, from thermodynamic considerations (see Eringen[1])

$$\left(\alpha\frac{\partial \nu_j}{\partial x_j}\frac{\partial \nu_i}{\partial x_i}+\beta\frac{\partial \nu_j}{\partial x_i}\frac{\partial \nu_i}{\partial x_j}+\gamma\frac{\partial \nu_i}{\partial x_j}\frac{\partial \nu_j}{\partial x_i}\right)\geq 0, \tag{6-c}$$

where, we introduced the vector notation

$$\tau_i = \begin{pmatrix} \sigma_{ij}(\vec{u},\vec{\nu})n_j \\ m_{ij}(\vec{\nu})n_j \end{pmatrix}, \qquad U_i = \begin{pmatrix} u_i \\ \nu_i \end{pmatrix} \tag{6-d}$$

From the above energy relation, it is clear that the present boundary value problem, with a given velocity and Micropolar rotation vector at each of the involved contours, cannot have more than one solution. In fact, it follows from eqns (6) that if $\vec{u}$ and $\vec{\nu}$ correspond to an homogeneous boundary condition, *i.e* $\vec{U}=0$, at each of the contours S_k, $k=0,1$, then

$$\begin{aligned} &(2\mu+\chi)\left(\frac{\partial u_i}{\partial x_j}+\frac{\partial u_j}{\partial x_i}\right)=0 \\ &(w_j-\nu_j)(w_j-\nu_j)=0 \\ &\left(\alpha\frac{\partial \nu_j}{\partial x_j}\frac{\partial \nu_i}{\partial x_i}+\beta\frac{\partial \nu_j}{\partial x_i}\frac{\partial \nu_i}{\partial x_j}+\gamma\frac{\partial \nu_i}{\partial x_j}\frac{\partial \nu_j}{\partial x_i}\right)=0 \end{aligned} \tag{7}$$

The above system has six linearly independent solutions, which are the following basic fluid motion vectors:

$$U_k^1 = \begin{pmatrix} \delta_{1k} \\ 0 \end{pmatrix} \qquad U_k^2 = \begin{pmatrix} \delta_{2k} \\ 0 \end{pmatrix} \qquad U_k^3 = \begin{pmatrix} \delta_{3k} \\ 0 \end{pmatrix} \tag{8-a}$$

$$U_k^4 = \begin{pmatrix} \epsilon_{1kj}x_j \\ -\delta_{1k} \end{pmatrix} \qquad U_k^5 = \begin{pmatrix} \epsilon_{2kj}x_j \\ -\delta_{2k} \end{pmatrix} \qquad U_k^6 = \begin{pmatrix} \epsilon_{3kj}x_j \\ -\delta_{3k} \end{pmatrix}, \tag{8-b}$$

corresponding to a rigid body motion velocity field and basic Micropolar rotation. Therefore, the only fluid motion compatible with the homogeneous boundary condition on the contours is the trivial solution $\vec{u}=\vec{\nu}=0$.

For the case of a flow field exterior to a closed surface S in an unbounded domain, Ramkissoon[16] proved that if we attempt to express $\vec{u}(x)$ and $\vec{\nu}(x)$ solely in terms of the following double layer potentials:

$$U_i(x) = W_i(x, \vec{\phi}) = \int_S K_{ij}(x,y)\phi_j(y)\, dS_y, \tag{9}$$

the corresponding second kind integral equation, for the unkonwn density $\vec{\phi}$, will have a solution, if and only if, the given boundary velocity and Micropolar rotational vectors satisfy certain orthogonality conditions, integral equation obtained by requiring that the double layer potential satisfies the given non–slip boundary condition, at the surface S. As we will see in the next section such cases are physically irrelevant, corresponding to dynamically neutral flow fields.

In eqn(9), we have introduced the vector and matrix

$$\phi_i = \begin{pmatrix} \phi_i^1 \\ \phi_i^2 \end{pmatrix} \quad \text{and} \quad K_{ik}(x,y) = \begin{pmatrix} F\sigma_{ij}^k n_j(y) & Fm_{ij}^k n_j(y) \\ L\sigma_{ij}^k n_j(y) & Lm_{ij}^k n_j(y) \end{pmatrix} \tag{10-a}$$

here $\vec{n}(y)$ denotes the unit normal to S at y directed outwardly from Ω_i, the domain interior to the surface S and $F\sigma_{ij}^k$, Fm_{ij}^k, $L\sigma_{ij}^k$, and Lm_{ij}^k are the components of the stresses associated with the fundamental solutions of the system (1)–(2), that is

$$\begin{aligned} F\sigma_{ij}^k(x,y) &= \sigma_{ij}(F\vec{u}^k, F\vec{\nu}^k)_y, \qquad & Fm_{ij}^k n_j(x,y) &= m_{ij}(F\vec{\nu}^k)_y \\ L\sigma_{ij}^k(x,y) &= \sigma_{ij}(L\vec{u}^k, L\vec{\nu}^k)_y, \qquad & Lm_{ij}^k n_j(x,y) &= m_{ij}(L\vec{\nu}^k)_y \end{aligned} \tag{10-b}$$

where the subscript y in the Cauchy and couple stresses denotes the derivative with respect to the y variables, and $(F\vec{u}^k(x,y), F\vec{\nu}^k(x,y))$, and $(L\vec{u}^k(x,y), L\vec{\nu}^k(x,y))$ are the velocity and Micropolar rotational vectors of the fundamental singular solutions due to a concentrated force, i.e. solution of the nonhomogeneous system with $-\delta(x-y)\vec{e}^k$ as a forcing function in eqn (1-a), and a concentrated couple, solution of the nonhomogeneous system with $-\delta(x-y)\vec{e}^k$ as a forcing function in eqn (1-b), respectively, defined as follows

$$
\begin{aligned}
F\vec{u}^k &= \frac{1}{4\pi\sigma^2(2\mu+\chi)}\{\frac{\chi}{\gamma}\frac{\vec{e^k}}{r} + \frac{\chi}{\gamma}\frac{(\vec{e^k}\cdot\vec{r})}{r^3}\vec{r} + \\
&\quad \frac{(2\chi-\gamma\sigma^2)}{\gamma\sigma^2}(\nabla\times\nabla\times\vec{e^k}\frac{(1-e^{-\sigma r})}{r}\} \\
F\vec{\nu}^k &= \frac{1}{4\pi(2\mu+\chi)}\vec{e^k}\times\nabla\frac{(1-e^{-\sigma r})}{r} \\
Fp^k &= \frac{1}{4\pi}\frac{(\vec{e^k}\cdot\vec{r})}{r^3}
\end{aligned} \tag{11}
$$

and

$$
\begin{aligned}
L\vec{u}^k &= \frac{1}{4\pi(2\mu+\chi)}\vec{e^k}\times\nabla\frac{(1-e^{-\sigma r})}{r} \\
L\vec{\nu}^k &= \frac{\vec{e^k}}{4\pi\gamma}\frac{e^{-\sigma r}}{r} + \frac{\nabla\nabla\cdot\vec{e^k}}{4\pi\gamma(\mu+\chi)(\alpha+\beta+\gamma)}\{\frac{\chi^2}{\sigma^2\theta^2}\frac{1}{r} + \\
&\quad \frac{\chi^2-\sigma^2(\alpha+\beta)(\mu+\chi)}{\sigma^2(\sigma^2-\theta^2)}\frac{e^{-\sigma r}}{r} + \frac{\chi^2-\theta^2(\alpha+\beta)(\mu+\chi)}{\theta^2(\theta^2-\lambda^2)}\frac{e^{-\theta r}}{r}\} \\
Lp^k &= 0
\end{aligned} \tag{12}
$$

where

$$
\sigma^2 = \frac{\chi(2\mu+\chi)}{\gamma(\mu+\chi)}, \qquad \theta^2 = \frac{2\chi}{\alpha+\beta+\gamma}, \qquad \text{and} \qquad r = |x-y|
$$

According to Ramkissoon's results, the Fredholm integral eqn (13) of the second kind, for the unknown density $\vec{\phi}$

$$
U_i(\xi) = \frac{1}{2}\phi_i(\xi) + \int_S K_{ij}(\xi,y)\phi_j(y)\,dS_y \qquad \text{for every} \qquad \xi\in S \tag{13}
$$

has solutions, if and only if, its right–hand side $\vec{U}(\xi)$ is orthogonal to each of the six eigenfunctions of the adjoint integral operator, where

$$
U_i(\xi) = \begin{pmatrix} a_i(\xi) \\ b_i(\xi) \end{pmatrix}. \qquad \xi\in S
$$

That is to say, it has solutions, if and only if

$$
\int_S U_i(y)\Psi_i^k(y)\,dS_y = 0 \tag{14}
$$

where $\vec{\Psi}^k$, $k = 1, 2, \cdots 6$ are linearly independent solutions of

$$\frac{1}{2}\Psi_i^k(\xi) + \int_S K_{ji}(y,\xi)\Psi_j^k(y)\, dS_y = 0 \qquad \text{for every} \qquad \xi \in S \qquad (15)$$

which are known to be six, since the following homogeneous adjoint integral equation

$$\frac{1}{2}\Phi_i^k(\xi) + \int_S K_{ij}(\xi,y)\Phi_j^k(y)\, dS_y = 0 \qquad \text{for every} \qquad \xi \in S \qquad (16)$$

has just six linearly independent solutions given by basic fluid motion (eqn (8)).

In the general case, where $U_i(\xi)$ is arbitrary, $u_i(x)$ and $\nu_i(x)$ cannot be represented in terms of double–layer potentials alone, but rather must be written as a linear combination of the six single–layer potentials with the eigensolutions $\vec{\Psi}^k$, as densities, plus a double–layer potential with unknown density $\vec{\phi}$ (see Ramkissoon[16]):

$$U_i(x) = W_i(x, \vec{\phi}) + \sum_{k=1}^{6} V_i(x, \vec{\Psi}^k), \qquad (17)$$

where

$$V_i(x, \vec{\Psi}^k) = \int_S J_i^j(x,y)\Psi_j^k(y) dS_y \qquad (18)$$

is a single layer potential with vector density $\vec{\Psi}^k$, and

$$J_i^k = \begin{pmatrix} Fu_i^k & F\nu_i^k \\ Lu_i^k & L\nu_i^k \end{pmatrix},$$

as we will see in the next section a double layer potential $\vec{W}(x, \vec{\phi})$ will not exert total force and torque upon its density carrying surface, and therefore it is dynamically neutral, however, a single layer yields total force and torque different from zero. Therefore, the representation formulae given by eqn (17) is able to exert the corresponding force and torque produced by a general Stokes' flow exterior to a closed surface.

As can be observed, Ramkissoon's approach for the exterior problem is not constructive, since the eigenfunctions, $\vec{\Psi}^k$, used as densities in the six single–layer potentials are not explicitly known. More recently, Power & Ramkissoon[15] explained how integral equations of the second kind can be obtained for general Micropolar flows exterior to a single close surface. They observed that, although the double layer representation that leads to a second kind integral equation coming from the jump property of its velocity and microrotation fields across the density carrying surface can represent only those fields corresponding to a force and torque free surface, the representation may be completed by adding terms that give arbitrary total force and torque in suitable linear combination.

For the case of a flow fields interior to a single closed surface S, where it is necessary that the given boundary velocity satisfies the following non–flux condition

$$\int_S a_i n_i dS = 0,$$

and the resulting interior flow field will not exert total force and torque upon its container, Ramkissoon[16] showed that the flow field can be given solely in terms of the double layer potential (9). In this case the resulting Fredholm integral equation of the second kind,

$$U_i(\xi) = -\frac{1}{2}\phi_i(\xi) + \int_S K_{ij}(\xi, y)\phi_j(y)\, dS_y \qquad \text{for every} \qquad \xi \in S \quad (19)$$

will have a solution, if and only if, the nonhomogeneous term is orthogonal to the only eigenfunction solution, $\vec{\Phi}$, of the homogeneous adjoint equation

$$0 = -\frac{1}{2}\Phi_i(\xi) + \int_S K_{ji}(y, \xi)\Phi_j(y)\, dS_y \qquad \text{for every} \qquad \xi \in S \qquad (20)$$

for which, it is known that its only nontrivial solution has a velocity field proportional to the normal vector and zero Micropolar rotation (Ramkissoon[17],

$$\Phi_i(\xi) = \begin{pmatrix} n_i(\xi) \\ 0 \end{pmatrix},$$

and therefore the previous orthogonality condition is always satisfied, since the given boundary velocity has to satisfy the required non–flux condition.

However, according to Fredholm's alternative the solution of the integral equation (19) is not unique, and hence $\phi_i(\xi) = \phi_i^p(\xi) + C\Psi_i(\xi)$, for any arbitrary constant C, here $\vec{\phi}^p$ is a particular solution and $\vec{\Psi}$ is the only eigenfunction solution of the homogeneous equation

$$-\frac{1}{2}\Psi_i(\xi) + \int_S K_{ij}(\xi, y)\Psi_j(y)\, dS_y = 0 \qquad \text{for every} \qquad \xi \in S$$

which is, in general, unknown, but the corresponding flow field at an interior point $x \in \Omega_i$, obtained by substituting the above density function on the double layer potential (9),

$$U_i(x) = \int_S K_{ij}(x, y)\phi_j^p(y)dS_y + C\int_S K_{ij}(x, y)\Psi_j(y)dS_y \tag{21}$$

is uniquely determined, since every double layer potential with the vector $\vec{\Psi}$ as a density function is identically equal to zero at points of the interior domain.

In spite of the complex form of the kernels in (10), their asymptotic form for small r, is:

$$\sigma_{ij}(F\vec{u}^k(x,y), F\vec{v}^k(x,y))n_k(y) = \left\{\frac{\chi\delta_{ij}}{8\pi(\mu+\chi)}\left(\frac{(x_k - y_k)}{r^3}\right) - \frac{\chi^2\delta_{jk}}{8\pi(\mu+\chi)(2\mu+\chi)}\left(\frac{(x_i - y_i)}{r^3}\right) + \frac{\chi(3\chi+4\mu)\delta_{ki}}{8\pi(\mu+\chi)(2\mu+\chi)}\left(\frac{(x_j - y_j)}{r^3}\right) - \frac{3(2\mu+\chi)}{8\pi(\mu+\chi)}\left(\frac{(x_i - y_i)(x_j - y_j)(x_k - y_k)}{r^5}\right) + O(r^{-1})\right\}n_k(y) \tag{22 - a}$$

$$m_{ij}(F\vec{v}^k(x,y))n_k(y) = O(r^{-1})n_k(y) \tag{22 - b}$$

$$\sigma_{ij}(L\vec{u}^k(x,y), L\vec{v}^k(x,y))n_k(y) = O(r^{-1})n_k(y) \tag{22 - c}$$

$$\sigma_{ij}(L\vec{u}^k(x,y), L\vec{v}^k(x,y))n_k(y) = \left\{\frac{(\alpha+\beta)(\beta+\gamma)\delta_{ij}}{8\pi\gamma(\alpha+\beta+\gamma)}\left(\frac{(x_k - y_k)}{r^3}\right) + \frac{(\alpha\beta+\beta^2-\alpha\gamma-\gamma\beta)\delta_{jk}}{8\pi\gamma(\alpha+\beta+\gamma)}\left(\frac{(x_i - y_i)}{r^3}\right) + \frac{(\alpha\gamma-\alpha\beta-\beta^2-\gamma\beta)\delta_{ki}}{8\pi\gamma(\alpha+\beta+\gamma)}\left(\frac{(x_j - y_j)}{r^3}\right) - \frac{3(\alpha+\beta)(\gamma+\beta)}{8\pi\gamma(\alpha+\beta+\gamma)}\left(\frac{(x_i - y_i)(x_j - y_j)(x_k - y_k)}{r^5}\right) + O(r^{-1})\right\}n_k(y) \tag{22 - d}$$

showing, that the integrals in the corresponding integral equations are singular in the sense of Cauchy principal value, with its most singular part of the form $(x_l - y_l)n_k(y)/r^3 \sim O(1/r^2)$, because the terms of the form

$$\frac{(x_i - y_i)(x_j - y_j)(x_k - y_k)}{r^5} n_k(y) = \frac{\partial r}{\partial x_i} \frac{\partial r}{\partial x_j} \frac{1}{r^2} \frac{\partial r}{\partial n_y}$$

are weakly singular, since it is known that in a Lyapunov surface the following inequality holds $|\partial r/\partial n_y| < Er^{\lambda}$, with $0 < \lambda \leq 1$, and then $(\partial r/\partial n_y/r^2 \sim O(1/r^{2-\lambda}))$.

The asymptotic limit of the kernels, for small r, implies that the corresponding systems of integral equations, discussed before, are systems of singular integral equations of Cauchy type, for which, in general, the theory of Fredholm integral equations does not apply, since the latter requires compact operators. In the original works of Rankisoon dealing with the interior and exterior problems, the application of Fredholm's theory has been taken for granted, and is non–trivial. However, the above asymptotic form of the kernels shows that our double layer potentials behave as the corresponding ones in the theory of elasticity, then, Kupradze's[8] regularization procedure to show that the fundamental Fredholm theorems and its alternative are valid for the singular system in elasticity, are still valid in the present case. Since, by adding and subtracting the terms (22) to the corresponding kernels in the integral equations, the above systems, for the interior and exterior problems, reduce to regular ones plus singular terms of elasticity type (for more details about this remotion of singularity see Power & Rankissoon[15]) .

In Section 4, we will present a constructive way of finding the solution for the problem of the slow motion of a Micropolar fluid exterior to a closed surface, but the fluid contained by an outer boundary. It is important to remember that this problem is a combination of interior and exterior ones. This is achieved by means of a second kind Fredholm integral equation without recourse to eigenfunctions. The approach presented in Section 4 is an extension to the Stokes flow of a Micropolar fluid of the completed double layer boundary integral equation method developed by Power & Miranda[14] for the equivalent problem of Newtonian fluid. In the present case, we will seek a solution for the flow field $(\vec{u}, \vec{\nu})$ as a double–layer potential, defined over the union of all the surfaces involved, with unknown vector density $\vec{\phi}$ plus singularities located in the domain interior to interior surface, Ω_i, singularities which give rise to force and torque, including the effect of the couple stress, with magnitude depending also on the unknown double–layer density $\vec{\phi}$ in a linear manner. In this way the boundary conditions (3) are satisfied, if and only if, a density $\vec{\phi}$ can be found satisfying a system of nonhomogeneous second kind Fredholm integral equations coming from the jump boundary properties of double–layer potentials. It will be shown that this system of integral equations possesses a unique continuous solution $\vec{\phi}$, and the proposed form of solution provides the unique solution of the present boundary value problem with arbitrary continuous datum $\vec{U}$ on Lyapunov boundary surfaces S_l, for $l = 0, 1$. In the following section we will analyze several dynamics properties of the fundamental singular solutions of the system of eqns (1)–(2), and the corresponding single and double layer potentials, relevant to the proof of existence and uniqueness of solution for the resulting system of Fredholm integral equations of the second kind. In the last section of this chapter the present formulation will be used to describe the motion of the cerebrospinal fluid through the subarachnoid space passing around the brain and the spinal cord.

3 Dynamic properties of the fundamental solutions and the corresponding single and double layer potentials

The starting point for the development of the singular solutions are the following inhomogeneous equations

$$\frac{\partial \sigma_{ij}(\vec{u}^l(x,y), \vec{\nu}^l(x,y))}{\partial x_j} = \alpha_i^l \delta(x-y) \qquad (23-a)$$

$$\frac{\partial m_{ij}(\vec{u}^l(x,y), \vec{\nu}^l(x,y))}{\partial x_j} + \varepsilon_{ijk}\sigma_{jk}(\vec{u}^l(x,y), \vec{\nu}^l(x,y)) = \beta_i^l \delta(x-y) \quad (23-b)$$

$$\frac{\partial u_i^l(x,y)}{\partial x_i} = 0. \qquad (23-c)$$

The i-th component of the total force exerted at the boundary S, by a Micropolar fluid in the interior domain Ω_i, is given by the following integral:

$$F_i = \int_S \sigma_{ij}(\vec{u}, \vec{\nu}) n_j dS_x. \qquad (24)$$

This integral, over a closed surface, can be transformed by the divergence theorem, to the following volume integral

$$\int_S \sigma_{ij} n_j dS_x = \int_{\Omega_i} \frac{\partial \sigma_{ij}}{\partial x_j} dx. \qquad (25)$$

Substituting eqn (23-a) into the above equation, with y located in-side Ω_i, and using the customary method to deal with the Dirac delt function, we obtain:

$$F_i^l = \int_S \sigma_{ij}(\vec{u}^l, \vec{\nu}^l) n_j dS_x = \alpha_i^l. \quad (26$$

Now, the i-th component of the total torque exerted at the boundar S, by a Micropolar fluid in the interior domain Ω_i, that includes the effec of the couple stress, is given by

$$T_i = \int_S (\varepsilon_{ijk} x_j \sigma_{kp}(\vec{u}, \vec{\nu}) n_p + m_{ij}(\vec{u}, \vec{\nu}) n_j) dS_x. \quad (27$$

where x_j is the j-th component of the position vector of the surfac element $\vec{n}\delta S$ relative to the origin, chosen inside the domain Ω_i. A before, this integral, over a closed surface, can be transformed by th divergence theorem, to the volume integral:

$$\int_S (\varepsilon_{ijk} x_j \sigma_{kp} n_p + m_{ij} n_j) dS_x = \int_{\Omega_i} \{\varepsilon_{ijk} \frac{\partial}{\partial x_p}(x_j \sigma_{kp}) + \frac{\partial m_{ij}}{\partial x_j}\} dx$$
$$= \int_{\Omega_i} \{\varepsilon_{ijk} x_j \frac{\partial \sigma_{kp}}{\partial x_p} + \varepsilon_{ijk} \sigma_{kj} + \frac{\partial m_{ij}}{\partial x_j}\} dx. \quad (28$$

Substituting eqns (23-a) and (23-b) into the above equation, with located inside Ω_i, we obtain

$$T_i = \int_S (\varepsilon_{ijk} x_j \sigma_{kp}(\vec{u}^l, \vec{\nu}^l) n_p + m_{ij}(\vec{u}^l, \vec{\nu}^l) n_j) dS_x = \varepsilon_{ijk} y_j \alpha_k^l + \beta_i^l. \quad (29$$

Therefore, a fundamental solution $(F\vec{u}^l(x,y), F\vec{\nu}^l(x,y))$, (given b eqn (11) and located at the point y) corresponding to the case of con centrated force, i.e. $\vec{\alpha}^l \neq 0$ and $\vec{\beta}^l = 0$, exerts a total force equal t $\vec{\alpha}^l$ and a total torque equal to $\varepsilon_{ijk} y_j \alpha_k^l \vec{e}^i$, then when the force funda mental solution, is located at the origin, i.e. $y = 0$, the correspond ing torque is identically zero. In the same way, a fundamental solutio $(L\vec{u}^l(x,y), L\vec{\nu}^l(x,y))$ (given by eqn (12) and located at the point y) co responding to the case of concentrated couple, i.e. $\vec{\alpha}^l = 0$ and $\vec{\beta}^l \neq$ (exerts a total force equal to zero and a total torque equal to $\vec{\beta}^l$.

Let us replace $(\vec{u}, \vec{\nu})$, in eqns (25) and (28), by some solution in th interior domain Ω_i of the system (1)–(2), that we will call interior regula solution. In this case, $\vec{\alpha}^l = 0$ and $\vec{\beta}^l = 0$, then any regular interior flow does not exert total force or torque on S, and the same will be true for any regular exterior flow that can be continued as a regular interior flow, with continuous Cauchy and couple surface tractions, $\sigma_{ij}n_j$ and $m_{ij}n_j$, across S. This will be certainly the case when such an exterior flow can be represented as a double layer potential with vector density $\vec{\phi}$, which is known to produce velocity and Micropolar rotational fields discontinuous across the surface S (see Ramkissoon[16]), whose limiting values are:

$$W_i(\xi,\vec{\phi})_{(i)} = -\frac{1}{2}\phi_i(\xi) + \int_S K_{ij}(\xi,y)\phi_j(y)\,dS_y \qquad \text{for every} \qquad \xi \in S \tag{30-a}$$

and

$$W_i(\xi,\vec{\phi})_{(e)} = \frac{1}{2}\phi_i(\xi) + \int_S K_{ij}(\xi,y)\phi_j(y)\,dS_y \qquad \text{for every} \qquad \xi \in S \tag{30-b}$$

where $\vec{W}(\xi,\vec{\phi})_{(i)}$ and $\vec{W}(\xi,\vec{\phi})_{(e)}$ denote the limiting values of the double layer $\vec{W}(x,\vec{\phi})$ on S, as S is approached from inside and outside Ω_i, respectively. In other words,

$$W_i(\xi,\vec{\phi})_{(e)} - W_i(\xi,\vec{\phi})_{(i)} = \phi_i(\xi). \tag{30-c}$$

On the other hand, it can be proven that the inner and outer limiting values of the Cauchy and couple surface tractions of the double layer potential coincide on S, i.e. $\tau_i(\vec{W}(\xi,\vec{\phi}))_{(i)} = \tau_i(\vec{W}(\xi,\vec{\phi}))_{(e)}$ (see Appendix A), where as before (i) and (e) denote the limiting values at the surface S, approaching it from inside and outside Ω_i, respectively, and

$$\tau_i(\;) = \begin{pmatrix} \sigma_{ij}(\;)_\xi \;\; n_j(\xi) \\ m_{ij}(\;)_\xi \;\; n_j(\xi) \end{pmatrix},$$

here the subscript ξ in the Cauchy and couple stresses denote the derivative with respect to the ξ variables.

Since the interior flow due to a double layer potential does not exert total force and torque on S, it follows that:

$$F_i = \int_S \sigma_{ij}(\vec{W}(y,\vec{\phi}))_{(e)}n_j(y)dS_y = \int_S \sigma_{ij}(\vec{W}(y,\vec{\phi}))_{(i)}n_j(y)dS_y \equiv 0 \tag{31-a}$$

and

$$T_i = \int_S \{\varepsilon_{ijk} y_j \sigma_{kp}(\vec{V}(y,\vec{\phi}))_{(e)} n_p(y) + m_{ij}(\vec{V}(y,\vec{\phi}))_{(e)} n_j(y)\} dS_y$$
$$= \int_S \{\varepsilon_{ijk} y_j \sigma_{kp}(\vec{V}(y,\vec{\phi}))_{(i)} n_p(y) + m_{ij}(\vec{V}(y,\vec{\phi}))_{(i)} n_j(y)\} dS_y \equiv 0. \quad (31-b)$$

Then, the exterior flow represented by a double layer potential, which is a regular exterior Micropolar flow, does not exert a total force and torque on S, and hence it is dynamically neutral.

On the other hand, a single layer potential with density $\vec{\phi}$ is well defined and continuous throughout the entire space, it has a Cauchy and couple surface tractions which experience a jump across S, given by (see Ramkissoon[16])

$$\tau_i(\vec{V}(\xi,\vec{\phi}))_{(i)} = \frac{1}{2}\phi_i(\xi) + \int_S K_{ji}(y,\xi)\phi_j(y)\, dS_y \qquad \text{for every} \qquad \xi \in S \quad (32-a)$$

and

$$\tau_i(\vec{V}(\xi,\vec{\phi}))_{(e)} = -\frac{1}{2}\phi_i(\xi) + \int_S K_{ji}(y,\xi)\phi_j(y)\, dS_y \qquad \text{for every} \qquad \xi \in S, \quad (32-b)$$

or equivalently:

$$\tau_i(\vec{V}(\xi,\vec{\phi}))_{(i)} - \tau_i(\vec{V}(\xi,\vec{\phi}))_{(e)} = \phi_i \qquad \text{when} \qquad \xi \in S \qquad (32-c)$$

where we have introduce the notation

$$\phi_i = \begin{pmatrix} \phi_i^1 \\ \phi_i^2 \end{pmatrix}, \quad K_{ki}(y,\xi) = \begin{pmatrix} \sigma_{kj}(F\vec{u}^i, F\vec{v}^i)_\xi \; n_j(\xi) & m_{kj}(F\vec{v}^i)_\xi \; n_j(\xi) \\ \sigma_{kj}(L\vec{u}^i, L\vec{v}^i)_\xi \; n_j(\xi) & m_{kj}(L\vec{v}^i)_\xi \; n_j(\xi) \end{pmatrix},$$

Therefore, since the interior flow due to a single layer potential does not exert total force or torque on S, the exterior flow represented by that single layer potential, which is a regular exterior Micropolar flow, exerts a total force and torque equal to:

$$F_i = \int_S \sigma_{ij}(\vec{V}(y,\vec{\phi}))_{(e)} n_j(y) dS_y =$$
$$\int_S \{\sigma_{ij}(\vec{V}(y,\vec{\phi}))_{(i)} n_j(y) - \phi_i^1(y)\} dS_y = -\int_S \phi_i^1(y) dS_y \quad (33)$$

and

$$\begin{aligned} T_i &= \int_S \{\varepsilon_{ijk} y_j \sigma_{kp}(\vec{V}(y,\vec{\phi}))_{(e)} n_p(y) + m_{ij}(\vec{V}(y,\vec{\phi}))_{(e)} n_j(y)\} dS_y \\ &= \int_S \{\varepsilon_{ijk} y_j (\sigma_{kp}(\vec{V}(y,\vec{\phi}))_{(i)} n_p(y) - \phi^1_k(y)) + \\ &\quad (m_{ij}(\vec{V}(y,\vec{\phi}))_{(i)} n_j(y) - \phi^2_i(y))\} dS_y \\ &= -\int_S (\varepsilon_{ijk} y_j \phi^1_k(y) + \phi^2_i(y)) dS_y. \end{aligned} \tag{34}$$

4 Integral equation formulation without recourse to eigenfunctions

Following Power & Miranda's[14] completed method, we will seek the solution $(\vec{u}, \vec{\nu})$ of the above boundary value problem, in the form of a double-layer potential, defined over the union of the sufaces involved, with unknown density $\vec{\phi}$ continuous on $S_0 \cup S$, plus a fundamental singular solutions due to concentrated forces, $(F\vec{u}^k(x,y), F\vec{\nu}^k(x,y))$, located in the domain bounded by the interior surface, singularity that exerts a total force and a total torque on any closed surface enclosing it, and plus a fundamental singular solutions due to concentrated couples, $(L\vec{u}^k(x,y), L\vec{\nu}^k(x,y))$, also located in the domain bounded by the interior surface, singularities that exerts a total torque and a zero total force on any closed surface enclosing it, then we have

$$U_i(x) = \int_{S_0 \cup S} K_{ij}(x,y)\phi_j(y)\, dS_y + F^k_i(x,y)\alpha_k + L^k_i(x,y)\beta_k \tag{35}$$

where U_i, ϕ_i, and K_{ij} are defined by expressions (6-d) and (10), and $\vec{F}^k$ and $\vec{L}^k$ are the fundamental solutions $(F\vec{u}^k(x,y), F\vec{\nu}^k(x,y))$ and $(L\vec{u}^k(x,y), L\vec{\nu}^k(x,y))$, located at the point y in the interior of surface S, respectively.

In comparison with Power & Miranda's original work for Newtonian fluid, where there is no fundamental solution due to a concentrated couple exerting pure torque upon any closed surface that encloses it, and therefore it was necessary to add besides a Stokeslets (fundamental solution of the Stokes equation for Newtonian fluid corresponding to a concentrated force) a Rotlets, high order singularity of the Stokes equation corresponding to the antisymmetric term of the Stokes doublet found by taking the gradient of the Stokeslet, and it is known that the Rotlet is the only high order singular solution that exerts torque upon any close surface enclosing it, here we do not need to use high order singularities, since the concentrated couples already complete the deficient torque of the double layer.

It will be convenient for later use to choose the strengths of the fundamental solutions, $\vec{\alpha}$ for the concentrated force and $\vec{\beta}$ for the concentrated couple, depending linearly upon the density ϕ of the double layer potential, in the following manner

$$\alpha_i = \int_S \phi_j(y) U_j^i(y)\, dS_y, \quad \text{for } i = 1, 2, 3 \tag{36-a}$$

$$\beta_i = \int_S \phi_j(y) U_j^{i+3}(y)\, dS_y \tag{36-b}$$

where $\vec{U}^k$, for $k = l, 2, \cdots, 6$, are the six linearly independent vector solutions of the homogeneous eqn (16), given by the expressions (8). Therefore, the strengths, $\vec{\alpha}$ and $\vec{\beta}$, are just the inner product between the vector density of the double layer and the basic motion of the fluid on the domain interior to the surface S. In the above expression, It can be observed that the component $\vec{\phi}^2$ of the unknown density of the double layer does not contribute to the final expression of the strength $\vec{\alpha}$, this is due to the zero components of the vectors U_j^1, U_j^2 and U_j^3.

From now on we will assume that S_k, for $k = 0, 1$, are closed Lyapunov surfaces, and that $\vec{U}$ is continuous on each of them. Applying the boundary condition (3) to the flow field defined by (35) to (36) leads to the following linear system of Fredholm integral equations of the second kind for the unknown vector density $\vec{\phi}$:

$$U_i(\xi) = -\frac{1}{2}\phi_i(\xi) + \int_{S_0} K_{ij}(\xi, y)\, \phi_j(y)\, dS_y + \int_S K_{ij}(\xi, y)\, \phi_j(y)\, dS_y + F_i^j(\xi, y)\, \alpha_j + L_i^j(\xi, y)\beta_j \quad \text{for all } \xi \in S_0 \tag{37-a}$$

and

$$U_i(\xi) = \frac{1}{2}\phi_i(\xi) + \int_S K_{ij}(\xi, y)\, \phi_j(y)\, dS_y + \int_{S_0} K_{ij}(\xi, y)\, \phi_j(y)\, dS_y + F_i^j(\xi, y)\, \alpha_j + L_i^j(\xi, y)\beta_j \quad \text{for all } \xi \in S \tag{37-b}$$

To show that (37) possesses a unique continuous solution $\vec{\phi}$, for a continuous datum $\vec{U}$, it is sufficient, according to Fredholm's alternative, that the following homogeneous system (38) for $\vec{\phi}^0$, admits only the trivial solution in the space of continuous functions:

$$-\frac{1}{2}\phi_i^0(\xi) + \int_{S_0} K_{ij}(\xi, y)\, \phi_j^0(y)\, dS_y + \int_S K_{ij}(\xi, y)\, \phi_j^0(y)\, dS_y + F_i^j(\xi, y)\, \alpha_j^0 + L_i^j(\xi, y)\beta_j^0 = 0 \quad \text{for all } \xi \in S_0 \tag{38-a}$$

and

$$\frac{1}{2}\phi_i^0(\xi) + \int_S K_{ij}(\xi, y)\, \phi_j^0(y)\, dS_y + \int_{S_0} K_{ij}(\xi, y)\, \phi_j^0(y)\, dS_y + F_i^j(\xi, y)\, \alpha_j^0 + L_i^j(\xi, y)\beta_j^0 = 0 \quad \text{for all } \xi \in S \tag{38-b}$$

with

$$\alpha_i^0 = \int_S \phi_j^0(y) U_j^i(y)\, dS_y \quad \text{for } i = 1, 2, 3 \tag{39-a}$$

$$\beta_i^0 = \int_S \phi_j^0(y) U_j^{i+3}(y)\, dS_y \tag{39-d}$$

From the uniqueness of solution of the present problem, and the system of homogeneous integral eqns (38-a,b), it follows that the pair of vector fields $\vec{V}^1$ and $\vec{V}^2$ defined below are identically equal in Ω

$$V_i^1(x) = \int_S K_{ij}(x, y)\phi_j^0(y)\, dS_y \tag{40-a}$$

and

$$V_i^2(x) = -(F_i^k(x, y)\alpha_k^0 + L_i^k(x, y)\beta_k^0) \tag{40-b}$$

On the other hand, since $\vec{V}^1$ yields zero total force and torque on each of the contours involved and $\vec{V}^2$ yields a total force and torque different from zero on each contour, it follows that both flow fields have to be the state of rest. Now, if $\vec{V}^2$ is equal to zero, it implies that a force fundamental solution has to be equal to minus a couple fundamental solution, but since the force fundamental solution yields non–zero total force and the couple fundamental solution yields zero total force and non–zero torque on any closed surface enclosing them, one can conclude that

$$\alpha_i^0 = \int_S \phi_j^0(y) U_j^i(y)\, dS_y = 0 \quad \text{for } i = 1,2,3 \tag{41-a}$$

$$\beta_i^0 = \int_S \phi_j^0(y) U_j^{i+3}(y)\, dS_y = 0 \tag{41-d}$$

Therefore, the system of integral eqns (38-a,b) reduces to

$$-\frac{1}{2}\phi_i^0(\xi) + \int_{S_0} K_{ij}(\xi, y)\, \phi_j^0(y)\, dS_y + \int_S K_{ij}(\xi, y)\, \phi_j^0(y)\, dS_y = 0 \quad \text{for } \xi \in S_0 \tag{42-a}$$

and

$$\frac{1}{2}\phi_i^0(\xi) + \int_S K_{ij}(\xi, y)\, \phi_j^0(y)\, dS_y + \int_{S_0} K_{ij}(\xi, y)\, \phi_j^0(y)\, dS_y = 0 \quad \text{for } \xi \in S \tag{42-b}$$

It can be proven that the above system of homogeneous integral equations has the following linearly independent vector solutions, $\vec{\phi}^0(y) = \vec{U}^k(y)$ for $k = 1, 2, \cdots, 6$ when $y \in S$, and $\vec{\phi}^0(y) = \vec{\Theta}(y)$ when $y \in S_0$ which is generally an unknown (see Appendix B for the analysis of the eigenfunctions of the system (42)). Hence any non–trivial solution of (38-a,b) is necessarily of the form $\phi_j^0(y) = \sum_{k=1}^{6} C_k U_j^k(y)$, when $y \in S$ and $\phi_j^0(y) = C_7 \Theta_j(y)$ when $y \in S_0$.

Substituting the general expression for points $y \in S$ of the vector density $\vec{\phi}^0$ into eqns (41), we found

$$C_k \int_S U_j^k(y) U_j^i(y)\, dS_y = 0 \quad \text{for } i = 1,2,3 \tag{43-a}$$

$$C_k \int_S U_j^k(y) U_j^{i+3}(y)\, dS_y = 0 \tag{43-d}$$

The above linear algebraic system for $C_1, C_2, \cdots C_6$ only has the trivial solution, implying that $\vec{\phi}^0 = 0$ on S, because the determinant of (43) has

$$\int_S U_j^l(y) U_j^q(y)\, dS_y$$

as element in the lth row and q^{th} column, and is thus the Gram determinant for the vector function $\vec{U}^k$, $k = 1, 2, \cdots, 6$, with a non vanishing value on account of the linear independence of $\vec{U}^k$.

To remove the remaining eigenfunction at points of the contour S_0, we will follow an idea given by Karrila & Kim[5] for the problem of Newtonian fluid analogous to the one we are dealing with. First, we will require that any possible solution of the system (37) satisfies the following orthogonality condition:

$$\int_{S_0} \phi_i(y)\Phi_i(y)\, dS_y = 0, \tag{44}$$

here

$$\Phi_i(y) = \begin{pmatrix} n_i(y) \\ 0 \end{pmatrix} \qquad \text{for every } y \in S_0,$$

or equivalently (see eqn (10-a) for the definition of the vector density function $\vec{\phi}$)

$$\int_{S_0} \phi_i^1(y) n_i(y)\, dS_y = 0,$$

this condition implies that the eigenfunctions solution of the system (38) will satisfy the following relation:

$$\int_{S_0} \phi_i^{01}(y) n_i(y)\, dS_y = 0, \tag{45}$$

This last equation forces the remaining eigenfunction $\vec{\phi}^0(y) = \Theta(y)$ to be zero when $y \in S_0$ (see Appendix B), and does not alter any of the previous statements about the other eigenfunctions. Therefore, after adding to the original double layer potential a force fundamental solution and a couple fundamental solution and requiring the above orthogonality condition, we have found a second kind Fredholm integral equation whose homogeneous form only admits the trivial solution.

To complete the actual system of integral equations, that possesses an extra equation coming from the required orthogonality condition, let us add to the velocity component of the eqn (37-a) an extra term which is multiple to the normal of the container surface, $Cn_i(\xi)$, for $\xi \in S_0$. In this way we get a complete system of Fredholm integral equations of the second kind, that has a unique continuous solution when the boundary of the internal surface and of the exterior container are Lyapunov surfaces and the velocity and Micropolar rotation data on each surface are continuous.

We will prove that the addition to the velocity component of the integral eqn (37-a) of a new constant variable multiple of the normal of the container surface, does not alter our final result, because due to the non–flux condition, this variable will end up being zero.

Multiplying the new integral eqn (37-a) by the vector $\vec{\Phi}$ at points of the surface S_0 and integrating with respect to $\xi \in S_0$, we obtain:

$$\int_{S_0} a_i^0(\xi)n_i(\xi)d\sigma_\xi = \int_S \phi_j^1(\xi)n_j(\xi)d\sigma_\xi + C\int_{S_0} n_i(\xi)n_i(\xi)d\sigma_\xi \tag{46}$$

doing the same to the velocity component of the integral eqn (37-b) at points ξ of the surface S, we obtain

$$\int_S a_i^1(\xi)n_i(\xi)d\sigma_\xi = \int_S \phi_i^1(\xi)n_i(\xi)d\sigma_\xi \tag{47}$$

where we have used in both equations, the fact that (Ramkissoon[17])

$$\int_{S_k} K_{ij}(\xi, y)\Phi_i(\xi)d\sigma_\xi = \begin{cases} 0 & y \in \bar{\Omega}_i^k \\ 1/2\delta_{jk}n_k(y) & y \in S_k \\ \delta_{jk}n_k(y) & y \in \Omega_i^k \end{cases} \tag{48}$$

and also the fact that a force fundamental solution and a couple fundamental solution yield zero net–flux through any closed surface, here Ω_i^k is the interior domain to S_k.

Substituting the integral relations (46) and (47) in the non–flux condition for an internal solenoidal flow, we obtain:

$$\int_S \phi_i^1(\xi)n_i(\xi)d\sigma_\xi + C\int_{S_0} n_i(\xi)n_i(\xi)d\sigma_\xi = \int_S \phi_i^1(\xi)n_i(\xi)d\sigma_\xi \tag{49}$$

consequently,

$$C\int_{S_0} n_i(\xi)n_i(\xi)d\sigma_\xi \equiv 0$$

implying that the constant C has to be equal to zero.

We have established that the non homogeneous integral eqn (38) has a unique continuous solution $\vec{\phi}$, for any arbitrary continuous datum $\vec{U}$, then the Stokes' velocity field of a Micropolar fluid defined by eqns (35), (36), with its corresponding pressure, provides the solution of the boundary value problem defined by eqns (1)–(5). In this way a constructive solution for the Stoke's flow of a Micropolar fluid exterior to a closed surface within a container, in terms of a well posed Fredholm integral equations of the second kind, has been obtained by completing the deficient range of a double layer potential, without recourse to eigenfuctions.

One of the advantages of completed double layer method is that the evaluation of the total force and torque upon the surfaces is found directly from the expressions of the strengths of the corresponding force fundamental solution and couple fundamental solution, respectively.

From the stated dynamic properties of the singularities considered here, as well as those of the double layer potential, it is found that the total force and torque exerted on the surface S, by the flow field represented by (35) are:

$$\begin{aligned} F_i &= \alpha_i = \int_S \phi_j(y) U_j^i(y)\, dS_y \\ T_i &= \varepsilon_{ijk} y_j \alpha_k + \beta_i = \varepsilon_{ijk} y_j \int_S \phi_l(y) U_l^k(y)\, dS_y + \int_S \phi_j(y) U_j^{i+3}(y)\, dS_y \end{aligned} \tag{50}$$

and upon the exterior container

$$\begin{aligned} F_i^0 &= -\left(\int_S \phi_j(y) U_j^i(y)\, dS_y \right) \\ T_i^0 &= -\left(\varepsilon_{ijk} y_j \int_S \phi_l(y) U_l^k(y)\, dS_y + \int_S \phi_j(y) U_j^{i+3}(y)\, dS_y \right) \end{aligned} \tag{51}$$

5 Modelling of the brain fluid dynamics

In this section, we propose a mathematical model for the flow of the cerebrospinal fluid in the subarachnoid space. This model is based on the solution found in the previous section for the micropolar flow between two closed surfaces. To the previous solution, we have added two terms. The first one represented the generation of the flow at the ventricles by means of a finite number N of potential sources located in Ω_i (the region modeling the space occupied by the brain), associated with constant pressure and zero microrotation vector, since every potential flow associated with such pressure and microrotational fields is solution of the creeping flow equation for a micropolar flow. The second term represents any possible motion of the enclosing system (modeling the craneal bone) as a rigid body, and then given by any linear combination of the basic flow field (8-a,b). Thus, our proposed flow representation formulae has the following form for the velocity and microrotational vector fields,

$$U_i(x) = \int_{S_0 \cup S} K_{ij}(x,y)\phi_j(y)\,dS_y + F_i^k(x,y)\alpha_k + L_i^k(x,y)\beta_k + \sum_{L=1}^{n} \frac{m_L x_j}{4\pi R_L^3(x)} + \sum_{p=1}^{6} C_p U_i^p(x) \tag{52}$$

for all $x \in \Omega$ (here Ω is the region which models the subarachnoid space).

In eqn (52) $\vec{\alpha}$ and $\vec{\beta}$ are given by equations (36–a,b), $m_L/4\pi$ represents the flux of the Lth potential source located at $x_L \in \Omega_i$, $(L = 1, 2, \cdots, n)$ which is controlled by the brain, and $R_L(x) = |x - x_L|$.

In what follows, Q will designate the total flux, *i.e.*

$$Q = \frac{1}{4\pi} \sum_{L=1}^{n} m_L.$$

In the representation (52), $\vec{\phi}$ stands for an unknown double layer density spread over $S_0 \cup S$, to be determined from the unique solution of eqns (37–a,b) with

$$a_i^0(\xi) = \frac{\delta^{av} Q}{A_{av}^m} - \sum_{L=1}^{n} \frac{m_L \xi_j}{4\pi R_L^3(\xi)} - \sum_{p=1}^{6} C_p U_i^p(\xi)$$

$$b_i^0 = 0.$$

where $\delta^{av} = 1$ for $\xi \in S_{av}^m \in S_0$, and $\delta^{av} = 0$ elsewhere, and $n = 1, 2, \cdots, M$, where we have assumed that there are M arachnoid villi. and

$$a_i^1(\xi) = \frac{\delta^f Q}{A_f^n} - \sum_{L=1}^{n} \frac{m_L \xi_j}{4\pi R_L^3(\xi)} - \sum_{p=1}^{6} C_p U_i^p(\xi)$$

$$b_i^1 = 0.$$

where $\delta^f = 1$ for $\xi \in S_f^n \in S_1$, and $\delta^f = 0$ elsewhere, and $n = 1, 2, 3$ corresponding to the three foramina.

Here S_{av}^m stands for a typical arachnoid villi location, *i.e.* surfaces where the flow leaves the subarachnoid space, and S_f^n stands for a typical foramina location, *i.e.* surfaces where the flow enters into the subarachnoid space. A_{av}^m represents the area of the $m-th$ arachnoid villi surface, and A_f^n the area of the $n-th$ foramina channel.

For simplicity, we have assumed that the fluid enters and leaves the subarachnoid space in a uniform manner over each surface S_f^n and S_{av}^m.

Under this assumption, the functions $\vec{a}^l$, for $l = 0, 1$, are piecewise continuous functions, but Fredholm's theory still holds for such data (see Goursat[4]).

It can be observed from the obtained representation formulae, eqn (52), that the brain has three possible ways of controlling the flow in the subarachnoid space: first, by controlling the total flux Q; second, for a given Q, by controlling the individual fluxes m_L of each potential source; and third, by changing the entrance and exit velocity by means of a corresponding change in the areas A_f^n and A_{av}^m in a valve-like type of mechanism. This control results from the change in density $\vec{\phi}$; resulting from the change in the data $\vec{a}^l, l = 0, 1$, which turn in a change of the total force and torque upon the brain surface. Another important feature that can be observed from the representation formulae obtained, is that the rigid body part of the nonhomogeneous terms $\vec{a}^0$ and $\vec{a}^1$ do not contribute to the dynamics of the flow, since it can be noticed from the previous proof of uniqueness of solution that for the case when $\vec{a}^0$ and $\vec{a}^1$ are only given in terms of the rigid body part, the unique solution of eqns (37) is given by $\vec{\phi}(\xi) = \sum_{k=1}^{6} C_k \vec{U}^k(\xi)$ when $\xi \in S_0$ and zero when $\xi \in S$. Therefore, it follows from eqns (36), (50) and (51) that the resulting force and torque, for such component of the total flow, are zero.

Under normal conditions, the brain selects an operating state of flow such that the total torque is zero, *i.e.* $\vec{\beta} = 0$, and the total force, $\vec{\alpha}$, is equal to the submerged weight of the brain. By controlling the flow the brain is able to change the total force and torque that the flow exerts upon itself, this is how the brain can restore its equilibrium position inside the cranial cavity after an extraordinary dynamic event, such as a punch on the head.

From the general theory of creeping flow, it follows that the present solution also holds for the case of non–permanent flow, as long as the phenomena occurs over a time scale of the order $O(U/L)$, so that the acceleration term is $O(R_e)$ smaller than the viscous term (quasi-static approximation). On the other hand, when the phenomena occurs over the much faster time scale, $O(L^2/\nu)$, the acceleration term is of the same order as the viscous term, then eqn (1) has to be equal to $\rho \partial u_i / \partial t$ instead of zero.

The numerical implementation of the present approach presents the difficulty of the apparent singularity in the numerical integration of the double layer potential kernel, due to the proximity between the two surfaces, because of the narrowness of the flow domain.

References

1. Eringen, A.C. Theory of Micropolar Fluids, *J. Maths. Mech.*, **16**,(1), (1966).

2. Davson, H. *A Textbook of General Physiology*, Churchill, London, (1970).

3. Goldberg, M.A. *Solution Methods for Integral Equations. Theory and Applications*, Plenum Press, New York, (1978).

4. Goursat, E. *Integral Equation Calculus of Variation, A Course in Mathematical Analysis*, Vol. III, Part Two, Dover Publication Inc., New York, (1964).

5. Karrila, S.J.; Fuentes, Y.O. & Kim, S. Parallel Computational Strategies for Hydrodynamic Interactions between Rigid Particles of Arbitrary Shape in a Viscous Fluid, *J. Rheolog.*, **33**, (1989).

6. Karrila, S.J. & Kim, S. Integral Equations of the Second Kind for Stokes Flow: Direct Solution for Physical Variables and Removal of Inherent Accuracy Limitations, *Chem. Eng. Commun.*, **82**, (1989).

7. Kim, S & Karrila, S.J. *Microhydrodynamics Principles and Selected Applications*, Butterworth–Heinemann Series in Chemical Engineering, Boston, (1991).

8. Kupradze, V.D. *Potential Methods in the Theory of Elasticity*, Israel Program for Scientific Translations, Jerusalem, (1965).

9. Ladyzhenskaya, O.A. *The mathematical theory of viscous incompressible flow*, Gordon y Breach, New York, (1963).

10. Lorentz, H.A. Ein Allgemeiner Satz, die Bewegung einer Reibenden Flussigkeit Betreffend, nebst einegen Anwendungen desselben (A General Theorem Concerning the Motion of a Viscous Fluid and a Few Consequences Dervived from it, *Versl. Kon. Akad. Wetensch.*, **5**, (1896).

11. Noback, C.R. & Demarest, R.J. *The Human Nervous System: Basis Principles of Neurobiology*, MacGraw-Hill, New York, (1967).

12. Odqvist F.K.G. Über die Randwertaufgaben der Hydrodynamik zäher Flussigkeiten (On the boundary value problems in hydrodynamics of viscous fluids), *Math. Z.*, **32**, (1930).

13. Power, H. & Miranda, G. Second Kind Integral Equation Formulation of Stokes Flows Past a Particle of Arbitrary Shape, *SIAM Appl.*, **47**, (1987).

14. Power, H. & Miranda, G. Integral Equation Formulation for the Creeping Flow of an Incompressible Viscous Fluid Between Two Arbitrarily Closed Surface, and a Possible Mathematical Model for the Brain Fluid Dynamics, *Journal of Mathematical Analysis and Application*, **137**, (1), (1989).

15. Power, H. & Ramkissoon, H. The Completed Double Layer Boundary Integral Equation Method for the Stokes Flow of a Micropolar Fluid, *Int. J. Engng. Sci.*, **32**, (4), (1994).

16. Ramkissoon, H. On the Uniqueness and Existence of Stokes Flows in Micropolar Fluid Theory, *Acta Mehanica*, **35**, (3-4), (1980).

17. Ramkissoon, H. Boundary–Value Problems in Microcontinuum Fluid Mechanics, *Quart. Appl. Math.*, **XLII**, (1984).

18. Youngren, M.A. & Acivos, A. Stokes flow past a particle of arbitrary shape; a numerical method of solution, *J. Fluid Mech.*, **69**, (1975).

Appendix A: Lyapunov Tauber theorem

The existence of either $\tau_i(\vec{W}(\xi,\vec{\phi}))_{(i)}$ or $\tau_i(\vec{W}(\xi,\vec{\phi}))_{(e)}$ for a double layer potential $\vec{W}(x,\vec{\phi})$ regular in Ω_i and Ω_e, implies the existence of the other limit and the following identity

$$\tau_i(\vec{W}(\xi,\vec{\phi}))_{(i)} = \tau_i(\vec{W}(\xi,\vec{\phi}))_{(e)} \qquad \text{for every} \qquad \xi \in S \tag{A.1}$$

as before (i) and (e) denote the limiting values at the surface S approaching from the inside and the outside, respectively (this theorem is the analog of Lyapunov Tauber theorem of potential theory).

Proof:

Let an external limit $\tau_i(\vec{W}(\xi,\vec{\phi}))_{(e)}$ to exist, we construct in Ω_i a single layer

$$V_i(x,\vec{\psi}) = \int_S J_i^j(x,y)\psi_j(y)dS_y, \tag{A.2}$$

such that

$$\tau_i(\vec{V}(\xi,\vec{\psi}))_{(i)} = \tau_i(\vec{W}(\xi,\vec{\phi}))_{(e)} \qquad \text{for every} \qquad \xi \in S. \qquad (A.3)$$

To prove that such single layer potential can be constructed, let us substitute eqn (32-a) into the left–hand side of the above expression

$$\frac{1}{2}\psi_i(\xi) + \int_S K_{ji}(y,\xi)\psi_j(y)\, dS_y = \tau_i(\vec{W}(\xi,\vec{\phi}))_{(e)} \qquad \text{for every} \qquad \xi \in S. \qquad (A.4)$$

The necessary and sufficient conditions for the solvability of the above equation are:

$$\int_S \tau_i(\vec{W}(y,\vec{\phi}))_{(e)} U_i^k(y) dS_y = 0 \qquad \text{for} \qquad k = 1,2,\cdots,6 \qquad (A.5)$$

where $\vec{U}^k$ are the six linear independent eigenfunctions given by eqn (8).

On the other hand, it is known that a single layer potential with a density $\vec{\Psi}^k$, $(k = 1,2,\cdots,6)$, solution of eqn (15), corresponding to the homogeneous form of eqn (A.4), has to be equal to the basic fluid motion, given by eqn (8) or any linear combination of them, in the bounded domain Ω_i, i.e. $\vec{V}(x,\vec{\Psi}^k) \sim U_i^k(x)$ for every $x \in \Omega_i$, since such flow field is the only interior flow with zero limiting value of the Cauchy and couple surface tractions. Therefore, (A.5) is equivalent to:

$$\int_S \tau_i(\vec{W}(y,\vec{\phi}))_{(e)} V_i(y,\vec{\Psi}^k) dS_y = 0 \qquad \text{for} \qquad k = 1,2,\cdots,6. \qquad (A.6)$$

Let us now write the external limiting value of the double layer potential $\vec{W}(x,\vec{\phi})$, given by (30-b), as:

$$W_i(\xi,\vec{\phi})_{(e)} = \frac{1}{2}\phi_i(\xi) + \int_S K_{ij}(\xi,y)\phi_j(y)\, dS_y \qquad \text{for every} \qquad \xi \in S. \qquad (A.7)$$

The solvability condition of the above equation is given by (14) with $\vec{U}(y) = \vec{W}(y,\vec{\phi})_{(e)}$, *i.e*

$$\int_S W_i(y,\vec{\phi})_{(e)} \Psi_i^k(y)\, dS_y = 0. \qquad (A.8)$$

Now, since $\vec{V}(x, \vec{\Psi}^k)$ and $\vec{W}(x, \vec{\phi})$ are regular solutions of the system (1)–(2) in the unbounded domain Ω_e, it follows from the corresponding Green's formulae for the exterior domain, that

$$\int_S \{\tau_i(\vec{W}(y, \vec{\phi}))_{(e)} V_i(y, \vec{\Psi}^k)_{(e)} - \tau_i(\vec{V}(y, \vec{\Psi}^k))_{(e)} W_i(y, \vec{\phi})_{(e)}\} dS_y = 0. \tag{A.9}$$

From the jump property of $\tau_i(\vec{V})$ and the way we constructed $\vec{V}(x, \vec{\Psi}^k)$, we obtain

$$\tau_i(\vec{V}(\xi, \vec{\Psi}^k))_{(e)} = \tau_i(\vec{V}(\xi, \vec{\Psi}^k))_{(i)} - \Psi_i^k(\xi) = -\Psi_i^k(\xi) \quad \text{for every} \quad \xi \in S. \tag{A.10}$$

In view of eqns (A.10) and (A.8) from eqn (A.9), we can write

$$\int_S \tau_i(\vec{W}(y, \vec{\phi}))_{(e)} V_i(y, \vec{\Psi}^k)_{(e)} dS_y = 0. \tag{A.11}$$

Since a single layer potential is continuous throughout the three dimensional space and since $\vec{V}(x, \vec{\Psi}^k)$, for any point $x \in \Omega_i$, is proportional to the basic fluid motion, i.e. $\vec{V}(x, \vec{\Psi}^k) \sim \vec{U}^k$ for every $x \in \Omega_i$, then, eqn (A.5) follows directly from (A.11).

Having thus established that we can construct a single layer $\vec{V}(x, \vec{\psi})$ satisfying the boundary condition given by equation (A.3), we can proceed with the demonstration. Let us consider the flow field

$$D_i(x) = W_i(x, \vec{\phi}) - V_i(x, \vec{\psi}) \qquad \text{for every} \qquad x \in \Omega_e \tag{A.12}$$

where $\vec{V}(x, \vec{\psi})$ is the single layer just constructed.

From the jump property of $\tau_i(\vec{V})$ and the boundary condition given by eqn (A.3), we have

$$\tau_i(\vec{V}(\xi, \vec{\psi}))_{(e)} = \tau_i(\vec{V}(\xi, \vec{\psi}))_{(i)} - \psi_i(\xi) = \tau_i(\vec{W}(\xi, \vec{\phi}))_{(e)} - \psi_i(\xi), \tag{A.13}$$

for every $\xi \in S$.

Then, from the above expression it follows that

$$\tau_i(\vec{D}(\xi))_{(e)} = \psi_i(\xi) \quad \text{for every} \quad \xi \in S. \tag{A.14}$$

Now, since $\vec{D}(x)$ is a regular micropolar flow in Ω_e, we have that it can be given in terms of the general Green's integral representation formulae as:

$$D_i(x) = -\int_S \{J_i^j(x,y)\tau_i(\vec{D}(y)) - K_{ij}(x,y)D_j(y)\}dS_y \quad \text{for every} \quad x \in \Omega_e \qquad (A.15)$$

the minus sign in the above integral representation formulae comes from the direction of the unit normal vector $\vec{n}$, taken here as directed outwardly from the domain Ω_i. Substituting (A.14) into the above equation, and using the definition of the single and double layer potentials, eqn (A.15) can be rewritten as:

$$D_i(x) = -V_i(x,\vec{\psi}) + W_i(x,\vec{D}) \quad \text{for every} \quad x \in \Omega_e. \qquad (A.16)$$

Subtracting (A.16) from (A.12), we obtain

$$0 = W_i(x,(\vec{D} - \vec{\phi})) \quad \text{for every} \quad x \in \Omega_e \qquad (A.17)$$

Letting x approach $\xi \in S$ coming from Ω_e, we find

$$0 = \frac{1}{2}(D_i(\xi) - \phi_i(\xi)) + \int_S K_{ij}(\xi,y)(D_i(y) - \phi_j(y))\, dS_y, \qquad (A.18)$$

for every $\xi \in S$.

It follows from eqns (16) and (8) that

$$(D_i(\xi) - \phi_i(\xi)) = \sum_{k=1}^{6} C_k U_i^k(\xi) \quad \text{for every} \quad \xi \in S \qquad (A.19)$$

where $C_1, C_2, \cdots, C_6$ are real constants.

If we let x lie in Ω_i, we find from eqn (A.15) that

$$0 = W_i(x,\vec{D}) - V_i(x,\vec{\psi}) \quad \text{for every} \quad x \in \Omega_i \qquad (A.20)$$

where, we have used (A.14) and the definition of the single and double layer potentials. Substituting (A.19) into the last equation, we obtain

$$0 = W_i(x,\vec{\phi}) + \sum_{k=1}^{6} C_k W_i(x,\vec{U}^k) - V_i(x,\vec{\psi}) \quad \text{for every} \quad x \in \Omega_i. \quad (A.21)$$

It can be proved that a double layer with density equal $\vec{U}^k$, is identica to $-\vec{U}^k$ at any point x belonging to Ω_i, because this double layer potentia has zero limiting values at point $\xi \in S$ as S is approached from the unbounded domain Ω_e (given by eqn (16)). Therefore, it follows from the jump property of the double layer potential, that the limiting values at point $\xi \in S$ of this double layer, as S is approached from the bounded domain Ω_i, are just $-\vec{U}^k$ i.e. $W_i(\xi, \vec{U}^k)_{(i)} = -U_i^k(\xi)$. Then, by the uniqueness of the interior problem, we have that $\vec{W}(x, \vec{U}^k)$ has to be equal $-\vec{U}^k$ (in this case the flow interior to the surface S moves with a velocity proportional to a rigid body vector and has a basic micropolar rotation). Using this behavior of the double layer $\vec{W}(x, U^k)$ in the domain Ω_i, we find from (A.21) that

$$0 = W_i(x, \vec{\phi}) - \sum_{k=1}^{6} C_k U_i^k(x) - V_i(x, \vec{\psi}) \quad \text{for every} \quad x \in \Omega_i. \qquad (A.22)$$

By taking the Cauchy stress and the couple stress tensors of the above null flow field in Ω_i, and taking their limiting values at the surface S, we obtain

$$\tau_i(\vec{W}(\xi, \vec{\phi}))_{(i)} - \tau_i(\vec{V}(\xi, \vec{\psi}))_{(i)} = 0 \quad \text{for every} \quad \xi \in S \qquad (A.23)$$

where we have used the fact that $\tau_i(\vec{U}^k) = 0$. Finally replacing $\tau_i(\vec{V}(\xi, \vec{\psi}))_{(i)}$ in the above equation by the expression given by (A.3), we obtain

$$\tau_i(\vec{W}(\xi, \vec{\phi}))_{(i)} = \tau_i(\vec{W}(\xi, \vec{\phi}))_{(e)}.$$

Appendix B: Eigenfunction solution of the system of integral eqns (42)

The homogeneous eqns (42) are

$$-\frac{1}{2}\phi_i^0(\xi) + \int_{S_0} K_{ij}(\xi, y)\, \phi_j^0(y) dS_y + \int_S K_{ij}(\xi, y)\, \phi_j^0(y) dS_y = 0, \quad \text{(B.1.a)}$$

when $\xi \in S_0$, and

$$\frac{1}{2}\phi_i^0(\xi) + \int_S K_{ij}(\xi, y)\, \phi_j^0(y) dS_y + \int_{S_0} K_{ij}(\xi, y)\, \phi_j^0(y) dS_y = 0, \quad \text{(B.1.b)}$$

when $\xi \in S$.

It is easy to prove that at least the vector functions $\vec{\phi}^0$ defined by $\vec{\phi}^0(\xi) = \vec{U}^k(\xi)$ for $k = 1, 2, \cdots, 6$ when $\xi \in S$ and $\vec{\phi}^0(\xi) = 0$ when $\xi \in S_0$ satisfy eqns (B.1). In fact, each $\vec{U}^k$ given by (8) is solution of

$$\frac{1}{2}U_i^k(\xi) + \int_S K_{ij}(\xi, y)\, U_j^k(y) dS_y = 0,$$

for any $k = 1, 2, \cdots, 6$ when $\xi \in S$, and every double-layer potential with the basic fluid motion as a vector density vanishes identically in the region exterior to the density carrying surface, so that:

$$\int_S K_{ij}(\xi, y)\, U_j^k(y) dS_y = 0,$$

for any $k = 1, 2, \cdots, 6$ when $\xi \in S_0$.

Therefore, (B.1.a) and (B.1.b) have at least six linearly independent solutions at the internal surface. By virtue of Fredholm's theorems, the vector solutions of (B.1.a) and (B.1.b) form a vector space of finite dimension N, the same being true for its adjoint integral equations. In fact, we will prove that the solutions of the adjoint integral equations to the system (B.1) form a vector space of dimension six, plus one coming from the exterior container:

$$-\frac{1}{2}\psi_i(\xi) + \int_{S_0} K_{ji}(y, \xi)\, \psi_j(y) dS_y + \int_S K_{ji}(y, \xi)\, \psi_j(y) dS_y = 0, \quad \text{(B.2.a)}$$

when $\xi \in S_0$ and

$$\frac{1}{2}\psi_i(\xi) + \int_S K_{ji}(y, \xi)\, \psi_j(y) dS_y + \int_{S_0} K_{ji}(y, \xi)\, \psi_j(y) dS_y = 0, \quad \text{(B.2.b)}$$

when $\xi \in S$.

To prove the above, we consider the following single layer potential:

$$V_i^1(x) = \int_{S \cup S_0} J_i^j(x, y) \psi_j(y)\, dS_y, \qquad \text{(B.3)}$$

where the density $\vec{\psi}$ is any non–trivial solution of (B.2.a,b)

Since the surface potential used in (B.3) is continuous across S_0 and S, it follows that (B.3) can express a continuous field at every point x belonging to the three-dimensional space. On the other hand, the Cauchy and couple surface tractions, $\vec{\tau}$, of (B.3) experiences a jump as a point x

crosses S_0 or S. The limiting value of the surface tractions of (B.3) when a point $x \in \Omega_e$, the domain exterior to the surface S_0, tends to a point ξ on the surface S_0 is given by the left-hand side of eqn (B.2.a), and the limiting value of these tractions when a point $x \in \Omega_i$, the domain interior to the surface S, tends to a point $\xi \in S$ is given by the left–hand side of eqn (B.2.b). Therefore, (B.3) for every $x \in \Omega_e$ represents a Stokes' flow with zero surface tractions $\sigma_{ij}n_j = 0$ and $m_{ij}n_j = 0$, at points on the surface S_0 and vanishing asymptotic velocity and micropolar rotation at infinity, and then has to be a stagnation flow. In the same way, it gives the motion of the fluid as the basic fluid motion, $\vec{V}^1(x) = \vec{U}^k(x)$ for $k = 1, 2, \cdots, 6$, with zero associated pressure, for every $x \in \Omega_i$, since for such zero traction conditions, we obtain from the energy relation for the domain in the interior of each surface S

$$\int_{\Omega_i} H \, dx = \int_S \tau_i(\vec{u}, \vec{\nu}) U_i dS_y = 0,$$

that $H = 0$, and hence the flow field $(\vec{u}, \vec{\nu})$ in the interior of each surface S has to satisfy the following system of partial differential equations

$$(2\mu + \chi)\left(\frac{\partial u_i}{\partial x_j} + \frac{\partial u_j}{\partial x_i}\right) = 0$$
$$(w_j - \nu_j)(w_j - \nu_j) = 0$$
$$\left(\alpha \frac{\partial \nu_j}{\partial x_j}\frac{\partial \nu_i}{\partial x_i} + \beta \frac{\partial \nu_j}{\partial x_i}\frac{\partial \nu_i}{\partial x_j} + \gamma \frac{\partial \nu_i}{\partial x_j}\frac{\partial \nu_j}{\partial x_i}\right) = 0,$$

which is known to have six linear independent solutions given by eqn (8), and then these flows are the only non–trivial interior Stokes' flows with zero tractions on the surface boundary, and therefore, no more than six of the $\vec{V}^1(x)$ are linearly independent inside the interior surface.

Now we can see that (B.3) for every $x \in \Omega$ represents a Stokes' flow, with the following boundary conditions at the surfaces S_0 and S, or any linear combination of them: $V_i^1(x) = U_i^k$ for all $x \in S$, $k = 1, 2, \cdots, 6$ and $V_i^1(x) = 0$ for all $x \in S_0$. This gives six degrees of freedom for the velocity and micropolar rotation boundary condition on the interior surface, and due to the indeterminacy in the base pressure, the interior Cauchy and couple stressess have a maximum of six degrees of freedom plus one. Due to the jump condition of the surface tractions of a single layer, the single layer density $\vec{\psi}$ is proportional to these tractions, and then the solution of the homogeneous system (B.2) form a vector space of dimension six plus one, i.e. $6 + 1$, and the same will be true for the solutions $\vec{\phi}^0$ of the system (B.1).

We will prove that the extra eigenfunction coming from the additional term, is restricted only to the exterior container S_0. We already noted that $\vec{\phi}^0(\xi) = \vec{U}^k(\xi)$, for $k = 1, 2, \cdots, 6$ and $\xi \in S$ and $\vec{\phi}^0(\xi) = 0$, for $\xi \in S = S_0$ are solutions of the system (B.1).

Now, the double layer $\vec{W}(x, \vec{\phi}^0)$ with density carrying contour $S_0 \cup S$, and density $\vec{\phi}^0$ as any non-trivial solution of (B.1) has to be the state of rest, $\vec{u} = \vec{\nu} = 0$, at any point $x \in \Omega$, and then the Cauchy stress, σ_{ij}, must be equal to a constant pressure C and the couple stress, m_{ij}, equal to zero. On the other hand, the surface tractions due to a double layer are continuous through its contours, so that the surface tractions on the inside of the contour S are also $\sigma_{ij}n_j = C\vec{n}$ and $m_{ij}n_j = 0$. Then, $\vec{W}(x, \vec{\phi}^0)$ has to be the basic fluid motion in the interior of the surface S, with constant associated pressure, and since $\vec{W}(x, \vec{\phi}^0) = 0$ for every $x \in \Omega$, we find from the jump property of the double layer potential that $\vec{\phi}^0(\xi)$ has to be linearly propotional to $\vec{U}^k(\xi)$, with $k = 1, 2, \cdots, 6$ for every $\xi \in S$ as before. Therefore, the additional eigenfunction, that we will call $\vec{\phi}^0(\xi) = \Theta(\xi)$, has to be restricted only to points on S_0.

The above double layer potential also defines a Stokes field in the domain Ω_e outside S_0, and by the previous argument the surface tractions on the contour boundary are also $\sigma_{ij}n_j = C\vec{n}$ and $m_{ij}n_j = 0$ outside S_0. The rate of dissipation outside S_0 is equal (see eqn (6-a))

$$D(\Omega_e) = \int_{S_0} (W_i(\xi, \vec{\phi}^0))_{(e)} \tau_i(\vec{W}(\xi, \vec{\phi}^0))_{(e)} dS$$

where $(\vec{W}(\xi, \vec{\phi}^0))_{(e)}$ denotes the limiting value of the velocity component of the double layer $\vec{W}(\xi, \vec{\phi}^0)$ on S_0, as it is approached from Ω_e. Because $\vec{W}(\xi, \vec{\phi}^0) = 0$ at any point $x \in \Omega$, it follows from the jump property of a double layer potential that $(\vec{W}(\xi, \vec{\phi}^0))_{(e)} = \vec{\phi}^0(\xi)$ for every $\xi \in S_0$, and since we have already shown that the exterior surface traction is given by

$$\tau_i(\vec{W}(\xi, \vec{\phi}^0))_{(e)} = \begin{pmatrix} Cn_i(\xi) \\ 0 \end{pmatrix}$$

then the above expression can be re–written as:

$$D(\Omega_e) = C \int_{S_0} \phi_i^{01}(y) n_i(y) \, dS_y.$$

It can be proven from the energy relation that the above inner product is different from zero, otherwise the velocity field in Ω_e has to be proportional to the rigid body motion vector and the micropolar rotation equal to the basic rotation (basic fluid motion), recalling that the double layer decays to zero at infinity, then $\vec{W}(x, \vec{\phi}^0)$ has to be zero outside S_0, since such null field is the only basic fluid motion compatible with the aximptotic behavior at infinity, from the discontinuity property of the double layer potential and the fact that $\vec{W}(x, \vec{\phi}^0) = 0$ for $x \in \Omega$, it follows that $\vec{\phi}^0(\xi) = \Theta(\xi) = 0$ at $\xi \in S_0$. This null density at points of S_0 contradicts our previous finding of the eigenfunction $\Theta(\xi)$.

From the above statement it follows that, if we require the homogeneous system (B.1) to satisfy the following additional orthogonal condition

$$\int_{S_0} \phi_i^{01}(y) n_i(y)\, dS_y = 0,$$

then the eigenfunction restricted to the exterior contour S_0 has to be identically equal to zero, $\Theta(\xi) = 0$.